THE FEEDING OF FARM ANIMALS IN INDIA

[*Frontispiece.*]

Open air experimental nutrition stalls for cattle at Lyallpur showing front and back view. The indoor stalls are in the background.

THE FEEDING OF FARM ANIMALS IN INDIA

P. E. Lander

BIOTECH
2015
Biotech Books
Delhi - 110 035

ISBN 81-7622-050-7
ISBN 978-81-7622-050-7

Published by : **BIOTECH BOOKS**
1123/74, Tri Nagar,
DELHI – 110 035
Phone: 27383999
e-mail: biotechbooks@yahoo.co.in

Showroom : 4762-63/23, Ansari Road, Darya Ganj,
NEW DELHI - 110 002
Phone: 23245578, 23244987

Printed at : **Chawla Offset Printers**
New Delhi – 110 052

PRINTED IN INDIA

FOREWORD

At a meeting of the Crops and Soils Wing of the Board of Agriculture and Animal Husbandry held in Delhi in February 1942, it was decided that the Imperial Council of Agricultural Research should proceed with the preparation of a series of Manuals dealing with all those subjects which are included in the comprehensive terms of Agriculture and Animal Husbandry and it fell to my lot to undertake the chief editorship of those Manuals dealing with Animal Husbandry.

The subjects chosen for treatment under Animal Husbandry are as follows :—

1. The Breeding of Farm Animals in India.
2. The Feeding of Farm Animals in India.
3. The Management of Farm Animals in India.
4. The Preparation of Animals and Animal Products for the Indian Market.
5. Dairying in India.

Arrangements for the preparation of each of the above Manuals are in train and they will appear in due course, but Dr. Lander has produced this Feeding Manual very quickly, for which we should be very grateful to him.

Not only has he been most expeditious in his methods, but he has treated the subject in a thoroughly comprehensive fashion and it is difficult to think of anyone, whether scientist or farmer, who will go to this publication for advice in regard to the feeding of farm animals under Indian conditions and come away empty.

In producing this, the first of the Animal Husbandry Manuals for Indian readers, Dr. Lander has set a very high standard and I believe this work will take a unique place amongst the literature on Animal Nutrition in the libraries of the future.

F. WARE, C.I.E., I.V.S.,

Animal Husbandry Commissioner with the Government of India.

PREFACE

In this book an attempt has been made to set forth the fundamental principles and practical aspects of Animal Nutrition with special reference to Indian conditions, in a manner which may appeal both to the scientist and the farmer.

Until comparatively recently the computation of rations in India for dairy cattle and other stock had to be made almost exclusively from data derived from feeding trials carried out in foreign countries. These served and continue to serve a useful purpose, as there are still many gaps in our knowledge of the true nutritive values and the best methods of using Indian feeding stuffs. This is much more the case with animals other than cattle, in which our knowledge of rationing values deduced from scientific feeding trials conducted in this country is largely a still unexplored field.

In writing this book I have drawn extensively on the results of my work on animal nutrition at Lyallpur during the last 22 years. I have also made considerable use of works, such as, Feeds and Feeding by F. B. Morrison, The Nutrition of Farm Animals by H. P. Armsby, The Scientific Feeding of Animals by O. Kellner, Animal Nutrition and Veterinary Dietetics by R. G. Linton, Feeding of Farm Live Stock by J. C. B. Ellis, and many others.

I am indebted for the photographs for Chapter I, on soils, to Mr. A. P. F. Hamilton, O.B.E., M.C., I.F.S., Chief Conservator of Forests, Punjab ; to Mr. W. Keventer, of Keventer's Dairies for those for Chapter XIV on pigs, and to Colonel Sir Edward Cole, Kt., C.B., C.M.G., of Coleyana Estate, Okara, Punjab, for permission to reproduce photographs of some of his horses, Chapter XII, and of silo pits. The photographs for Chapter XIII were taken by kind permission of the officer Commanding No. 41, Camel Coy. Lahore Cantonment.

The graphical method for determining the cash value of feeding stuffs, described in Chapter XVI, is largely the work of Pandit Lal Chand Dharmani, B.Sc. (Agri.), my Chief Research Assistant (now Agricultural Chemist to Government, East Punjab). I am especially indebted to him for invaluable assistance throughout the preparation of the book, particularly in the computation of rations, and the preparation of the appendices. The laborious task of Proof-reading was also done by him. I express my gratitude to my collaborators, and also to Mr. F. Ware, C.I.E., F.R.C.V.S., I.V.S., Animal Husbandry Commissioner with the Government of India, for many valuable suggestions in the final stages of the work.

P. E. LANDER.

CONTENTS

LIST OF ILLUSTRATIONS

LIST OF ILLUSTRATIONS

LIST OF ILLUSTRATIONS

LIST OF ILLUSTRATIONS

CHAPTER I

THE SOIL

The soil is mankind's most precious material heritage, as it is the source from which all living organisms ultimately derive a considerable part of the substance of which they are composed, and on whose healthy and balanced condition all plant and animal welfare ultimately depend.

It is desirable that the owner of livestock should take into account the nature and condition of the soil on which he grows his crops, as well as of the crops themselves, as the two are very intimately related, and adverse soil conditions will give rise to crops of defective nutritive value which will, eventually, be reflected in the condition of the livestock. Conversely, the manure produced will be valuable, if properly used, in maintaining the fertility of the soil.

The practice of agriculture has undergone considerable changes in recent years, particularly in western countries, and traditional customs in many parts of the world are giving place to improved methods of scientific agriculture with increased yields as the goal and frequently directed to 'money crops' rather than to 'food crops'.

Unless, however, the proper balance between the soil and the plant is maintained by preserving the fertility of the former by judicious manuring, the scientific farmer may in time discover that he is unable to maintain his increased yields.

Under pressure of modern conditions, the rise in human population and the continually increased demand for more food, the old-time farmer in many Western countries who employed animal labour and traditional methods of cultivation is becoming converted into the scientific agriculturist employing machinery instead of animal labour, and artificial fertilizers to a greater extent than natural ones, with the result that the biological equilibrium between the soil and the plant may tend to be upset.

In India, China and the Orient generally this tendency is not yet so pronounced, and time honoured methods of cultivation are still largely in vogue, though as the population increases there will inevitably be greater demands on the soil for more food crops, as is indeed the case at the present time of War (1943).

The soil is a very complex product, brought into existence by many natural agencies operating over millions of years during which it was not interfered with by the hand of man. Compared with the time taken in its preparation, the time during which the human race has existed, and still more so, the time during which modern methods of scientific agriculture have been employed, are negligible. The soil must not be regarded as a 'dead' inert mass, but rather as a 'living' entity teeming with lowly organisms whose interaction with the organic and inorganic parts of the soil largely determines its healthy condition. Natural conditions of cultivation assist this mutual relationship, and a judicious combination of organic with artificial manures where the latter are necessary to supply necessary plant food nutrients, is a good method of ensuring healthy soil conditions.

Modern agriculturists today realize that if artificial fertilizers are used alone they may adversely affect the natural equilibrium in the soil and that organic manures, preferably farmyard manure, are valuable, in fact essential, for the long-term maintenance of soil fertility; they emphasize the need to use both organic manures and artificial fertilizers together. In this way soil fertility can be maintained, while the further crop increases needed by the growing population of the world can be secured.

The soil is a depository of many minerals which the plant requires, and which become available under healthy soil conditions, and some of the more important of these from a quantitative aspect, such as calcium, phosphorus, iron and so forth may have their natural availability disturbed by exclusively artificial manures. Apart from these, there are also many minerals in the soil, present in only infinitesimally small quantities, but which are carefully selected by the plant and no doubt exert a great, but largely unknown, influence on plant life.

For example, copper is present in only minute traces in the soil. A copper sulphate solution of 1 : 1,000,000,000 is definitely injurious to Algae such as Spirogyra, and in a dilution of

1 : 700,000,000 it hinders the development of wheat sprouts, and at a dilution of 1 : 800,000 completely stops their growth. In the soil a dilution of this strength will act as a poison to bacteria, yet various crops contain considerable amounts of copper. Oats contain up to 0.9 per cent, barley up to 0.01 per cent, and wheat and rye about the same; potatoes contain four milligrammes per kilogramme, hay from six to twelve milligrammes per kilogramme, whereas lentils contain 0.015 per cent, peas 0.01 per cent, soybeans 0.01 per cent, calculated on the weight of the ash. [Pfeiffer, 1938.]

Again, sugar beet contains an amazing collection of 'rare' elements with over 50 per cent of its ash consisting of sodium, lithium, manganese, titanium, vanadium, strontium, caesium, copper and rubidium. [Pfeiffer, 1938.] Many other plants show similar phenomena as regards mineral content, and exhibit great selective capacity in extracting the mineral salts from extremely low soil concentrations.

Our knowledge is still scanty concerning the part played by many of the minerals in plant and animal welfare, but it might be possible that a heavy dressing of artificial fertilizers with no organic manure may alter their usual concentrations in the soil solution, or otherwise render the plant less capable of utilizing them.

The human race and the plant kingdom suffer from many diseases, the origin of which is inexplicable with our present knowledge; many are 'diseases of civilization', so called, but it might be found, if our knowledge could extend so far, that some of these diseases could be traced, among other factors, to alterations in the composition of food crops due to disturbances in the composition of the soil brought about by the natural equilibrium between the soil and the plant being upset.

Susceptibility to disease, in human beings, animals and plants may thus be increased by unnatural treatment of the soil or by allowing it to lose condition.

Again, it is well known that the losses which occur in crops as a result of insect pests and plant diseases are colossal as is also the amount of labour and money spent in preventive measures. Insect pests and diseases may be regarded by some people as an inevitable natural phenomena, but it is quite likely that when soil conditions deteriorate from whatever the causes may be, a state of affairs ensues favourable to the development of pests and diseases. On the other hand if the soil is kept in a healthy condition it is quite conceivable that the incidence of loss from pests and diseases might be considerably reduced.

The preservation of soil condition, and its reclamation where it has deteriorated, can only be achieved if a comprehensive view is taken of the nature of the soil and the close relationship of the inorganic and organic parts of which it is composed with the living organisms. Natural methods of cultivation and natural manures help to maintain this harmony. Unnatural methods which may be called for by an ever increasing demand for greater yields may disturb it, and eventually upset the natural balance in the soil between the inorganic and the organic constituents.

In this connection the results obtained over a short period, or even the span of human lifetime may be deceptive, and it is the long time effects as judged by soil deterioration and defective nutritive properties of the crops produced which ultimately matter.

Viswanath and Suryanarayana [1927] have made an interesting study on the relative effects of natural and artificial manuring under Indian conditions. They found important qualitative differences in the seeds of millets, *Eleusine corcocan* (*Mandal*), *and Panicum milaceum* (*Cheena*), and of wheat, and used pigeons to demonstrate these differences in feeding trials.

In the Madras climate these grains gave greater yields with organic fertilizers than with chemical ones or without any fertilizers. Chemical fertilizing gave an increased yield over the no-fertilizer plots of 32.8 per cent, but the organic fertilizer gave an increase of 100.7 per cent in the case of *Panicum milaceum.* The same strain of seed was used again and again for the same fertilizer test, and thus an increase was obtained in the various qualitative aspects.

Feeding tests with Eleusine gave the following results:

Average percentage of loss of body weight during the days of the test:

Group with basic ration	37.7
Group with basic ration plus plants grown on stable manure basis	22.4

Group with basic ration plus chemically fertilized plants	37·4
Group with basic ration plus unfertilized plants ..	40.9

Even with manifold changes in the conditions of the experiment, a better result was evident in the case of the seed raised with organic fertilizers than with seed raised with chemical fertilizers. In the case of wheat, the seed raised with chemical fertilizers reacted less favourably than the seed raised without fertilizers.

The same workers carried out a series of feeding trials with pigeons and found that stable manured barley gave better results than barley to which mineral fertilizers were given.

Experiments with rats, with a basic ration of meat residue, refined starch flour, olive oil and salt, cod liver oil (or Marmite or Vegex) for a vitamin supplement, and in addition, either organically fertilized wheat, or minerally fertilized wheat, both sorts having been grown on adjacent land, gave the following results:

Percentages of gain in body weight:

Basic ration plus stable manured wheat	114
Basic ration plus chemically fertilized wheat plus vitamin supplement	104
Basic ration plus chemically fertilized wheat alone ..	89

That is, the stable manured wheat was better despite the added influence of the vitamin supplement to the chemically fertilized soil.

Somewhat similar results have been recorded by Professor Boas of Munich [1932] who has shown that pasture grasses organically fertilized possess a higher albumin content than when fertilized by mineral manures, and in the latter case the grasses contained more peptones which are one of the first disintegration products of proteins.

It is interesting in this connection to recall the famous Broadbalk wheat experiments at Rothamsted, which have been continued for many years with different manures, and with no manure whatsoever. The Rothamsted report for 1938 records the following:

"It is sometimes stated that wheat grown without organic manure has less nutritive value than wheat grown with it. The Broadbalk experiments afford no evidence of this claim. Samples of grain from the different plots were sent to the Dunn Nutritional Laboratories at Cambridge and examined by Dr. Harris, but no consistent differences in their content of vitamin B1 could be found. Nor have the milling and baking tests ever shown any superiority of organic, over inorganic manures. The claim is also made in regard to other crops; fruits, vegetables, tea, etc., but no good experiments have shown any difference. Bad misuse of artificial fertilizers may *of course* lead to loss of quality of produce and it is well known that farmyard manure has various beneficial effects on the soil."

These results were obtained on English soils which are richer in organic matter than the average Indian soils, which are notoriously deficient in this respect. Organic manure is the one fertilizer above all others which Indian soils need, but in certain parts of the country there are conspicuous deficiencies in essential plant food material; for example, the soils of Bihar and Orissa are deficient in phosphates, in many of the sub-montane tracts and the lower Himalayan foot-hills where heavy rainfall occurs, calcium is deficient, whilst in the north-western corner of the Punjab, soils respond to phosphatic fertilizers to a greater extent than in other parts of the Province.

More long period work is still required on artificial manures versus organic ones, but the general consensus of the scientific opinion to-day is in favour of a judicious combination of the two.

It may well prove to be the case that the best policy in India will be to develop the greatest possible use of organic manures and supplement these with artificials to supply actual soil deficiencies.

SOIL DETERIORATION FROM HUMAN AND ANIMAL AGENCY

The history of civilization has been associated with the destruction of large tracts of fertile land by the activities of human beings and animals. The process of the transformation of fruitful land into desert has been going on since very ancient times, but it is not necessary to go far back in history for evidence of soil destruction, as the process is going on at present in many countries of the world. In India, and to a greater degree in South Africa and the United States of America, large tracts of

FIG. 1. Khud village land in the Hoshiarpur District of the E. Punjab, showing destructive erosion due to deforestation and over-grazing.

FIG. 2. A dreadful picture of erosion at Talwara in the Himalayan foothills (Hoshiarpur District, E. Punjab). Hill slopes have been washed away and the entire well lies exposed.

land are being lost to cultivation by soil erosion. In India erosion is most conspicuous in the Himalayan foothills and submontane tracts due to the denudation of forest and hill lands of their protective vegetation. In the American middle west, that vast prairie lands which originally supported enormous herds of wild cattle were built up by nature over long periods of time and nature preserved what she had so laboriously achieved. With the coming of civilization, these natural lands which acted as a sponge for the absorption and gradual release of rainwater to other levels, were broken up for farm crops and the surface of the soil became exposed to winds. Concurrently, forests were cut down and over-grazing practised on the remaining grazing lands with the result that the protective soil covering was destroyed. Drought and wind then played their part and the American West 'Dust Bowl' covering many hundreds of thousands of acres of land, replaced what were formerly fertile plains. Jenny [1933] has recorded that the nitrogen and humus content of the soil, when contrasted with the original prairie, have fallen about thirty-five per cent in about thirty years, and the only thing that has increased is soil acidity. Dust storms which are the natural sequence are completing the destruction, and it has been computed that a third of the cultivated area of the United States of America is on the way to becoming useless. This scientific commercializing of Agriculture —a policy seriously lacking in foresight—has resulted in one of the greatest grain producing areas of the world—the United States—importing cattle feed in recent years and actually importing wheat (1936) instead of exporting it. Drastic measures are now being taken in the United States to cope with the evil and large reafforestation measures are being undertaken, but the damage done affords a valuable lesson which is not without significance to India.

The writer has drawn attention in a recent publication [Lander, 1942] to the widespread destruction of cultivated and grazing lands in this country as a result of overgrazing. The results of excessive and uncontrolled grazing, *viz.*, soil erosion, are more conspicuous in the hills and submontane tracts, but their ill effects are felt far beyond the immediate localities in which they occur. (Plate I, Figs. 1 & 2.)

The semi-arid foothills of the Western Himalayas, and the

low dry hills of peninsular India support an excessive human and animal population, and it is in such areas of light rainfall that the greatest damage from erosion occurs. Many of the hilly tracts show large areas of bare rock which in comparatively recent times were covered by a protective cover of vegetation. The original soil and herbage acted like a sponge, holding rain water as it fell, and enabling it to percolate gradually downwards on its journey to the plains where it fed the subsoil water and smaller streams and rivers. The soil and its covering thus served a double function, *viz.,* the production of more vegetation and fodder, and a deepening of the soil on the rocky hills, and by acting as a valve, guarding and controlling the water supply of the lower levels. As in other parts of the world, the removal of the protective herbage exposes the soil to the corrosive action of rain, and destroys its porosity and its profile. Hence when rain does occur, it washes away the surface soil in the currents which are formed, so that not only is the soil lost, but the rain water is also lost, both locally and in the lower areas to which it should have found its way by slow natural percolation. The inevitable result is the incidence of floods during rains, and drought soon afterwards. (Plate II, Fig. 1.)

Gorrie [1933] has drawn attention to this evil, and in a note to the Animal Husbandry Wing of the Board of Agriculture (1936) says: "In the Punjab the outstanding need is for an immediate and very drastic reduction in the total of grazing animals. The grazing for the whole provincial forest area is 1.2 acres per animal, but even this indication of heavy stocking gives no accurate picture of the real state of affairs. The whole of the foothills of the outer Himalayas, the Siwaliks, the Salt Range and many smaller ranges of low hills are literally crawling with livestock of every description. The east of the province gets enough rainfall to allow of fairly rapid recovery, but the west is becoming more arid as the ground cover becomes less efficient for catching and storing the already inadequate rainfall. Both through fresh legal powers, and through a widely organized series of practical demonstration areas a really drastic reduction of livestock must be aimed at, beginning with the destruction of useless scrub bulls and bullocks and complete exclusion of goat browsing". (Plate II, Fig. 2.)

The problem of soil deterioration and destruction in India

FIG. 1. Stages in erosion showing exposed roots of Anogeissus in undemarcated Chulari, Kutlehar Jagir, E. Punjab.

FIG. 2. The last stage of erosion in a Chir pine forest. The mature forest at a low altitude has been over-grazed and has almost disappeared. (Nadaun State, Hoshiarpur District, E. Punjab.)

FIG. 1. Erosion of fields of the Khattar tract near Kherimar forest, Attock District, W. Punjab.

FIG. 2. Some badly eroded land near mile 19 on Chakwal-Sohawa Road (Jhelum District, W. Punjab). This was previously a highly fertile plateau.

has not yet received the attention it deserves. The vast Indo-Gangetic alluvium stretching almost unbroken from the N. W. F. P. and Sind to Assam, the Red Soils of the C. P. and Southern India, and the Black Soils of the Deccan have been cultivated from time immemorial, and appear to contain inexhaustible reserves of plant food material. Situated midway, as India is, between the 'Far East' and the 'Far West' the preservation of her soils in a healthy condition should be a matter of prime importance to all connected with the feeding of men and animals.

The preservation of the soil and the production of crops of high nutritive quality is a matter of more than mere agricultural or scientific concern. Ultimately it has a very direct bearing on national stability and national defence, and this applies to India no less than to other countries. The standing committee of the Council of Agriculture for England under the chairmanship of Sir Merrick Burrell [1936] stated that it was seriously concerned over the position of the National Food Supply which at that time was regarded as one of the weakest links in the chain of national defence. The Council pointed out that both the fertility of the soil and the means for the increased production of foodstuffs were less in 1936, than they were in 1914 and subsequent years when the shortage of food supply placed the country in a most perilous position. The Council also warned the Government that the lowered reserves of fertility in the greater part of the soils of the country and the fewer men employed, made a rapid expansion of foodstuffs absolutely impossible, and that only a carefully thought out long-term agricultural policy embracing all sections of the Industry would be likely to prove adequate. Under the stress of a Second World War, England has set to work to remedy her agricultural defects, and appears to have done this with considerable success (1943). At the beginning of the Second World War, England was only producing sufficient food to feed her population for about one and a half days a week; by the autumn of 1943 food production in England, as a result of unprecedented efforts by the farmers was sufficient to feed the population for about five days a week, and it is expected that food production will be still further increased. India has in turn adopted a 'Grow More Food,' (Plate III Figs. 1 & 2) campaign; but as demands on the soil increase, means must be taken to see that her soils do not lose their fertility.

As far back as 1926 the writer drew attention to the loss which the soils of the Punjab alone incur through the practice of burning dung for fuel instead of using it as manure. [Lander 1926] The loss was calculated both on a manurial, and a cash basis, and it was shown that:

(1) Based on the market value of Farm Yard Manure, the loss amounted to 1.5 crores of rupees per annum and

(2) When based on the value of Farm Yard Manure at Rs. 7 per ton, calculating this value from the increased yields obtained from land manured with Farm Yard Manure as against controls with no manure, the monetary loss comes to no less than 8 crores per annum.

Those figures were obtained by combining manurial values from chemical analyses and from increased crop yields. Even considering (1) alone, the picture is sufficiently impressive to indicate that every endeavour should be made to tackle the problem of the preservation of this valuable plant food for the soil. Similar data if computed for the whole of India would be much greater, and a matter for alarm at the losses taking place from the soil, and if to those be added the fertilizer values represented by exported products they would be astronomical in size. Yet inspite of such losses the soil continues to produce crops, but these might be much greater and of higher food value if all the natural plant food removed from the soil could be returned to it.

A decade, or even a lifetime is too short a period to pronounce judgment on the ultimate degree of, or effects, of soil deterioration, but some illuminating facts in this connection may be seen in China, a country renowned for its intensive cultivation and strict devotion to manurial and compost economy, carried out with almost religious zeal. In China, everything that can be composted and returned to the soil is put back; all sorts of refuse, night soil, etc., etc., are collected in layers, properly treated and turned into humus earth,—all by manual labour, so that it has been possible to keep much of the land in its original state of fertility by the oldest cultural methods of humanity—humus conservation and manual labour. The more heavily cultivated regions of China show, according to Pfeiffer [1938], from 1,750 to 2,000 people to the square mile (the average for China as a whole is considerably less), against 41 per square mile for the U.S.A., 225 for Switzerland, 343 for Germany and 660 for England. The figure for India is approximately 100. The extreme density of population in China has resulted in the natural equilibrium between the soil, the crops and the human population being disturbed, and this is assisted by the animal population which helps to convert hills and mountain chains which should be covered with vegetation, into more or less barren and impoverished land which in many places is approaching the characteristics of a desert.

It is doubtful, however, whether any large scale use of night soil as is customary in China, is to be recommended for India or Western countries. There is considerable prejudice against its use and its scientific value as a manure still remains to be vindicated. On the other hand there is very great scope for increasing the production of compost manures by utilizing the vast quantities of town waste available, and steps have already been taken to do this on a large scale.

It is not difficult to realize that in Oriental countries where the human population is abnormally dense, or rapidly increasing year by year, as in the case of India, any factors which conduce to deterioration in soil conditions and the consequent reduction in the soil's capacity to produce food, may in time result in human catastrophes of the first magnitude.

Mention has been made previously of the serious conditions prevailing in some parts of India and a good, although more extreme parallel may be found in a vivid description of similar conditions in China, by Prof. G. Wegener [1936].

"On the slopes of the hills we see either a dusty poor stand of plants of secondary quality, or the complete and fearful barrenness and emptiness of stony ground. The higher portions are badly furrowed and torn by the rain while the lower are always newly covered by streams of mud which destroy the fruitful soil at the bottom of the hills. The levels and valley bottoms are cultivated as gardens right upto their extreme edges. The great lack of fuel drives the inhabitants to a ruthless interference with the natural reforestation, even on the uncultivated lands. In some parts of the country grass roots, dung, etc., already serve as the only source of fuel.

In south China, although this was settled much later by the Chinese, and is today still much more thinly populated, the exhaustion of the woods is already very far advanced and goes forward at a gigantic pace, and with all its devasting effects. This is especially true in the areas of thickly populated sections. The Chinese culture which in its conservatism is still today very largely a culture of wood, needs this wood in large amounts—and so the consumption of the forests goes ceaselessly on. It has become clear that even for the inexplicably cheerful Chinese peasant the minimum standards of existence are being rapidly left behind, and that precisely from these provinces comes the main stream of the tremendous Chinese emigration which Manchuria has in recent years absorbed. The denuded hills and mountainous regions, about 50 per cent of the total area, remain practically unutilized."

"From here, as from the similarly oversettled regions of China", says Prof. Wegener "there comes forth a type of man who, because of his low standards of living, and the toughness of his body and nerves for the performance of his traditional labour, eliminates every competitor from the field both at home and in the lands in which he settles. Herein lies perhaps the greatest 'yellow peril', when the spreading of the Chinese over the earth becomes more extensive than today."

These words have a significance for India, where in many parts, soil conditions, although not presenting quite such a sombre picture as the above, nevertheless already constitute a problem of the greatest national concern. Europe, America, China, South Africa, India and other parts of the world, all have to face the problem of the disturbance of the natural equilibrium between the soil and the human and animal populations, brought about by increasing numbers, industrial and agricultural development and the competition between nations for a 'place in the sun.'

The preceding discussion indicates that one of the greatest problems confronting the world at present is that of food supply. The economist tells us that if the world's food supply were properly distributed there would be enough for all, but it is questionable if this is really so, or if it is, how long it will continue to be so with increases in human and animal populations, and deterioration in soil conditions? The modern Agriculturist tries to solve the problem by intensive farming combined with an increasing use of artificial fertilizers. This no doubt produces more food—for a time—but it is necessary, taking a long view, to guard against the natural balance between the soil, the livestock and the human population being upset, with the result that, in course of time, all may suffer. When a country produces crops, not only for its own needs but also for export, it is natural that it will tend to concentrate on those which are likely to give the biggest cash returns, or in other words, produce 'money crops' rather than 'food crops'. The result is that a considerable part of the land which might otherwise have been put under food crops is devoted to other purposes, and livestock are liable to suffer more than human beings. This is well illustrated in India where natural pastures are diminishing, the fodder supply tends to become more and more inadequate, and the majority of cattle hardly get a subsistence ration.

The large numbers of weak, emaciated, ill fed and badly cared for cattle in India support this contention. According to the figures given in the census for 1937-38 the number of livestock in British India and Indian States was roughly 300,000,000. Yet in spite of these numbers the needs of the people for dairy produce cannot even be remotely met. India is predominantly an Agricultural country, yet the *per capita* production of milk and milk products is only 6.6 oz., as against two to four pounds in many foreign countries. One of the main reasons for this is that, although India has so many cattle, a large proportion of them produces little or no milk, and the number of cattle is far too large for the fodder resources of the country. The area of cultivated land per head of cattle is approximately 1.2 acres, compared with 3.4 acres in Great Britain, 31.4 in Canada and 24.9 in the United States of America.

It is a very easy matter to tell the people what they should feed to their cattle to increase milk production, but quite another to produce the crops to enable this to be done. India is up against a vicious circle. The scientists are continually advocating limiting the number of nondescript and useless cattle and feeding the remainder better rations, but social and religious customs do not permit weeding out the useless stock. Instead they are kept alive and fed with fodder crops which, from a social and agricultural improvement point of view, could be more profitably employed in increasing dairy products for the people.

Whatever may be the immediate difficulties, the ultimate solution of the problem appears to lie in restoring the proper equilibrium between the soil and the human and cattle population.

Better conservation and utilization of the soil and the food which it produces is the only means of building up robust and healthy livestock. Some knowledge of the soil from the point of view of livestock requirements is thus a very important aspect of all problems of animal nutrition.

SOIL TYPES, COMPOSITION AND FUNCTIONS

Geological ages have been required for the formation of the soil by the various weathering and other agencies which have occurred from the time when primitive or crystalline rocks represented the earth's original crust. These agencies, operating through the ages, gradually covered the original exposed rock with layers of 'debris' which acted as a protection for the lower layers, and guarded them against further rapid corrosion, although the percolation of water still went on. The finer particles in the upper layer became slowly washed away to lower levels and so, in course of time, the surface of the earth became so modified under the influence of meteorological conditions that life in its lowest forms became possible. Vegetation appeared and this in turn again reacted with the weathering rocks to form the first beginning of what eventually became 'soils'.

It is not proposed to enter here into any detailed review of the origin and composition of soils but to confine this review to such salient features as have a practical bearing and interest for the animal husbandman.

An important consideration to be kept in mind is that the soil is by no means 'just dirt', or 'earth' or 'something in which plants grow', but the seat of many complex chemical and biological operations in which the bacteria and protozoa play a fundamental part. It is by means of these changes that the mineral and organic matter of the soil are rendered available—*i.e.*, capable of being taken up in the soil solution by the plant for its sustenance and growth.

A plant is the produce of the particular soil on which it grows, and its value as a feeding stuff for animals is governed to a considerable degree by the nature and composition of that soil, and the efficiency with which the living organisms in it are able to function.

Untold millions of bacteria exist in every handful of normal soil, and it is this living population which, considered against the background of the other soil constituents, is foremost among the various factors which determine the value of a soil as a medium for plant growth. The soil may therefore be reviewed as a medium for producing feeding stuffs for animals under the following captions:

1. The main types of soil found in India.
2. The mechanical nature of the soil.
3. The chemical composition of the soil.
4. The living organisms of the soil.

Main Types of the Soil Found in India

The chief soil types found in India are:

(*a*) *Alluvial soils*

These soils which cover about 300,000 square miles and stretch from Sind through the Punjab, the United Provinces, Bengal and Assam to the Eastern coastal regions, constitute by far the largest and most important of the soil groups of India. They are derived from the deposition of silt over long periods of time by the numerous tributaries belonging to the three chief river systems of India, *viz.*, the Indus, the Ganges and Brahmaputra. In superficial appearance they appear to be very uniform, but in reality they vary considerably from drift or blown sand, through intermediate loams and silts to very stiff clays. They are argillaceous (clay), and as a rule their nitrogen content is low, their phosphorus and potassium content is adequate and the amount of lime they contain variable. They contribute the largest share of the agricultural wealth of the country and are suitable for growing almost all kinds of crops except those found exclusively in tropical regions.

(*b*) *Black soils*

These, sometimes known as Black Cotton Soils, are found in the western part of the Central Provinces and Hyderabad State, interspersed with deeper black soils, and a deep black alluvial soil running from the North of Hyderabad State, towards the coast near Kathiawar. They are highly argillaceous, somewhat calcarious and extremely sticky when wet, and they retain moisture for a considerable time. Conspicuous among their constituents are calcium and magnesium carbonates, iron and a large admixture of organic matter, sometimes amounting to 10 per cent. It is owing to the latter two constituents that the prevailing dark or black colour is due. Black Cotton Soils are credited with an extraordinary degree of fertility by the people, and in some cases they are known to have been cultivated for centuries without manuring or being left fallow, and without showing any apparent signs of exhaustion.

(*c*) *Red soils lying on metamorphic rocks*

These soils are found chiefly in the eastern parts of the Central Provinces, Hyderabad State and the Deccan, and have a characteristic colour due to the presence of moderate amounts of limonite fairly evenly distributed. Such soils vary from poor, thin, gravelly and light coloured types on the uplands, to much more fertile, deep dark varieties in the plains and valleys. They are generally poor in nitrogen content, phosphorus and humus, and are mildly alkaline. Some are comparatively rich in potash, iron, lime and magnesia, but they are as a rule poorer in lime, potash and iron oxide than the black soils, and are also uniformly low in phosphorus content.

(*d*) *Laterite soils*

These soils appear to be peculiar to India and certain other tropical climates with a heavy rainfall, and are derived from the atmospheric weathering of several types of rocks under conditions of alternating wet and dry seasons, the siliceous matter of the rocks being almost completely leached away during weathering.

They are found only to a small extent in India, and occur in Eastern Assam, South Eastern Bengal and along the Malabar Coast. Laterite soils are widely divergent in character. A true Laterite is very infertile and contains but little plant nutrients, but good tillage can effect considerable improvement. These soils when found at high altitudes are generally thin and gravelly, very porous and agriculturally unimportant.

Low level laterite soils are dark, and comprise heavy loams and clays and are red brown or dark in colour, and retain moisture well.

All laterite soils are poor in lime and magnesia, and deficient in nitrogen and potassium, but the phosphorus content is occasionally high, and they may sometimes show a high content of humus.

(*e*) *Desert soils*

These comprise a large tract in eastern Sind, extending over the whole length of the province along the border of the Indus alluvium, Rajputana and the south Punjab. The sands of the desert are mainly derived from the old sea coast and have been transported and deposited in the interior by south westerly winds. As their name suggests, these soils are poor, but by no means entirely barren, for there are towns within the regions containing them, even in the driest parts of Rajputana, where underground water and irrigation enable a certain amount of cultivation to be carried on.

It may be expected that fodders and natural grasses obtained from soils so different in their geologic origin and character, and exposed to such widely different climatic conditions, will show significant variations in their quality and chemical composition. In this respect data regarding soils and crops in India is still somewhat scanty and no very clear picture can be given of any precise correlation between soil composition and the composition of crops grown on them. It is well known, however, that conspicuous soil deficiencies can be reflected in the composition of crops grown on them, and these in turn may effect animals and human beings. Thus, for example, the soils of some of the hill tracts along the Himalayas which are subjected to heavy rainfall are lacking in calcium, and so are the crops, and rickets and osteomalacia are wide-spread among the population. Again, the soils of Bihar are notoriously deficient in phosphates, while those in the Hissar district in the Punjab are, by contrast, rich in phosphates, and well balanced generally. This is reflected in the excellent cattle indigenous to this region—the Hariana breed. Dub grass grown at Hissar is the richest of its kind in India and may contain up to 20 per cent of protein. As a contrast the Bihar cattle are notoriously poor.

For further details concerning the relationship between the minerals in the soil and in crops, the reader may consult Orr's 'Minerals in Pastures', which deals comprehensively with the problem in different parts of the Empire. Most of the data at present available in India [Lander, 1942] on the composition of natural and cultivated fodders have been published. Some general features of the composition of soil may now be considered.

(2) Mechanical Nature of the Soil

The response of a soil to the water supply, and its suitability for cultivation depend to a large extent on its physical texture or composition, and the determination of this constitutes the mechanical analysis.

In determining the mechanical composition, the inorganic particles are divided into groups according to the size of the particles as determined by special analytical methods.

This grouping is shown below:

	Maximum size in mm.	Minimum size in mm.
Stone and gravel	—	3.0
Fine gravel	3.0	1.0
Coarse sand	1.0	0.2
Fine sand	0.2	0.04
Silt	0.04	0.01
Fine silt	0.01	0.002
Clay	0.002	—

A soil is designated sandy or light, a light, medium or heavy loam, or a heavy soil, according to the relative proportions of the different groups contained in it. The category in which a soil is placed will also be governed to some extent by the amount and kinds of salts which are present. For example, lime or calcium salts tend to keep a soil open and lighten it, while sodium salts will do the reverse, and if the latter are present to a high degree, a hard uncultivable soil will result, as is seen in the *bara* soils of parts of the Punjab. The growth of different types of crops will be governed to a large extent by these factors—among many others. Thus, though the size of the various particles is an important factor in determining the physical nature of a soil, this can be modified considerably by the composition of the finer particles, and their ability to form a colloid system with water, and by the presence or absence of certain salts in the soil. For example, clays differ considerably in their chemical composition, and a sodium clay will give a much

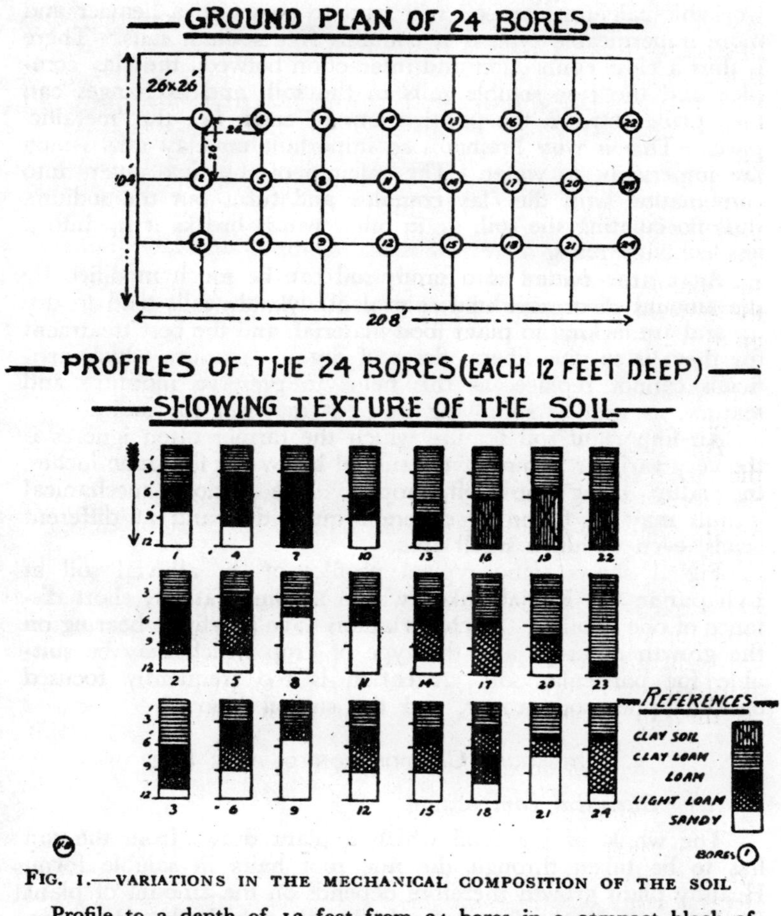

FIG. I.—VARIATIONS IN THE MECHANICAL COMPOSITION OF THE SOIL

Profile to a depth of 12 feet from 24 bores in a compact block of about half an acre (104′ x 208′). Each bore is situated at a distance of 26 feet from the adjacent bore.

heavier and intractable soil than a calcium clay, and a readily workable calcium clay soil will tend to pass into a heavier and more impermeable type if it contains free sodium salts. There is thus a close connection and interaction between the clay complex and the free soluble salts in the soil, and exchanges can take place between the positive ions of each, *i.e.*, the 'metallic' parts. This is why liming is so important on clay soils which are impervious to water. The calcium of the lime enters into combination with the clay complex and turns out the sodium, thus flocculating the soil, or in other words breaks it up into a less cohesive mass.

Again, the nature of a sandy soil can be much modified by the amount of organic matter present. Sandy soils tend to dry up and are lacking in plant food material, and the best treatment for them is to give liberal doses of organic manure which artificials cannot replace, as this helps to preserve moisture and texture, as well as supplying nutrient material for crops.

An important soil feature which the farmer often ignores is the very variable nature of the subsoil below the first nine inches, the latter being the 'soil proper'. The various mechanical groups may be found in different quantities, and at different levels, even within a small area.

Fig. 1 shows some typical profiles of an alluvial soil at Lyllapur in the Punjab taken within a comparatively short distance of one another. Such variations have a definite bearing on the growth of crops and the type of crop which may be suitable for particular soil. Attention is too frequently focused on the upper soil proper, and the subsoil ignored.

3. Chemical Composition of the Soil

(*a*) *Chemical composition*

The whole of the food which a plant draws from the soil has to be taken through the fine root hairs in soluble form. Healthy plant growth therefore depends on the amount of plant food present in the form of soluble salts which the plant can absorb, and the amount present as reserves which can be made available progressively under the influence of the various activities which go on in the soil, largely as a result of bacterial activity. Chemical analysis can tell something about the reserves of plant food material in the soil when there are any conspicuous deficiencies, but it is a matter of extreme difficulty to say with certainty that a particular crop will do well on a certain soil from chemical analysis alone. Many imponderable factors come into play which chemical analysis cannot reveal. Although the chemical analysis of a soil resolves itself chiefly into determining the nitrogen, calcium, phosphorus, potassium and iron, and to a less degree the magnesium, manganese, aluminium, chlorine and sulphur, there are many other minerals present only in extremely small quantities in the soil such as copper, iodine and many of the rarer elements, whose importance is too frequently not recognized. These can only be accurately determined by spectrographic analysis, and our knowledge of the part many of them play in regard to plant and animal health is still scanty.

(*b*) *Humus content*

A conspicuous feature of most Indian soils is their lack of humus—the dark brown material produced by bacterial action from organic matter of vegetable origin, and which is so valuable to plants as a source of food. Its chemical composition is indefinite as it consists of a variable mixture of different substances, themselves of very complex constitution. Humus constitutes nature's own fertilizer and is far better for the soil than artificial fertilizers. It helps to bind light soils together, and open out heavy ones, and in the decomposition of the carbohydrate portion, carbon dioxide is liberated which aerates the soil. Humus also forms a store of food for the bacteria, which thus derive energy for the important work they have to do in breaking down the humus into soluble food material in combination with the salts, as inorganic nitrates, etc., and for the many other groups of bacteria, some of which will now be briefly described.

4. Living Organisms of the Soil

Vast numbers of micro-organisms exist in the soil—bacteria—protozoa—fungi—and it is through their manifold activities, chiefly of the bacteria—that vegetable matter is broken down and converted into humus, and ultimately into carbon dioxide, nitric acid and other simple products. Without these changes there would be a vast accumulation of 'dead' matter in the world, the soil would be sterile through lack of nutrients, and there would be a break in the biological cycle of construction, destruction and reconstruction. The changes effected in the organic matter in the soil are associated with different groups of organisms, each with its specific functions, but all are of fundamental importance in the nutrition of the plant.

Certain groups of organisms attack the carbonaceous parts of plant residues such as soluble carbohydrates, cellulose, etc., converting them into a variety of organic acids, carbon dioxide and water, while the proteins are attacked by certain special groups of bacteria and the lower fungi.

Perhaps the most important groups of soil bacteria, called amonifiers are those responsible for the breaking down of nitrogenous plant residues into amino-acids and then ammonia. The ammonia is then acted upon by other groups called the nitrifiers (nitrosomas and nitrobacter) and oxidized to nitric acid, which combines with the bases of the soil to form nitrates. Under certain conditions, in the presence of large amounts of carbonaceous matter, and in the absence of air a group of anaerobic bacteria called the denitrifiers comes into play and causes actual losses of nitrogen by changing nitrates and other nitrogenous compounds into free nitrogen which is lost in the air.

It is important to remember that bacteria are present and flourish only in the upper crust of the soil, mostly in the top four inches, and that their numbers decrease with depth. Their numbers also vary according to the amount of food material, moisture and the reaction of the soil. Hence while aeration of the soil is necessary for the optimum activity of the bacteria, deep ploughing which turns up the sub-soil on to the surface helps to render the soil unproductive for a considerable time due to sterile soil being brought to the region where the young plants will start to grow.

Nitrogen Fixation by Bacteria

There are two groups of bacteria which possess the power of taking free nitrogen from the air and synthesizing it into their own body proteins which are later mineralized for the use of plants.

The members of one of these groups, the Azotobacter and Closteridium, live free in the soil and derive their energy from carbonaceous matter, fixing free nitrogen from the air in the oxidation processes of their life cycle. These organisms are recorded to be capable of fixing as much as 50—100 lb. of nitrogen per acre per annum.

The members of another group, *Rhizobium leguminosarum*, live in symbiotic relationship with leguminous plants which possess small round nodules on their roots in which the bacteria live. They utilize the sap of the host for food, and in return fix nitrogen from the air and pass it on to the host plant. This capacity of leguminous plants to fix nitrogen from the air is of great importance in nutrition problems. The legumes are themselves rich in nitrogenous products, and they also enrich the soil for succeeding crops, which is why they are used so commonly in crop rotations.

Sulphur and Iron Bacteria

Two other interesting groups of soil bacteria are:—

(*a*) the sulphur bacteria which oxidize sulphur to sulphuric acid, thus helping to render soluble certain other minerals needed by the plant, and

(*b*) the iron bacteria which secrete hydrated ferric oxide in badly drained soils whose water contains ferrous bicarbonate in solution, a condition made possible only if calcium carbonate is absent.

GROWING CROPS WITHOUT SOIL

After having stressed the importance of the soil and the necessity of maintaining it in proper condition for the growth of plants, it may appear strange to state that plants can be grown to healthy maturity without any soil whatever. Yet this is so provided they have the proper salt solutions and other environmental conditions necessary for good growth. This is not so strange as it appears to be at first sight if it is remembered that, apart from the carbon dioxide which plants absorb through their leaves for the manufacture of carbohydrates, the whole of the food they require is taken in through the fine root hairs situated near the tips of the roots in the form of a dilute solution of salts comprising what is known as the 'soil solution'. If, therefore, a plant is provided with a solution in which to grow approximating as nearly as possible to that which constitutes the soil solution, it might be expected that corresponding results, namely, normal growth would follow. This is what actually happens when plants are grown in nutrient solutions.

The elements which are necessary to promote normal growth in nutrient solutions consist of nitrogen, phosphorus, potassium, calcium, magnesium and sulphur. All of these with the exception of sulphur are required by plants in large amounts but each is absorbed by plants in amounts different from the others, and they can be given in considerable excess without producing harmful results. Sulphur in the form of sulphate is generally absorbed to less varying degree by plants, and can also be supplied in excess without producing toxic effects.

In addition to the above, plants require certain other elements known as trace elements in only very small quantities. The optimum concentration of these may be only a fraction of one part in a million, and if this is exceeded toxic effects are likely to follow. Iron, manganese, copper, zinc and boron fall into this category, and in addition certain other elements may also play an important part in plant growth under natural conditions.

Potassium nitrate, calcium phosphate and magnesium sulphate are the chief commercial chemicals from which the major elements required in nutrient solutions can be obtained. The trace elements can be furnished by boric acid, manganese sulphate, zinc sulphate, copper sulphate, ferric chloride or nitrate or ferric sulphate (see below).

It is well known that plants vary very considerably in chemical composition, and consequently a considerable range in the composition of the nutrient solution in which plants are grown is possible, although an excessive absorption of any one element in the early stages of plant growth may have unfavourable effects later.

The chief principles underlying the possibility of growing plants in nutrient solutions, or culture solutions as they are sometimes called, have been known for a long time but their development on a commercial basis is largely due to Gericke

[1929; 1936] in the U.S.A. who developed the methods for growing plants on a large scale in specially prepared containers filled with nutrient solution. The development of the commercial production of plants by this method created great popular interest in the U.S.A., and the practice has attained considerable dimensions in that country. The chief drawbacks are partly the cost of the chemicals, but more particularly the cost of the containers and other equipment. Refined glass or glazed earthenware containers which resist the solvent action of salt solutions are expensive, and the basins now generally used are constructed of concrete, wood, sheet iron and various preparations of asphalt.

These basins or trays range from 2½ feet to 6 feet in width, and 50 feet in length. Wooden troughs 1 foot wide and 40 feet long, and even large tanks 12 feet wide and of sufficient length to provide a water surface area of 1/200 of an acre have also been employed.

Many types of plant can be grown by this method, and apart from the actual containers, certain other accessories are needed. For example, seeds must be started off on a seed-bed which may consist of a mat of some vegetable litter such as straw or sawdust properly mounted over the surface of the nutrient solution on wire netting and fixed to a portable frame which rests on the top of the basin. Wire supports will also be needed, the size and cost of which will depend on the type of plant being grown.

Naturally, different types of plant require different treatments, and the beginner who wishes to take up the culture of plants in nutrient solutions either as a hobby or on a more ambitious scale should carefully study the details of the technique required from one of the excellent textbooks available on the subject. A suitable book for this purpose is 'Soilless Growth of Plants' by Carleton Ellis and Miller W. Swaney, Reinhold Publishing Corporation, New York, U.S.A.

The culture of plants by this method makes a fascinating hobby but it is questionable whether it will ever supplant the normal production of crops from the soil except in certain countries and in localities where defective soil conditions may render its commercial exploitation an economic possibility.

Arnon and Hoagland [1939] have described the results of some of their experiments on tomatoes, but the yields they obtained were not markedly different from those which could be obtained from soil, and they state that their results do not justify the contention that the potential crop yields from suitable nutrient solutions are higher than from fertile soil.

These investigators also studied the chemical composition of tomatoes grown in fertile soil in greenhouses, and the same varieties grown in water culture solutions under the same climatic conditions. It was found that there was no significant difference in the calcium, phosphorus, magnesium, potassium, nitrogen and sulphur content of fruit grown in the nutrient solutions and in soil, neither could any significant difference be found in the contents of vitamin A (carotene) and vitamin C.

The author has grown tomato plants in culture media in glass jars and found no difference in the vitamin content of the ripe tomato from the vitamin content of tomatoes grown on soil.

It is sometimes contended that plants grown in nutrient solutions are not as susceptible to plant diseases or insect pests as are plants grown in soil, but Arnon and Hoagland point out that, while strictly soil-borne diseases are eliminated, recent observations suggest that plants grown in nutrient solutions may be attacked by diseases specific to these solutions. The general consensus of opinion appears to be that the potentialities for crop production in suitable nutrient solutions do not exceed those of a good fertile soil.

Hosain [1942] has recorded the results of trials in which he grew maize and barley in water culture solutions at Izatnagar in the U.P. where very considerable ranges of temperature and humidity occur. Detailed observations were made in regard to attack by fungi, the yields obtained, the quality of the crop as revealed from chemical analyses and the cost of production. The results of these investigations show that the climatic conditions of Izatnagar which are typical of those of a large part of northern India preclude the possibility of growing fodder in the type of cabinet used during the summer months. Furthermore, the high humidity was found to favour the development of fungi which could not be easily suppressed by chemical treatment. More favourable results were obtained with fodder grown under controlled conditions in winter, but Hosain concluded that it seems highly improbable that the cost of production involved will justify this method of growing fodder in India. There is no doubt, however, that most excellent crops can be produced in water culture media under controlled conditions and provided proper equipment is installed.

Preparing the Nutrient Solutions

The nutrient solutions required for soilless growth are of two kinds; firstly, a solution of trace elements, *i.e.*, those elements which the plant requires in only very small quantities; and secondly, the main nutrient solution containing the salts needed in considerable amount. Ellis and Swaney [1929] recommend the following two stock solutions for providing the trace elements.

Stock Solution A

One teaspoonful (3.2 grammes) each of boric acid, manganese sulphate and zinc sulphate are simultaneously dissolved in ½ gallon of water, to which ⅛th of a teaspoonful of copper sulphate is then added. Stock solution A may be added to the main culture solution at any time before use.

Stock Solution B

One quarter of a teaspoonful (0.8 gramme) of ferric chloride or ferric nitrate is dissolved in one pint of water. Ferrous sulphate may also be used to provide the iron, but this salt has a tendency to be precipitated from solution on standing. Iron salts are inclined to precipitate when brought into contact with the main culture solution, consequently stock solution B should only be added to the latter just before it is required for use with plants.

Stock solution A may be added to the main culture solutions in the proportion of two teaspoonfuls (10 c.c.) of the former to 5 gallons of the latter if pure chemicals are used, or one teaspoonful (5 c.c.) of the former to 5 gallons of the latter if commercial chemicals are employed.

Stock solution B may be used in the proportion of 4 teaspoonfuls (20 c.c.) to each gallon of the main culture solution.

STANDARD CULTURE SOLUTIONS

No. 1 Culture Solution

Unit of measure	Mono-potassium phosphate KH_2PO_4	Calcium nitrate $Ca(NO_3)_2 \cdot 4H_2O$	Magnesium sulphate $MgSO_4 \cdot 7H_2O$	Ammonium sulphate $(NH_4)_2SO_4$ (Dry)
Grammes per 5 gallons of solution	5·9	20·1	10·7	1·8
Teaspoonfuls per 5 gallons of solution (approximate)	1¼	4	2½	½

Each of the salts should be dissolved separately in about a pint of water, and the separate solutions mixed and made up to 5 gallons with ordinary water. The trace elements provided in stock solutions A or B may then be added as indicated above.

No. 2 Culture Solution

Unit of measure	Monopotassium phosphate KH_2PO_4	Sodium nitrate $NaNO_3$	Magnesium sulphate $MgSO_4 \cdot 7H_2O$	Calcium chloride $CaCl_2$ (Dry)
Grammes per 5 gallons of solution	3·9	6·4	10·3	3·2
Teaspoonfuls per 5 gallons of solution (approximate)	1	1	2½	1

This solution should be prepared exactly as directed for No. 1.

No. 3 Culture Solution Prepared from Commercial Grade Chemicals

Unit of measure	Superphosphate (Mono-calcium, $CaHPO_4$	Sodium nitrat $NaNO_3$	Magnesium sulphate $MgSO_4 \cdot 7H_2O$	Potassium chloride KCl
Grammes per 5 gallons of solution.	5·8	6·4	10·3	3·9
Teaspoonfuls per 5 gallons of solution (approximate)	2	1	2½	1

Superphosphate usually contains a certain amount of insoluble matter which should first be allowed to settle and the clear solution then poured off. Only half the quantity of stock solution A or B should be added to culture solution No. 3.

The above solutions have been recommended and developed by the New Jersey Agricultural Experiment Station, U.S.A., and have been found to produce excellent results with vegetables (tomatoes, potatoes, radish, lettuce, etc.) and flowers (begonia, gladioli, tulip, rose, gardenia, carnations, snapdragons, etc.).

For producing flowers and vegetables on a small scale as a hobby, good quality glass containers may be used, of size and dimensions to suit the investigator, or they may be made of any of the materials mentioned above. The beginner will experience many practical difficulties before he has perfected his technique and will have to find out from experience when to replenish the solutions and give them the periodic aeration they need.

BIBLIOGRAPHY

Arnon, D. I. and Hoagland, D. R. 1939. *Science,* 89, 512-14.

Boas, F. F., 1932, "*Untersuchungen fur eine dynamische Grunland Biologie*" Praktische Blatter fur Pflanzenbau, 9, 173.

Burrel, Sir Merrik, 1936, J. of the Ministry of Agric. 43, 372.

Ellis C. and Swaney, M. W. 1929. *Soilless growth of Plants.* Reinhold Publishing Corporation, New York, U.S.A.

Gericke, W. F. 1929. *Amer. J. Bot.,* 16, 862.

————and Tavernetti, J. R. 1936. *Agric. Engineering,* 17, 141-43.

Gorrie, R. M. 1933. *The Better Utilization of Forests for Grazing,* Indian Forester.

————1936. *Proceedings of Meeting of the Anim. Husb. Wing of the Board of Agric. and A. H. in India,* Imperial Council of Agricultural Research, India, New Delhi.

Hosain, M. F., 1942. *Production of Fodder in a Sprouting Cabinet,* Indian Farming 3, 329-32.

Jenny, H., 1933, *Soil Fertility Losses under Missourie Conditions,* Columbia Mo. Bul. 324, Agric. Experiment Station.

Lander, P. E. and Mukand Lal. M. 1926, Agric. J. of India, 21, 115-24.

——— 1942. *Indian Grazing Conditions and the Mineral Content of Indian Fodders.* Imperial Council of Agricultural Research, India, Misc., Bul. 16.

Pfeiffer, E. 1938. *Biodynamic Farming and Gardening.* 1938 (Rudolf Steiner Publishing Co., London).

Viswanath and Suryanarayana 1928, *Memoirs of the Dept. of Agric. in India,* 9, No. 4, 85.

Wegener, G., 1936. *Handbuch der Geographischen Wissenschaft* (as quoted by Pfeiffer page 13 above).

CHAPTER II

CONSTITUENTS OF FEEDING STUFFS

Different types of farm stock are fed a variety of feeding stuffs which vary considerably in composition, but have the common feature that they are all composed of a comparatively small number of main food groups or nutrients. These may be classified as follows:

Feeding stuff

- **Organic**
  - **Nitrogenous substances**
    - **True Proteins**
    - **Non-protein nitrogenous substances**
  - **Non-nitrogenous substances**
    - **Carbohydrates**
      - **Soluble carbohydrates**
      - **Crude fibre**
    - **Fats**
- **Inorganic**
  - **Mineral matter** e.g. Calcium, Phosphorus, Iron, Iodine, Sodium, etc.
  - **Water**
- **Vitamins**

The material of which plant food material is composed consists in the first place of water and dry matter. If the material is heated to 120°C for some time, depending on the nature of the material, the water is driven off and the residue is the dry matter. It is the dry matter which forms the basis on which chemical analyses are performed in order to express the percentage of the different groups of nutrients present in a feeding stuff. For example, if a pasture grass contains 90 per cent. of water, and chemical analysis reveals that it contains 1 per cent. of calcium on the oven dried material then the grass as it stands in the field would obviously contain 0.1 per cent. of calcium by weight.

No physical manifestations of life are possible without water, so it may be considered as *primus inter pares* among the constituents comprising the plant. The amount of water present in feeding stuffs varies considerably. Wheat grown in a hot dry region is drier than wheat from moister areas, but contains on an average about 10 per cent. of water; succulent feeding stuffs such as root crops may contain anything above 75 per cent. of water and are called succulent in distinction to those of a water content below this figure, *viz.*, the non-succulent feeds. Water plays an important mechanical part in supporting the plant structure by virtue of the osmotic pressure set up in the plant cells by the salts in solution. It also acts as a medium for transporting food material, and is essential for the proper functioning of cellular life.

ORGANIC

Nitrogenous Substances

Feeding stuffs contain a great variety of nitrogenous substances of varying degrees of chemical complexity; these have been divided into the proteins proper, and those substances which, although containing nitrogen in their composition are not true proteins. Proteins contain on an average 16 per cent. of nitrogen, so when the chemist wishes to determine the protein content of a feeding stuff he finds out the amount of nitrogen it contains, and then multiplies that figure by 6.25 ($6.25 \times 16 = 100$).

Such an estimation reveals what is called the total or crude protein. A more refined analysis subdivides the total or crude protein into true protein and the non-protein nitrogenous substances, sometimes referred to as the amides.

True protein has a greater nutritive value than the amides, and consequently an analysis which calculates all the nitrogen as though it were protein proper, will give a somewhat higher nutritive figure than is justified. The relative proportion of true to crude protein in feeding stuffs varies considerably, and in some young succulent fodders may be as high as 50—50. The amount of non-protein nitrogen, however, is usually very much smaller than this. Reference to this subdivision will be made again in Chapter V.

What Proteins Are ?

The farmer is continually having to take into account the nature of the food he has to give his animals, and particularly the protein foods, and it is important that he should have some knowledge of the relative values of different protein foods, and why some proteins are so much more valuable from a feeding standpoint than others, as this has a close relationship to the desirability of feeding mixed concentrates instead of a single one.

The chemistry of the proteins is very complex and has by no means been fully worked out yet, although the chemical structure of some of the ultimate constituents of which the protein molecule is composed is well known. The study of the proteins may be made either from an analytical aspect, *i.e.*, breaking them down and discovering the nature of their component parts, or by the synthetic method, *i.e.*, endeavouring to build up substances resembling proteins from much simpler substances of which they are known to be composed.

In order to illustrate briefly the chemical nature of the proteins, examples will be taken from the synthetic method of attempting to elucidate their ultimate structure. A few simple examples of chemical substances and their formulae will serve as a useful prelude both to an understanding of the chemical nature of the proteins and also of the other constituents of feeding stuffs.

One of the simplest chemical formulae is that for water, *viz.*, H_2O, which tells us that a molecule of water, that is the smallest particle of water that can exist as such, contains two atoms of hydrogen and one atom of oxygen. Another way in which the formula for water could be written is H.OH, which shows how the various atoms are linked together. This is called the structural formula and shows in this case, that one of the hydrogen atoms and one of the oxygen atoms act so to speak as a group, which can be displaced by certain other atoms or groups of atoms. For example if this OH group is turned out of the water molecule and replaced by chlorine, a yellow poisonous gas, hydrochloric acid (HCL) results.

A slightly more complex substance may now be taken before proceeding to study the foodstuffs proper.

Acetic acid is familiar to everyone as the acid ingredient in vinegar. The smallest particle of this acid which can exist as such, *i.e.*, a molecule, has a formula $C_2H_4O_2$, which means that one molecule of acetic acid contains two atoms of carbon, four atoms of hydrogen and two atoms of oxygen, all totally different substances which when chemically united in a certain way and in this proportion give a colourless liquid, acetic acid. Another way in which this formula can be written, as in the case of water, is $CH_3.COOH$, which shows its structure or division into groups, each of which can act as an entity as though it were a single atom.

Acetic acid is one of a group of related acids each of increasing complexity and formed by adding an additional CH_2 group to its predecessor. These are known as the fatty acids and have to be considered both in regard to the chemistry of the proteins, and the fats.

The first member of the group is formic acid (H.COOH), well known as the substance in stinging nettles which sometimes causes trouble. The next is acetic acid, ($CH_3.COOH$), mentioned above—this is clearly H.COOH with one added CH_2 group,—the next is propionic acid, C_2H_5COOH, and so on for a whole series until we come to the higher fatty acids, which will be referred to in greater detail in connection with the fats.

	Palmitic Acid,	$C_{15}H_{31}COOH$
and	Stearic Acid,	$C_{17}H_{35}COOH$

are two such examples, and their relationship with acetic acid and the added CH_2 groups is quite clear.

We may now return to acetic acid for an introduction to the proteins. It is possible to replace one of the hydrogen atoms in the CH_3 group by a group of atoms, NH_2, called the amino group, by treating acetic acid with ammonia, NH_3, or as it may be written, $N \leftarrow \begin{matrix} H \\ H \\ H \end{matrix}$ with the formation of an amino acid and liberation of hydrogen as a result.

This reaction is shown in equation form as follows:

$$\underset{\text{(acetic acid)}}{CH_3.COOH} + \underset{\text{(ammonia)}}{NH_3} = \underset{\text{(amino acetic acid)}}{CH_2.(NH_2)COOH} + \underset{\text{(hydrogen)}}{H_2}.$$

Another name for the amino acetic acid thus produced is glycine. Some of these amino acids are very simple, like glycine, others are very complex; some contain only carbon, hydrogen, oxygen and nitrogen in their molecules, while others contain in addition, sulphur or phosphorus. Glycine, however, is the simplest of them all, and they are important because they constitute the actual substances of which all proteins are composed.

We may now consider another amino acid of slightly greater complexity. The next fatty acid in the series after acetic acid, is propionic acid, $CH_3CH_2.COOH$. If now an NH_2, or amino group, is introduced into this as was done for acetic acid $CH_3.CH(NH_2).COOH$ known as amino propionic acid, or alanine, which is another and very important amino acid present in all proteins is obtained.

The amino acids are of fundamental importance in stock-feeding in connection with protein metabolism and some more complicated examples may now be given to illustrate their nature more fully.

It would have been noticed that alanine, unlike glycine, has its CH_3 group intact. If one of the hydrogen atoms of this CH_3 or methyl group as it is called, is replaced by benzene (C_6H_6) the following equation shows what happens:—

$$\underset{\text{(Alanine)}}{CH_3.CH(NH_2).COOH} + \underset{\text{(Benzene)}}{C_6H_6} = \underset{\text{(Phenyl alanine)}}{C_6H_5.CH_2.CH(NH_2).COOH} + H_2$$

Phenyl alanine is yet another amino acid, a little more complicated than the former, but not yet very formidable.

It will be noticed in the above equation that one of the six hydrogen atoms in benzene has been removed and also one of the hydrogen atoms in the CH_3 group of amino propionic acid. To make what follows clear, it may be suggested that benzene C_6H_6 can be written thus:—

$$CH.CH.CH.CH.CH.CH.$$

This is not actually so and benzene behaves as though the above chain formula was taken and closed in the form of a circle or ring, such as this:—

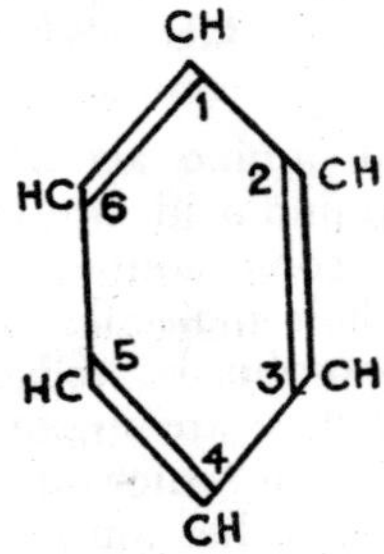

in much the same way as though the two ends of a piece of string were curved round and joined together. We can now at once see that if two or more of the hydrogen atoms in benzene are replaced by other atoms or groups of atoms, the resulting product will depend on which of the hydrogen atoms in respect to any other are so replaced, that is to say,

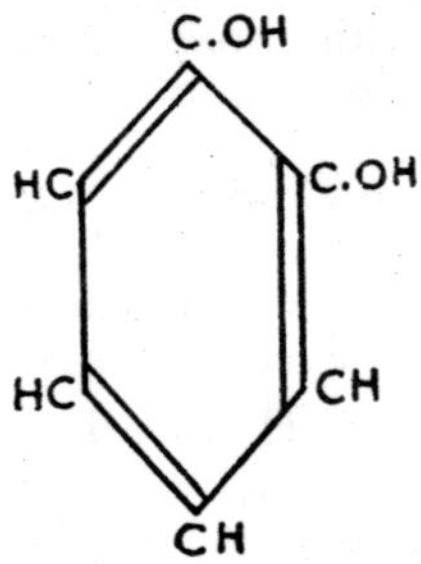

will be quite a different substance from

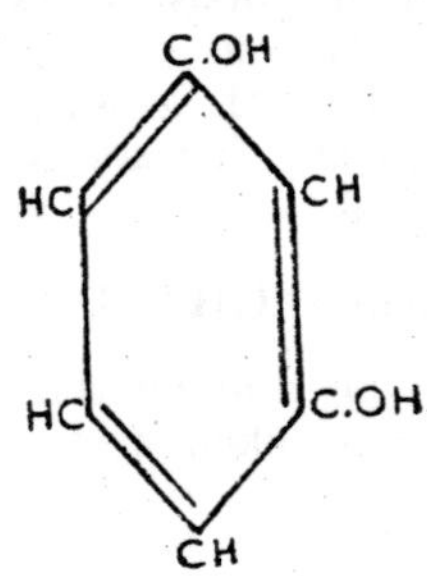

It will now be seen that phenyl alanine can be written thus:

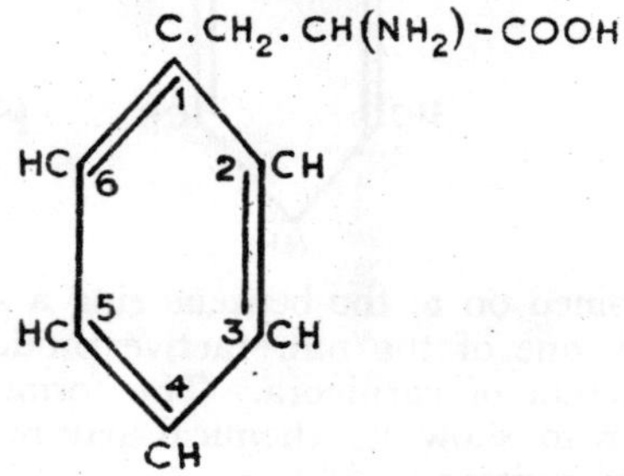

Still another amino acid can be formed from phenyl alanine if the hydrogen atom in the No. 4 CH group is replaced by an OH or hydroxyl group. If this is done tyrosin results, *viz.*,

$$C_6H_4.OH.CH_2.CH.(NH_2).COOH$$

or, as described above, this may be written thus:—

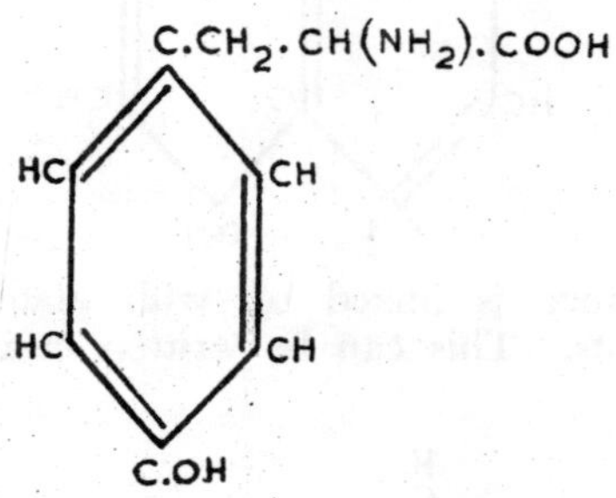

Tyrosin is an amino acid of fundamental importance in the proteins of feeding stuffs and is indispensable to growth; the same remarks apply to the last example of amino acids which will now be given, *viz.*, tryptophane, which again is more complex than any of the former.

In order to understand the nature of tryptophane we must refer to pyrrole, a substance related to the alkaloids, with a formula C_4H_5N, and which, like benzene, has to be written in the form of a closed ring to show its chemical nature, thus:—

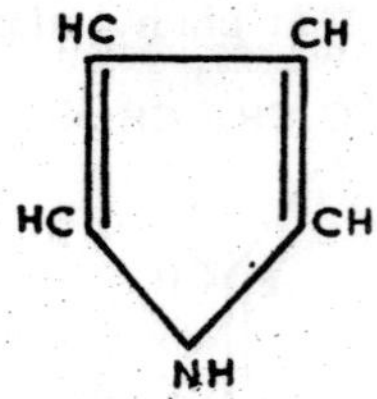

Pyrrole

If pyrrole is joined on to the benzene ring a substance called indole is obtained, one of the putrefactive products of proteins present in the excreta of carnivora. The formula of indole is C_8H_7N. In order to show its chemical nature and structural formula it may be written:

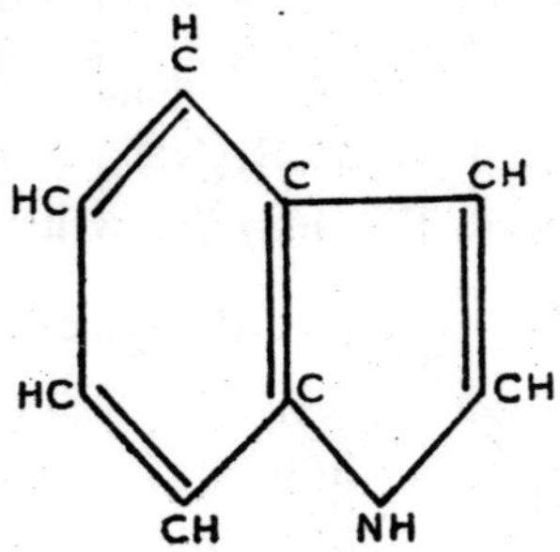

Indole

If this substance is linked up with alanine, tryptophane $C_{11}H_{12}O_2N_2$ results. This can be written in its full structural form as follows:—

Tryptophane

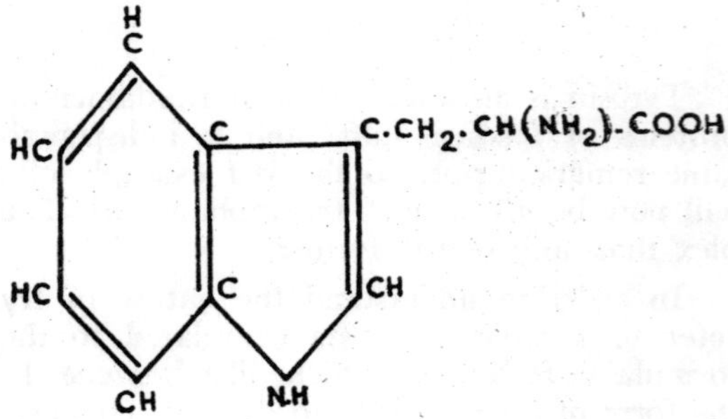

Another way of regarding it is to consider one of the hydrogen atoms of the CH_2 group in propionic acid as being replaced by the amino group NH_2, and one of the hydrogen atoms of the CH_3 group as being replaced by indole. Thus the chemical name for tryptophane is ζ amino β indole propionic acid. In the fatty acids the CH_2 group nearest the COOH group is called the ζ group, the next the β group, then the γ group and so on, using Greek letters as shown below for propionic acid.

$$\underset{\beta}{CH_3}.\underset{\zeta}{CH_2}.COOH$$

So far the origin and chemical nature of only five amino acids have been given, *viz.*,—glycine, alanine, phenyl alanine, tyrosin and tryptophane. There is a large number however, formed by permutations and combinations of different substances of varying degrees of complexity. Chemists have been able to combine some of the amino acids together to form other substances in a regular series as will now be briefly described.

Two molecules of glycine can be combined together with the loss of one molecule of water, to form a diamino acid or dipeptide as it is called. When a number of amino acids are thus combined the resulting substance is called a polypeptide. More complex bodies built up on the same system are known as peptones. Then as complexity increases the proteoses are obtained. The final protein has never yet been synthesized by this joining together of simpler bodies, although the famous German chemist, Emil Fischer, succeeded in building up in the laboratory a substance containing many hundreds of atoms from simple amino acids, and although this substance did not conform to any known protein, it nevertheless possessed the properties of a protein.

The significance of what has been said above in regard to the complexity and variability of the proteins will be apparent. All proteins are built up from amino acids, and differ among themselves according to the number and complexity of the amino acids they contain, and although proteins have not actually been synthesized in the laboratory, they have been broken down and their individual amino acids studied. Once it is appreciated how complex and varied the proteins are by virtue of the various amino acids they contain, the importance of the quality of the proteins fed to animals will be understood and the necessity for a mixed ration.

For example, the protein zein of maize is lacking in two of the amino acids essential for growth, *viz.*, tyrosine and tryptophane so that while maize is an excellent food when properly balanced by others it would be disastrous to feed a growing animal with maize as the sole source of protein. These simple facts can be put differently, thus. Different proteins contain different amino acids. The flesh of animals also consists largely of protein, or in other words, amino acids. The animals' food, therefore, must contain proteins which in the course of digestion will furnish those amino acids necessary for building or maintaining the animals' tissues. Hence the term 'biological value' as applied to proteins and the necessity to take the quality of proteins into account as well as the quantity.

We have thus seen that proteins are built up from amino acids, some being more complex than others, but all different. Some lack essential amino acids many contain them. Some supply phosphorus or sulphur in addition to carbon, hydrogen, oxygen and nitrogen. The different types of protein will now be reviewed.

Classification of the Proteins

Owing to our imperfect knowledge of their structure, and the extremely complicated nature of the proteins, it has not been possible to classify them according to any scheme based on chemical composition, such as is the case with the fats and carbohydrates. A somewhat arbitrary scheme, however, has been adopted, depending partly on physical and chemical properties, and partly on their source. This classification, adopted by the British and American Physiological Societies is as follows:

The Simple Proteins

A. The Albumins

These (with the globulins) are among the most widely distributed proteins both in the animal and vegetable kingdoms. They are found in the lymph and blood stream, in eggs and milk as egg-albumin and lact-albumin, and they are also present in small quantities in many seeds, such as wheat, barley, gram, peas, soybeans, etc. They are soluble in pure water and are coagulated by heat.

B. The Globulins

These are insoluble in distilled water but soluble in dilute salt solutions and are coagulated by boiling. They are distributed similarly to the albumins, but are much more prevalent in seeds, in different species of which Osborne [1909] has enumerated 39 different globulins.

C. The Glutelins

The most typical members of this group are the glutelin of wheat and orzyenin of rice. Others exist in other seeds and vegetables but they are not well defined. The glutelins are insoluble in neutral solvents, but are soluble in acids and alkalis. When a globulin changes its solubility and is changed into a protein such as the myosin of muscle it is classified as a 'derived protein', although nothing is known of the chemical changes which have come about, and although the resulting protein still has all the properties of the glutelins.

D. The Prolamines (Gliadins)

These are the characteristic proteins of the cereal grains, the chief of which are the gliadin of wheat and rye, the zein of maize and the hordein of barley. They are also found in vegetables, and are distinguished from the glutelins by their solubility in 75 per cent alcohol. They are, however, insoluble in water, absolute alcohol and other neutral solvents.

E. The Protamines

These are basic substances which contain a high percentage of nitrogen but no sulphur, and are the characteristic proteins of ripe sperm cells and ova, and are apparently associated with nucleic acid in chromatin. They are easily soluble in water, insoluble in alcohol and ether, do not coagulate, and cannot pass across semi-permeable membranes such as parchment paper. Pepsin does not affect them but they are readily digested by trypsin and erepsin.

One of the best known of the protamines is salamin of the ripe salmon sperm. It is said to have a formula, $C_{81}H_{155}N_{45}O_{18}$

i.e., with a molecular weight of 2045 or some multiple of this, and contains over 30 per cent nitrogen.

F. The Histones

The histones are found in unripe spermatozoa and are the characteristic proteins of cellular tissues such as the glands. They are also found in the red blood corpuscles, the protein of heamoglobin being a histone combined with a non-protein radicle, haemin. They are precipitated by ammonia, form salts with the strong mineral acids, and combine readily with nuclein or nucleic acid.

G. The Albuminoids (Scleroproteins)

These belong to a heterogenous class of proteins found in the skeletal structure and connective tissues of animals and are characterized by their great insolubility in all neutral solvents. They are generally resistant to chemical reagents and to peptic and tryptic enzymes. Included in the group are gelatin, collagen, which forms the basis of cartilage and bone, elastin found in ligaments and the keratins of epidermal tissues such as hair, wool, etc. Some of them appear to contain but few amino acids.

For example, raw silk is a mixture of two proteins belonging to this class, silk gelatin and silk fibroin. The former is peptized by water under slight pressure and by dilute alkali solutions and constitutes from 15 to 20 per cent. of the weight of the silk. Silk fibroin has a very high content of three particular amino acids, *viz.*, glycine, alanine and tyrosin, which make up about 70 per cent of the raw silk. Silk gelatin is low in glycine content.

An interesting member of this group is spongin which constitues the skeletal protein of sponges and coral. Spongin contains no tyrosin, but a di-iodo salt of tyrosin, the iodine content being from 1 to 1.5 per cent. It is interesting to record in this connection that the ancient Greeks considered that ground sponges had a beneficial action on goitre. This, like many other crude beliefs of long ago, was ridiculed in scientific quarters till it was finally discovered that sponges contain a compound of iodine and tyrosin, which has proved to be an effective preventive of goitre, and dried ground sponges have found a place in the French Pharmacopoea as an official remedy. Many of the keratins contain a high percentage of cystein, a sulphur containing amino acid, and human hair which is richer in this than wool, is generally used for preparation of cystein.

Gelatin is worthy of special note as it appears to be the only protein which does not give specific immunological reactions, and can accordingly be injected intravenously without producing deleterious effects common to other proteins, such, for example, as snake venoms which are protein in nature. Thus in the first world war gelatin was used in cases of hæmorrhage, and was responsible for saving many lives, as it was capable of replacing the blood proteins which had been lost and helping to hold liquid in the blood vessels, and thus keep the volume of the blood within normal limits.

The Conjugated Proteins

A. The Nucleo-proteins

The members of this group consist of a simple protein such as a histone or protamine combined with nucleic acid, and are especially characteristic of both plant and animal cell nuclei. Their special point of interest is the phosphorus containing nucleic acid which, on being broken down, yields phosphoric acid with purine bases, pyrimidine bases and a pentose carbohydrate. They are soluble in dilute alkalies but are precipitated by acids.

B. The Phosphoproteins

These are combinations of proteins with a phosphorus containing substance other than nucleic acid, and must be distinguished from the nucleoproteins. Typical examples are the caseinogen of milk and vittelin of the yolk of eggs.

C. The Gluco or Glycoproteins

The glucoproteins are found chiefly in mucins and the white of eggs, and are combinations of a protein with a carbohydrate group. They are soluble in dilute alkalies but not in acids.

D. The Chromoproteins or haemoglobins

These are compounds of protein with haematin or some similar substance and constitute the red colouring matter of blood-hæmoglobin.

THE NON-PROTEIN NITROGENOUS SUBSTANCES

Although all proteins contain nitrogen, not all nitrogen containing substances are proteins, and the existence is known of a large class of bodies in feeding stuffs which may sometimes account for one third to one half of the total nitrogen present; hence it is necessary to distinguish between these and the true proteins. They are much less complicated in structure than the true proteins and of considerably less nutritive value. Many are crystalline and readily soluble in water.

Our knowledge of their exact nature is still incomplete but their more important groups are as follows:—

A. Nitrogenous muscle extractives

Such as creatin, creatinin and the purin bodies, xanthin and hypoxanthin, which form the basis of meat extractives and possess little or no nutritive value.

B. Nitrogenous Lipoids

These contain phosphorus, of which the lecithins are an example. The actual amount in plants and animals is small.

C. Amino acids and amides

These constitute the most important part of the non-protein nitrogenous bodies. Typical examples of the amides are, asparagin, ($COOH.CH.NH_2.CH_2CO.NH_2$), and glutamin, ($COOH.CH_2.CH_2.CH.NH_2CO.NH_2$), which are respectively the amides of aspartic acid ($COOH.CH_2.CH.NH_2.COOH$), and glutamic acid ($COOH.CH_2.CH_2.CH_2.CH.NH_2.COOH$), both of which are constituents of the protein molecule. Although the non-protein nitrogenous substances are sometimes referred to as amides, this is not strictly correct, as the term amide refers to a special type of compound of which asparagin and glutamin are the only two found to any extent in feeding stuffs. Most of the other substances in this category are not true amides at all, but are comprised of many amino acids and purine bases which make up the greater part. The non-protein nitrogenous substances are especially abundant in young growing plant tissues such as fresh pasture, green fodders and soiling crops generally, which accordingly contain a considerable part of their nitrogen in the form of non-proteins. As the plant matures this proportion decreases progressively so that hay is relatively poor in non-protein nitrogen. Roots and tubers, however, are an exception and contain a considerable part of their nitrogen in the non-protein form as a reserve for future growth. Silage which has undergone fermentation also shows an increase of non-protein nitrogen over the original material from which it was made.

D. Alkaloids and organic bases

Alkaloids are comparatively rare in plants although organic bases are more widely distributed.

E. Glucosides

These bodies are very common in the vegetable kingdom and contain nitrogenous compounds combined with simple sugars. They have little or no nutritive value, but under the influence of enzymes they may release hydrocyanic acid (prussic acid), which, as is well known, is a deadly poison. The sorghums when grown under drought conditions are particularly liable to develop glucosides to such an extent as to be fatal to cattle which eat them. (See also Chapter VIII.)

Carbohydrates

The carbohydrates, as their name suggests, are composed of carbon, hydrogen and oxygen, the latter two being in the same proportion as in water; hence on complete combustion only water and carbon dioxide result.

In analysing feeding stuffs the carbohydrates are divided into two main groups, *viz.*, the crude fibre which remains after treating the material with dilute acid and alkali, and the soluble carbohydrates, or 'nitrogen free extract' as it is called, which pass into solution. The former is generally supposed to be indigestible although a certain proportion is actually digested, or more accurately fermented by bacteria in the intestinal tracts of animals, particularly sheep and other ruminants.

Foods are classified as concentrates or roughages mainly on the basis of their crude fibre content, concentrates containing less than 15—20 per cent, and roughages more than this. Obviously the nutritive value of a food will largely depend on the relative proportions of these two groups. Crude fibre consists mainly of celluloses, lignin, pentosans, etc., and performs a useful function in giving bulk to a ration and facilitating the digestive processes.

Soluble Carbohydrates or Nitrogen Free Extracts

This part of the foodstuff is not directly estimated in a chemical analysis, but is obtained by subtracting the sum of the crude fibre, fat, crude protein, ash and water from the whole. The carbonaceous foods such as the cereals, maize, root crops, etc., contain a high percentage of nitrogen free extract (N.F.E.) which varies considerably in different foodstuffs, and comprises the following main groups:

1. *The Monosaccharides*

These are simple sugars and include the pentoses, ($C_6H_{10}O_5$), and the hexoses ($C_6H_{12}O_6$). The pentoses are widely distributed in nature and differ from the hexoses in not being fermented by yeast. The hexoses comprise:

(a) Dextrose (or glucose or grape sugar) widely found in plant juices and root crops. It possesses the property of turning a ray of polarised light to the right.

(b) Laevulose (or fructose or fruit sugar), which turns a ray of polarised light to the left, but to a greater extent than dextrose turns it to the right, and is especially abundant in honey and plant juices where it occurs along with dextrose.

(c) Galactose and mannose are isomers of dextrose and occur in nature only in combination as di- or polysaccharides.

2. *The Disaccharides* ($C_{12}H_{22}O_{11}$)

The most important of these are:

(a) Cane sugar (or sucrose) found chiefly in sugar cane and beet juice, in the sorghums, and in plant juices generally. Under the influence of acids or alkalies, or adverse biological conditions in the plant it becomes split up into dextrose and laevulose, thus:

$$\underset{\substack{\text{cane sugar}\\\text{(dextro rotatory)}}}{C_{12}H_{22}O_{11}} + \underset{\text{water}}{H_2O} = \underset{\substack{\text{dextrose}\\\text{(dextro rotatory)}}}{C_6H_{12}O_6} + \underset{\substack{\text{laevulose}\\\text{(laevo rotatory)}}}{C_6H_{12}O_6}$$

(b) Lactose or milk sugar is characteristic of the milk of mammals, but is less soluble and less sweet than sucrose. During digestion or acid hydrolysis it yields galactose and dextrose:—

$$\underset{\text{lactose}}{C_{12}H_{22}O_{11}} + H_2O = \underset{\text{galactose}}{C_6H_{12}O_6} + \underset{\text{dextrose}}{C_6H_{12}O_6}$$

(c) Maltrose is produced by the action of certain ferments on starch during the germination of seeds, and hence is abundant in malt as an intermediary between starch and alcohol. It is also produced in the course of digestion of starch by animals. (See Chapter III.) During the fermentation of starch the first products are dextrins, which in turn are converted into maltose, and finally the latter is changed into dextrose thus:

$$\underset{\text{maltose}}{C_{12}H_{22}O_{11}} + H_2O = \underset{\text{dextrose}}{C_6H_{12}O_6} + \underset{\text{dextrose}}{C_6H_{12}O_6}$$

3. *The Polysaccharides*

These comprise a wide variety of different substances of unknown molecular composition but all conforming to the general formula $(C_6H_{10}O_5)n$, where n is unknown. Many of them such as the dextrins, glycogen (liver sugar), and inulin found in dahlia roots, are tasteless amorphous substances, and are soluble in water, while some, such as the celluloses and starches, are insoluble in water.

(a) The starches. These constitute one of the most important reserves of food material stored in the seeds, leaves and roots of plants, and each plant produces its own characteristic type, distinguishable from others by the particular conformation and markings of its granules. This readily enables flours and foods produced from different starchy materials to be identified. Seeds possess enzymes which under favourable conditions of temperature and moisture convert the starch into sugar, *viz.*, maltose, and eventually into glucose, for the nutrition of the young seedling. This is a parallel to the digestion in the animal alimentary tract to be described later. (Chapter III.)

(b) Cellulose and Lingin. Cellulose forms the basis of the supporting walls of plant cells and the general skeleton of the plant. Cotton fibre is an elongated plant cell of almost pure cellulose, from which most or all extraneous protein matter, etc., has disappeared. As the plant matures the cell wall thickens and develops more cellulose intermingled with a group of numerous other substances of tougher nature allied to cellulose and known generally as lignins. These contain more carbon than cellulose and may be separated from the latter by oxidizing agents.

Thus, as the plant matures the cellulose becomes changed in character and interspersed with tougher and less digestible material. In the analysis of a young plant, therefore, the major part of the cellulose may be accounted for in the nitrogen free extract, and but little in the crude fibre portion, while later the reverse is the case.

Pure paper is almost pure cellulose, while wood in which lignin predominates may be reconverted into cellulose for paper making by appropriate treatment.

(c) The Hemicelluloses. These bodies differ from true cellulose in that they are hydrolysed when boiled for only a short time. This results ultimately, not only in the production of dextrose (as with cellulose) but also of a variety of other hexose sugars such as galactose, mannose and laevulose and pentose sugars such as arabinose and xylose ($C_5H_{10}O_5$) in addition. While the celluloses proper constitute the framework of the cell, the hemicelluloses which contain both hexosans (producing hexose sugars) and pentosans (producing pentose sugars) may be considered as furnishing a reserve of plant food material. In an ordinary analysis of feeding stuffs they appear partly in the crude fibre and partly in the nitrogen free extract.

(d) The Dextrins. When starch is hydrolysed a series of intermediate bodies of varying composition called the dextrins is formed. They may be considered as intermediate between starch and the simple sugars, and are also produced in the plant by the action of cellular enzymes. They are also produced during the cooking of starchy foods, and the brown crust of bread is largely dextrin, and hence more digestible than the white undextrinized portion.

(e) Glycogen is an approximate counterpart in the animal body to dextrins in the plant. It is a substance derived from starches and sugars and is stored in the liver as a reserve sugar to be called upon by the body when needed.

FATS AND RELATED BODIES

In performing an analysis of a feeding stuff the material after preparation is extracted with ether or other suitable solvent, and when the solvent is evaporated from the extract, the residue is recorded as ether extract or simply 'fat'. The fat content of a feeding stuff is of importance to the stock breeder, but the fat extracted during analysis and recorded as such contains other substances, which are not true fats, such as the lipoids, gums, waxes, etc., which are of little or no value as food. The proportion of these, however, is not usually large enough to be of great importance. In order to understand the nature of a fat it is necessary to revert to one or two simple chemical illustrations. Earlier in this Chapter there was mentioned in connection with the proteins, the series of bodies known as the fatty acids, starting with formic acid (H.COOH)), then acetic acid (CH_3COOH), and so on till the higher fatty acids such as palmitic acid ($C_{15}H_{31}COOH$) and stearic acid ($C_{17}H_{35}COOH$) are arrived at. The two latter enter into the constitution of hard fats, while soft fats and oils are characterized by a different acid, *viz.*, oleic acid ($C_{17}H_{33}COOH$), which is called unsaturated because it is capable of having more hydrogen atoms put into its molecule. The process of hydrogenation of oils by which liquid vegetable oils are converted into hard vegetable fats performs this operation. The soft oleic acid containing oils can thus be changed into hard stearic fats.

If now a fat is taken and treated with soda (sodium hydroxide) or potash (potassium hydroxide), the resulting product consists of a mixture of the sodium or potassium salts of fatty acids, depending on the fat, and glycerine. Glycerine is known as a trihydric alcohol because it contains three hydroxyl or OH groups. Its formula is $C_3H_5.(OH)_3$ or $C_3H_5 \leftarrow \begin{matrix} OH \\ OH \\ OH \end{matrix}$

This operation on the fat can therefore be expressed as follows:

$$C_3H_5(C_{17}H_{35}COO)_3 + 3\ KOH = C_3H_5 \leftarrow \begin{matrix} OH \\ OH \\ OH \end{matrix} + 3\ C_{17}H_{35}COOH$$

fat or tristearin potash glycerine potassium stearate or soap

The equation shows that an alkali salt of the fatty acid, or soap, and glycerine are formed by the action of alkalies on fats.

The oils from oilseeds such as cotton seed, linseed, rape, etc., contain a high proportion of oleic acid in contrast to animal fats and some vegetable fats which contain more stearic and palmitic acids.

Butter is characterized by its high content of those fatty acids which are soluble in water, *viz.*, butyric, capric and caprylic, and also oleic acid. It is of importance to stock breeders to note that excessive feeding of feeds rich in the soft fats or oils, tends to produce in animals a soft fat which is specially undesirable.

In addition to the ordinary fats present in plants and foods, free fatty acids may also be present in varying degree specially in the early stages of growth. These are also likely to develop in feeding stuffs which have become damp and mouldy, due to an enzyme, lipase, present in tissue cells, which splits the fat into free fatty acid and glycerine. These free acids impart the characteristic musty odour and rancid taste to such materials.

The oilseeds are particularly rich in fat or oil which contains a far greater variety of fatty acids than most animal fats. Linseed may contain over 36 per cent oil, cotton seed over 20 per cent, while the oil content of palm nut kernels may be as high as 50 per cent. It is these high percentages which render large quantities of native oilseeds unsuitable for rations, and it is more usual to feed the oilseed cakes which contain only about 5-6 per cent oil. This will be referred to again in Chapters IX-XV.

The vegetable oils can be divided into three main groups according to their capacity for or rapidity in drying. The drying oils which dry readily include linseed and sunflower oils, the former being particularly rich in the unsaturated linolic and linolenic acids. The non-drying oils include olive oil, some rape oils and others, while the semi-drying oils include cotton seed oil and sesame oil.

Related substances

In addition to the fats and oils present in plant tissues there are also present other substances of a fatty nature, including the waxes, and lipoids or sterols.

Waxes are the fatty acid esters of alcohols other than glycerine, that is to say compounds of glycerine and alcohols. Spermaceti is a typical example and consists chiefly of the palmitic ester of cetyl alcohol $CH_3 (CH_2)_{14} CH_2 OH$.

Beeswax is a true wax containing the palmitic ester of myricyl alcohol $CH_3 (CH_2)_{29} CH_2OH$. Lipoids, although fatty in physical properties are chemically alcohols, such as cholesterol ($C_{27}H_{45}OH$), found in animal tissues, and its counterpart the phytosterols found in the vegetable kingdom, such as sitosterol ($C_{27}H_{45}OH$).

An important member of the sterol group is ergosterol ($C_{28}H_{43}OH$), closely allied chemically to cholesterol, and which, on irradiation by sunlight or ultra violet light, constitutes vitamin D.

INORGANIC

MINERALS

That part of the plant or feeding stuff which is left behind after careful combustion during which all the water is driven off and the organic matter completely burnt away is called the ash.

In the past, attention has been chiefly focused on the fat, carbohydrates and protein, or energy yielding parts of the feed, and little heed paid to the minerals which constitute a considerable percentage of the oven dried material, and about 3 per cent of an animal's body weight.

A considerable number of minerals is present in feeding stuffs, varying in quantity according to their nature. These include calcium, phosphorus, sodium, potassium, iron, magnesium, manganese, sulphur, chlorine, silicon and also to a lesser extent zinc, copper, iodine and no doubt microscopic amounts of rarer elements. It is not possible to say to what extent any or all of these exist freely as salts in solution in the cell sap, or as a part of complex organic molecules. It has, however, come to be realized in recent years that minerals play a fundamental part in the life of the plant, and what is perhaps more important, in the life of the animal. The owner of stock is too prone to concentrate attention on the protein and energy requirements of his animals and neglect the quantity and quality of the minerals. This onesided outlook has in the past no doubt contributed much to ill-health and disease among stock, and much more attention should be given to mineral requirements if animals are to be maintained in health and productive capacity free from the numerous and often ill-defined, malaises associated with the term malnutrition.

It is known that a considerable number of minerals is essential for animal health and that these must not only be adequate in quantity, but well balanced. Naturally, these factors will vary with the purpose for which feeding stuffs are fed. For example, they will be different for a young growing animal, or a fully grown one and so forth. This is an important point to be taken into account in regard to mineral mixtures which are so prolific on the market. These mixtures are not necessarily made up with regard to any specific crop deficiencies or special mineral adjuncts which may be needed, and unless such factors are taken into consideration, the indiscriminate use of unspecified mineral mixtures may possibly be deleterious.

It is likely that mineral deficiencies or defects in crops may not be reflected in adverse conditions of animal health for some considerable time, and perhaps too late to remedy them, although deficiencies in iron, calcium and phosphorus are likely to be among the earliest to become evident.

A most important matter is the relative proportion of various minerals—or their 'balance'. For example, certain minerals are basic in character, such as lime and potash, while others are acidic, *e.g.,* phosphoric acid, and an excess of the basic over the acidic elements or *vice versa* in the ration may have deleterious effects. It is essential that the pH of the blood (*i.e.,* the hydrogen ion concentration) should be kept within narrow limits, and a satisfactory acid-base mineral equilibrium in the rations is necessary for this purpose. Again, apart from the ratio of total acids to total bases the inter-relationship between various minerals is of great importance. This is particularly so in regard to calcium and phosphorus, and sodium and potassium. Should potassium be in excess in the food there is a tendency for sodium to be excreted in excess in the urine. It follows therefore that what might otherwise be a satisfactory sodium content may become a deficiency in the presence of an excess of potassium.

It is not easy to say what constitutes the most suitable balance between the acid and basic groups, but a reasonable guide is to take the ratios as found in milk. These are as follows:

Acid equivalents	: Base equivalents	= 1.1 : 1
Phosphoric acid (P_2O_5)	: Lime (CaO)	= 1.5 : 1

Ratio between soda and potash

Soda (Na_2O)	: Potash (K_2O)	= 0.7 : 1

As an example, taking these figures as a standard for reference, the corresponding figures for berseem grown at Lyallpur are:—

P_2O_5	: CaO = 0.20	: 1
Na_2O	: K_2O = 0.082	: 1

Corresponding ratios can be worked out for other rations or combinations of rations and by a judicious admixture a satisfactory balance may be obtained. It is important in this connection to note that the relative amounts of the different minerals vary with the stage of growth of a crop, and the climatic features of the locality in which it is grown. This also applies to successive cuttings for a particular fodder. Examples illustrating this will be given in Chapter VI, and it will be noted that the calcium and phosphoric acid (and also the protein content) of fodders vary very considerably from month to month during growth.

Potassium

Potassium is found in all parts of the plant but especially in active growing parts and is indispensable for plant growth.

Sodium

Sodium is not indispensable, but is usually found associated with potassium in the stems and leaves of plants, but to a lesser extent.

Calcium

Calcium is found chiefly in the leaves and stems of plants but sparsely in the seeds. Legumes, leafy salads and some vegetables are particularly rich in calcium which is essential for plant growth. It occurs in all parts of the animal body but especially in the skeleton, the bones of which are composed almost entirely of calcium phosphate. The proper utilization of calcium is intimately bound up with vitamin D as will be described later under Vitamins.

It is worthy of note that the calcium in the ration is generally utilized more efficiently than the phosphorus. An interesting point, which will be enlarged on in Chapter VII, is the considerable variation which occurs in the calcium-phosphorus content of plants during growth, the calcium and phosphorus on a dry basis being greater during the early stages than later on. Also the variations in calcium are far greater than with phosphorus. This has an important bearing on the relative merits of controlled grazing versus stall and hay feeding.

Non-legume roughages contain less calcium than legumes, and both are affected by calcium deficient soils; cereal straws contain even less. Maize and sorghum silages approximate to ordinary grass hays in calcium-phosphorus content, while roots and tubers are invariably low, especially in calcium. In fact it may be said that none of the common roughages are rich in phosphorus as is shown in Appendix I.

All cereal grains are very low in calcium, varying from 0.06 per cent for maize, and 0.19 per cent for wheat, to 0.31 per cent for oats, but they are richer in phosphorus—maize 0.84 per cent, wheat 0.70 per cent and oats 0.81 per cent.

Thus most of the protein rich concentrates of plant origin are much higher in phosphorus than the grains and roughages, but none of them are rich in calcium. Wheat bran is specially rich in phosphorus containing 1.97 per cent, cotton seed meal 1.2 per cent and linseed meal 1.0 per cent. The legume seeds are not as a rule very rich in phosphorus, soybeans having 1.64 per cent, soybean meal 0.66 per cent and groundnut oil meal 0.99 per cent.

Corn gluten meal and distillers' grains are relatively low in phosphorus containing 0.7 per cent and 0.4 per cent respectively. Milk being a natural food is rich in both calcium and phosphorus, while fish meals are among the richest of all common feeding stuffs in both calcium and phosphorus. White fish meal has 10.50 per cent. calcium and 9.0 per cent phosphorus.

The approximate percentages of calcium and phosphorus in common feeding stuffs are given in Appendix I.

Ratio of Calcium to Phosporous

The importance of sufficient calcium and phosphorus in feeding stuffs has been stressed, but the proportions in which they occur is also of significance. If there is a great excess of one over the other, detrimental effects may be produced, even though the smaller moiety be present in otherwise adequate amounts. It is difficult to say precisely how this adverse effect is produced but it appears that the efficiency with which vitamin D can carry on its controlling function in calcium metabolism appears to be upset with an unsatisfactory calcium-phosphorus ratio. A satisfactory ratio is 1 to 2 parts of calcium to 1 of phosphorus, although much wider ratios have been found not to produce bad effects when vitamin D is adequate.

Mention will be made later in this chapter of the occurrence of rickets in young animals, and osteomalacia in adults, in cases of calcium-phosphorus deficiency, either alone or combined with vitamin D deficiency. In rickets the blood calcium-phosphorus content is deficient, with the result that normal bone formation is prevented and the bones become weak and deformed. Young cattle are very liable to suffer from this disease, the general symptoms of which are stiffness, bent and swollen knees and hocks and curvature of the leg and backbones, and there is usually a craving for such material as bones, and chalk of walls, which provide the deficient minerals. Pigs are particularly liable to suffer from rickets unless properly fed.

In osteomalacia, the causes and symptoms are similar, the main difference being that calcium and phosphorus are drawn from the skeletal structure to provide the minerals which the body demands. Whenever such cases occur attention should be given to the diet, or a skilled veterinarian consulted and remedial measures taken according to the principles above described.

Magnesium

Magnesium is essential for plant growth, and is found in all plant tissues, though in smaller amounts than calcium. It is more abundant than calcium in seeds, and appears to have a specialized function in connection with seed formation.

Manganese

This occurs in plant tissues to a small extent and appears to play some as yet unknown rôle in maintaining the health of the plant. Plants of the N. O. Solonaceae appear to be particularly susceptible to manganese deficiency in the soil which manifests itself in poor and stunted growth and liability to virus diseases.

Iron

Iron is a *sine qua non* of all plant growth, as it is indispensable for the formation of the green chlorophyl by means of which carbon dioxide is assimilated from the air. A corresponding function in the animal is the formation of the red haemoglobin of the blood which is the agency for conveying oxygen to the tissues. Iron is thus one of the most fundamentally necessary minerals for all life processes.

Iron and Copper

Iron and copper although present in the body in extremely small quantities—iron being present in the proportion of 1 : 25,000 by weight in the body—perform functions of fundamental importance and appear to work in unison. Thus while iron enters into the composition of the red haemoglobin of the blood, which constitutes the mobile oxidising agent of the body—copper does not. Yet copper is necessary to enable the body to produce this compound.

In cases, therefore, where the ration is deficient in iron and copper or both, a condition known as nutritional anæmia will result. It is interesting that copper which is a violent poison to cellular tissues, above a certain amount, is nevertheless a fundamental necessity in ultra minute quantities.

As a rule the ordinary farm rations are not lacking in iron or copper but cases are on record in which cattle would not thrive in certain areas and became weak and emaciated, and many died. This disease, known as salt sick, is due to an iron-copper deficiency and can be cured by free access to a mineral mixture composed of 100 pounds of common salt, 25 pounds of red oxide of iron and 1 pound of finely ground copper sulphate —very thoroughly ground and mixed.

Milk which is the natural food of young animals is exceptionally low in both iron and copper, but the young are generally born with enough of both in their bodies to enable them to carry on for a certain time.

In cases of prolonged suckling, however, trouble is likely to ensue owing to iron-copper deficiency, and a condition of anaemia may result. This may be cured by feeding about one-eighth of an ounce of a mixture made as follows:

135 grammes of ferric sulphate
22½ grammes of copper sulphate
dissolved in 600 ozs. of water.

Phosphorus

Phosphorus is another indispensable mineral; not only does it play an important part with calcium in building up bone tissues in the animal, but it appears to be necessary for the proper formation and transportation of proteins, and also plays an important part in regulating fermentation processes. It is found chiefly in seeds, but also in lesser quantities in all parts of the plant. While a small amount of phosphorus does occur in the inorganic state, phosphorus is chiefly present in organic combination in the form of phosphatides or phosphorised fats, the nucleo-and phosphoproteins, and as phytin, which is the chief phosphorus compound of seeds.

Sulphur

Sulphur is taken up by the plant in the form of soluble sulphates, but is largely transformed in the plant into organic combination in the formation of the sulphur containing proteins, in which form it is also found in the tissues of animals.

Chlorine

Chlorine is not apparently an essential of plant life though it is usually found in plants as sodium chloride. In the animal it is essential for the formation of the gastric juice (pepsin hydrochloric acid).

Fluorine

Fluorine occurs in varying proportions in feeding stuffs and is regarded as necessary for healthy plant growth [Susuki and Aso 1903]. It is also claimed by Gautiar and Clausmann [1919] that a dressing of five kilogrammes of sodium fluoride per acre was followed by an increase in cereal crops of 5 to 18 per cent and sometimes considerably more in the case of roots. These observations, however, lack confirmation. Very few data are available on the amount of fluorine present in crops but it is known that fluorine is a violent poison when taken into the body in considerable amounts or when very small amounts are steadily consumed over a long period of time. For instance, Mazumdar, Ray and Sen, of the Animal Nutrition Section of the Veterinary Research Institute at Izatnagar [1943], have recorded cases simulating rheumatoid arthritis, which have occurred chiefly in places situated at altitudes between 600 to 1,200 feet above sea level in the vicinity of mountain ranges in Hyderabad, Madras, Bihar, Kangra, etc., near the sites of old volcanic regions or sulphur springs, which on investigation proved to be due to ingestion of excess of fluorine. This malady which affects chiefly the teeth and bones may be distinguished from osteomalacia, as in the latter case, the calcium and magnesium content of the blood and bones shows a considerable reduction, but this does not occur when the attendant symptoms are caused by excess of fluorine.

The toxic effects of fluorine in producing this condition are of great importance because rock phosphates, or lime phosphates as they are sometimes called, which are used to some extent as a mineral supplement for farm animals, may contain as much as 4 per cent of fluorine.

Reed and Huffman [1930] have found that when 1.5 per cent of rock phosphate was used in the concentrate mixture fed to dairy cattle the teeth became soft and wore down to such an extent within two years, that the cattle were unable to chew their food properly, and were so worn and painful that they could not even drink cold water.

In this connection a matter of great importance both to the stock owner and human beings in general, is that during the manufacture of super-phosphate for fertilisers, most of the fluorine is not removed, and consequently long continued use of super-phosphate on land on which feeding stuffs are grown, may tend to an eventual accumulation of fluorine in the soil, which may be sufficiently great to affect adversely the teeth of animals or human beings using the crops for food. The results produced will be much the same as when rock phosphates are used.

Very little definite information is available on this subject but it is quite possible that many symptoms of disease or malaise or defective teeth formation which are still undiagnosed and unexplained may have their origin in the use of super-phosphate or other artificial fertilisers as has been pointed out in Chapter I.

Iodine

All plants contain minute quantities of iodine necessary for the body for the production of thyroxine, an iodine containing compound secreted by the thyroid gland in the neck, which controls the general rate of metabolism in the body. When the iodine content of a ration is inadequate the thyroid gland enlarges to a great extent in an endeavour to manufacture the thyroxine needed by the body and the ordinary type of goitre is produced characterised by a heavy swelling in the neck.

Young dairy cattle, pigs, lambs and foals are most likely to be affected, and in cases where iodine deficiency in the ration is prevalent the young may be born dead or extremely weak—and in the case of pigs, hairless.

A safe preventive for goitre in localities and conditions in which it occurs, is to supply iodine to the mother sometime before parturition occurs. This may best be done by giving her access to iodised salt instead of ordinary rock salt.

Iodised salt may be made by mixing one ounce of potassium iodide with three hundred pounds of rock salt. The iodised salt will thus contain 0.02 per cent of potassium iodide, and enough should be available to supply the mother with about two grains of potassium iodide per week.

Excepting cases where goitre is known to be prevalent there should be no necessity to add iodine in any form to the rations of farm animals. Goitre is very prevalent in the Kangra district of the Punjab and parts of South India.

In addition to the foregoing, certain other minerals occur in the soil and are found in traces in plant tissues. Some, such as copper, zinc, cobalt and boron, have not been thoroughly investigated and their functions are imperfectly known. Experimental evidence, however, indicates that they have an important role to play in plant and animal metabolism. Boron certainly appears essential for the tobacco plant, and copper in trace doses with iron proves beneficial in cases of anaemia in pigs. There are undoubtedly other elements present in the soil and in plant tissues in traces only which can only be determined by spectrographic analysis, but which may nevertheless have some unknown function to perform. Much research is still needed to throw further light on this problem and also on the possible effects which artificial fertilisers may have in rendering them unavailable to the plant by precipitating them in the soil in insoluble form.

The quantitative distribution and the biological significance of the elements as yet unrecognised as of value in the biologic cycle and of the so called 'rare elements' is still as unexplored a field, as was the domain of vitamins a brief half century ago.

THE VITAMINS

What are Vitamins?

At the beginning of this century nobody had heard of vitamins, and even to-day there is much popular confusion as to their nature, even among farmers, and there is a tendency to regard them as rather mysterious entities distributed in foodstuffs rather like currants in a bun. The argument is sometimes heard that animals got on very well before they were discovered, so why all the bother about them now! The answer is that before the existence of vitamins was known a vast number of human beings and also animals suffered from ill health or actual disease as a result of vitamin deficiency without anyone knowing the reason why. Correspondingly vast numbers were fit and healthy because their diets or rations happened to be sufficiently natural and well selected. The same argument applies with no less force to-day in spite of the advancement of knowledge and the availability of vitamins in medicated form.

Let it be stated at the outset that vitamins are chemical substances, just as water, or fats or mineral salts are, the main differences being that some of them are very complicated in chemical nature, and that they are distributed in foodstuffs in extremely small quantities. For example, it has been seen that lipoids or sterols are associated with fats in plant and animal tissues, and that one of them is ergosterol, of a known empirical chemical composition, *viz.*, $C_{27}H_{41}OH$. When this substance is exposed to ultra violet light it takes on new properties and the activated substance is known as vitamin D. Similarly, vitamin C is a comparatively simple organic acid known by the popular name of ascorbic acid. Its chemical name is hexuronic acid and its formula is $C_6H_8O_6$ or

$$CH_2OH.CHOH : CH.C(OH) = C(OH).CO$$
$$\llcorner\_\_\_\_\_\_ O \_\_\_\_\_\_\lrcorner$$

The existence of these substances, which later became known as vitamins, was first brought to light by the discovery of the fact that when the materials of which ordinary feeding stuffs are composed, *viz.*, fats, carbohydrates and proteins, were very carefully purified so that nothing else could be present, and were fed to animals together with pure mineral salts, growth did not occur and the animals ultimately died. Hence they were at first called accessory food substances, because they were necessary as adjuncts to the then ordinarily accepted feeding stuffs. Paradoxical as it may appear, although most of the experimental work on vitamins has been carried out on small animals, animals in general are less susceptible to vitamin deficiencies, than are human beings, or perhaps it would be more accurate to say that their rations are on the whole less likely to exhibit vitamin deficiency. The only serious deficiencies which are likely to occur in the usual rations of farm animals are in vitamin A and vitamin C.

With the advance of knowledge a considerable number of vitamins is now known, but for simplicity and convenience they may be designated as A, B, C, D and E, although B is really a complex of two (or more) separate vitamins called B_1 and B_2, the latter being sometimes referred to as vitamin G. Vitamin D is also now known to be a complex. These five best known vitamins are divided into two groups known as the water soluble vitamins and the fat soluble ones, because of their association with or solubility in water and fat respectively. To the latter class belong A, D, and E and to the former B and C.

Vitamin A

Sources

Vitamin A is found in both animal and vegetable tissues. Chief among the former are oils and fats, particularly halibut-

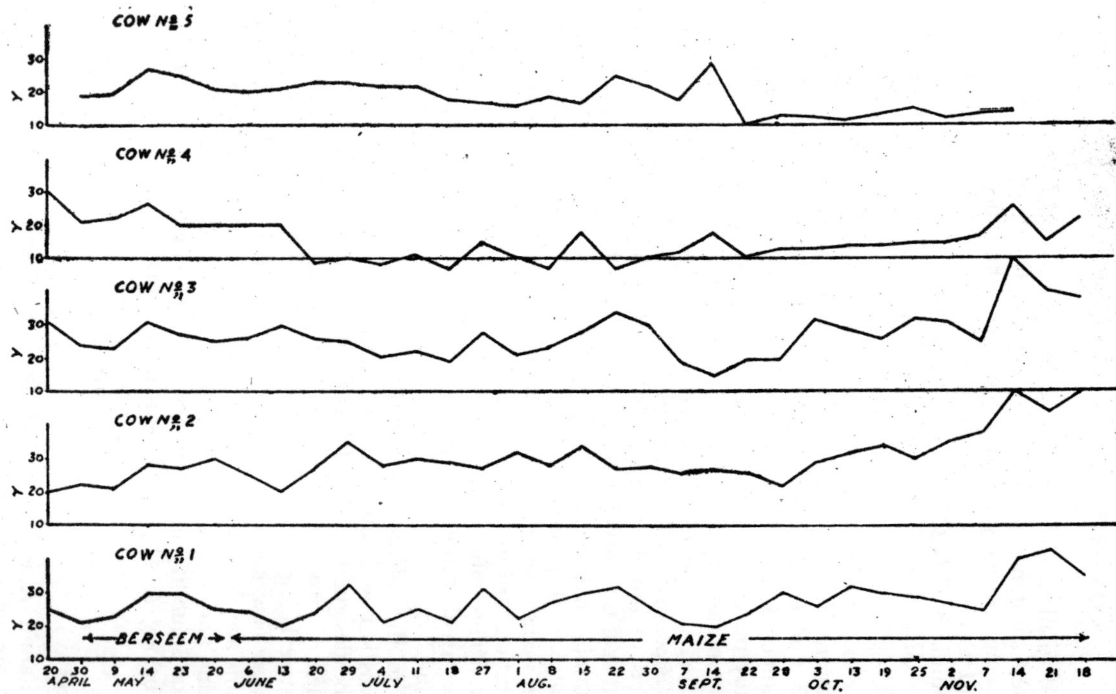

FIG. 2. Graphs showing the amount of carotene in gammas, γ, ($1\gamma = 0.000001$ gram.) per 100 c.c. milk from 5 experimental Montgomery Cows at Lyallpur during a lactation from April to November 1942.

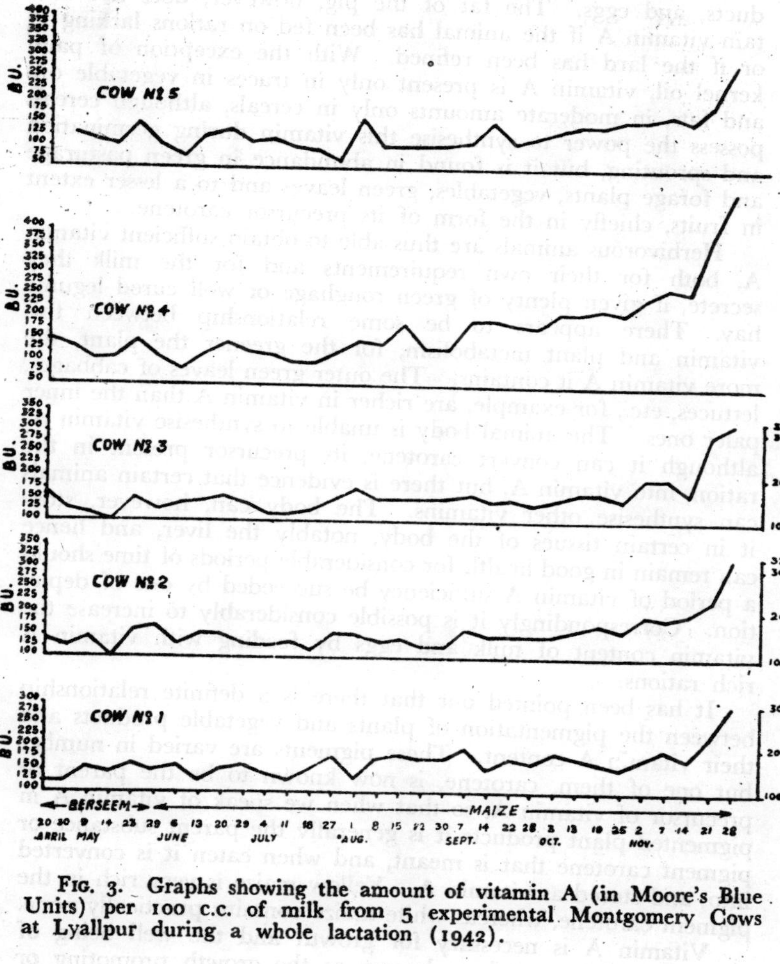

FIG. 3. Graphs showing the amount of vitamin A (in Moore's Blue Units) per 100 c.c. of milk from 5 experimental Montgomery Cows at Lyallpur during a whole lactation (1942).

liver oil and cod liver oil, the livers of certain other varieties of fish, butter, the cellular tissues of animals, milk and its products, and eggs. The fat of the pig, however, does not contain vitamin A if the animal has been fed on rations lacking it, or if the lard has been refined. With the exception of palm kernel oil, vitamin A is present only in traces in vegetable oils and fats, in moderate amounts only in cereals, although cereals possess the power to synthesise this vitamin during germination and sprouting, but it is found in abundance in green pasturage and forage plants, vegetables, green leaves and to a lesser extent in fruits, chiefly in the form of its precursor carotene.

Herbivorous animals are thus able to obtain sufficient vitamin A, both for their own requirements and for the milk they secrete, if given plenty of green roughage or well cured legume hay. There appears to be some relationship between this vitamin and plant metabolism, for the greener the plant, the more vitamin A it contains. The outer green leaves of cabbages, lettuces, etc., for example, are richer in vitamin A than the inner paler ones. The animal body is unable to synthesise vitamin A, although it can convert carotene, its precursor present in the ration, into vitamin A, but there is evidence that certain animals can synthesise other vitamins. The body can, however, store it in certain tissues of the body, notably the liver, and hence can remain in good health for considerable periods of time should a period of vitamin A sufficiency be succeeded by one of depletion. Correspondingly it is possible considerably to increase the vitamin content of milk and eggs by feeding with vitamin A rich rations.

It has been pointed out that there is a definite relationship between the pigmentation of plants and vegetable products and their vitamin A content. These pigments are varied in number but one of them, carotene, is now known to be the parent or precursor of vitamin A, so that when we speak of vitamin A in pigmented plant products it is generally the parent substance or pigment carotene that is meant, and when eaten it is converted into, and stored as vitamin A. Yellow maize is very rich in the pigment carotene, whereas white maize contains practically none.

Vitamin A is necessary for growth and the well being of animals, and is sometimes known as the growth promoting or anti-infective vitamin. If a young animal is deprived of it for any length of time it becomes restive and ceases to grow and develops a characteristic disease of the eyes known as xerophthalmia. In the adult animal similar conditions develop and general malaise sets in. This can be cured if treated in time, but long continued deprivation of the vitamin is likely to produce permanent effects.

The explanation of eye and other corresponding diseases in vitamin A deficiency is as follows:—

There are two main types of nerve fibre and nerve cells in the peripheral and central nervous system:

(a) The motor nerve cells which originate, and the motor or efferent nerve fibres which conduct motor impulses from the central nervous system of the various muscles of the body which cause movement. For example, in walking the motor impulse starts unconsciously in a part of the cerebral cortex known as the motor area and is conducted along the motor nerve fibres to the muscles of the legs. These cells and fibres are not affected by vitamin A deficiency.

(b) The afferent or sensory nerve fibres are connected with sensory surfaces such as the skin, the conjunctiva of the eye, the retina of the eye and similar tissues. These fibres conduct sensory impulses from the exterior to the various cells with which they are connected in the central nervous system. In vitamin A deficiency these nerve fibres and nerve cells become damaged, and in serious conditions become completely destroyed. The fibres and cells govern the nutrition, tone and general health of the sensory surfaces with which they are connected. Thus in vitamin A deficiency the sensory surfaces of the eye become damaged because the nerves which govern them are injured. Xerophthalmia is a typical instance of such damage.

Another well-known example of the result of mild vitamin A deficiency is night blindness. The retina of the eye contains certain cells which govern the capacity to see in feeble light, and in vitamin A deficiency their efficiency becomes impaired and they do not function properly in feeble light or semi-darkness. In serious conditions of vitamin A deficiency the whole retina, and in due course, the optic nerves and the cells in the cerebral cortex governing sight may be damaged and ultimately destroyed and blindness may result.

All epithelial tissues, of which the alimentary tract is another example, are similarly weakened by vitamin A deficiency, and may ultimately show lesions which permit the ingress of pathogenic organisms. It is on this account, and by virtue of the function of vitamin A in maintaining the health of these tissues and the body generally, that it has been called the anti-infective vitamin.

Vitamin A is destroyed in the presence of air or oxygen. This fact has an important bearing on the storage of dairy produce such as butter and cheese, whose outer layers if exposed to air, even in cold storage, may become devoid of vitamin A and take on a bleached tint. Any agency which tends to facilitate oxidation will also destroy vitamin A. Quick heating of milk and other foods, however, is not harmful, provided air is excluded.

Poultry and dairy cattle are very susceptible to vitamin A deficiency in their rations. Cotton seed cake is deficient in this vitamin and the so-called cotton seed poisoning said to ensue after feeding the cake may be due to vitamin A deficiency combined with an otherwise unbalanced ration.

In chemical constitution vitamin A is supposed to be an alcohol closely related to cholesterol, and the following formula which has been assigned to it illustrates its chemical nature:

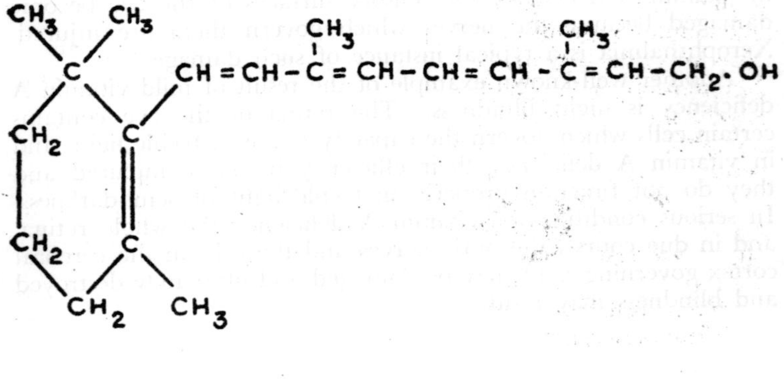

VITAMIN B

The vitamin B complex. (Vitamins B_1, and B_2 or G.)

There is sometimes confusion in the use of different European and American forms of terminology in regard to this vitamin. It may be explained that when the original vitamin B, as it was called, was shown to be a complex consisting of one entity known as the anti-neuritic vitamin, because it prevents beri beri in humans and polyavian neuritis in birds, and another known as the anti-pellagric factor which prevents pellagra, European workers designated these respectively B_1 and B_2 although they are not necessarily related. It was then proposed by the Americans that as the part called B_1 was discovered by Funk, and B_2 part by Goldberger, they should be called respectively vitamins F and G (a natural patriotic sentiment!). This would have left no B's in the vitamin nomenclature and so to-day, although the previous B_1 and B_2 are now recognised as B and G in America, these terms are not always used in European literature which still sticks to B_1 and B_2, or B_1 and G. In this discussion they will be referred to as B_1 and B_2, but for reference it may be remembered that:

B_2, (or G) European = B, American.
B_1, European = G, American.

Vitamin B_1 (Anti-Neuritic)

Beri beri is a very common disease in the East and is found chiefly among people whose staple article of diet is polished rice, *i.e.*, the grain with its outer cuticle and germ removed. The symptoms of the disease are a type of paralysis which ultimately affects the whole body and causes death. Similar symptoms can be induced in birds and animals when fed polished rice or other rations deficient in vitamin B_1. Fowls, for example, can be afflicted with the disease when kept on such rations for as short a time as a fortnight. The addition, however, of a small quantity of rice polishings or yeast, or other material containing vitamin B_1, will effect a most remarkable recovery and the birds can be restored to health in a very short time. Prolonged affliction, however, is more difficult to cure whether in humans, birds or animals as the deficiency causes degenerative changes in the nervous system, responsible for the paralysis and other structural and metabolic disturbances particularly of the heart, suprarenal glands and other organs of the body. There is also general impairment of appetite, when animals are fed vitamin B_1 deficient food, which has led to its being called a growth promoting vitamin, although B_1 and B_2 have both been shown to be necessary for normal growth. Vitamin B_1 is of special interest as it is found in considerable quantities in yeast, indicating that very low forms of life can synthesise it. Certain bacteria present in the rumen of the cow can also synthesise it, which renders that animal to a considerable degree independent of its rations for its vitamin B_1 supply.

Vitamin B_1 is found extensively in cereals, egg yolk, rice and maize, pulses, nuts, peas, lentils, milk, and in fresh vegetables and fruit. The embryos of grains usually contain much more than the rest of the seed, and hence any milling operation which removes the germ produces a vitamin deficient flour. This applies also to vitamin E. Although such manufactured products have other important nutritive values, it should always be remembered that they are deficient in vitamins and that natural foods are always better than manufactured ones. Another important point to remember is that vitamin B_1 is destroyed by heat and by the action of alkalis. Even natural foods, particularly grains and cereals, are known to lose their anti-neuritic properties to a considerable degree when heated to the temperature of boiling water or above, for half an hour. Hence a certain quantity of natural uncooked foods should always form part of the dietary. Unlike vitamin A, however, vitamin B is not affected by aeration.

Vitamin B_2 (Antipellagric)

It was discovered that when yeast was autoclaved or heated, it still possessed a water soluble vitamin which had the property of stimulating growth and appetite, although the autoclaved yeast no longer possessed the power to ameliorate neuritic conditions. It was also discovered by Goldberger and co-workers that this residual vitamin, which had been unaffected by heat had the specific property of preventing and curing human pellagra and a corresponding disease in dogs known as black tongue.

These maladies occur most commonly among the poor in many countries, especially among people who subsist largely on maize or rice, and are characterised by loss of hair, hyperaemia and ulceration with haemorrhage, gastro-intestinal disturbances and changes in the central nervous system. This connection between the disease and the heat resistant vitamin left in the yeast endows the latter with very specific properties—hence its name anti-pellagra vitamin (or vitamin B_2). It is difficult to dogmatise on the functions of this vitamin in nutrition though its growth inducing aspects appear to be connected with the stimulation of appetite, but as far as is known the whole B complex is necessary for normal growth, and young growing animals should be fed liberal amounts of these vitamins to ensure good health and maximum growth.

Although vitamin B_2 appears to be distributed in nature with vitamin B_1, yet there is considerable variation in their relative distributions. Its chief sources are milk (which is richer in B_2 than in B_1) yeasts, cereals (wheat and maize are poor in B_2), animal tissues, vegetables and fruits. In general, however, seeds, grains and roots are relatively poor in vitamin B_2, but rich in vitamin B_1, while the reverse is the case for foods of animal origin such as meat, milk and eggs. Yeast, is rich in both vitamins, but like other plant materials is relatively richer in vitamin B_1.

Very little is yet known concerning the chemical nature of the vitamins B_1 and B_2, although they appear to be chemically related and contain the element sulphur in their molecules.

Vitamin C

It has frequently been the case that some simple remedy has been adopted to cure a disease in the absence of any accurate knowledge about the actual curative substance involved. An example of this is the chewing of cinchona bark by the natives of South America as a remedy for fever, although the fact that the alkaloid quinine in the bark was responsible, was of course, a sealed book to them. Similarly in the old days of sailing boats when fresh fruit and vegetables were difficult to get, and long voyages were common, sailors developed scurvy, a disease characterised by loss of appetite and weight, accompanied by pain in the joints, sores and capillary haemorrhages. It was found, however, that a small amount of lemon juice, or better orange juice, effected a miraculously quick recovery.

It is now known that the prophylactic and curative agent in the juice is vitamin C or antiscorbutic acid (hexuronic acid). Susceptibility to scurvy is apparently confined to man, monkeys and guinea pigs, and the disease appears to be largely one, involving capillary haemorrhage. Other animals appear to be immune, and this immunity apparently takes the form of an inherent capacity of these animals to manufacture vitamin C in their bodies. For example, it has been found that if rats are fed for many months on a vitamin C deficient diet—which would have killed guinea pigs—and they are then killed and their livers fed to guinea pigs suffering from scurvy, the guinea pigs recover in sufficient numbers to indicate that the rats must have been able to synthesise vitamin C in their body tissues.

Farm animals undoubtedly also possess this power, as the presence of vitamin C in the livers of calves and chickens, and in the milk of cows fed vitamin C deficient diets, may be taken to show that these animals are also able to synthesise this vitamin from its precursor in grains and seeds, but under natural conditions of feeding they should obtain an adequate amount of this vitamin for their needs.

Vitamin C is essentially a product of the vegetable kingdom and is present in immature seeds, in seeds during germination, fruits and vegetables, water cress and green leafy vegetables. The freshly pressed juices from all these plants are also rich in vitamin C.

Oranges, lemons and tomatoes are particularly rich in vitamin C, but it has been shown that alkalis and oxygen tend to destroy it; so the use of alkalis and undue exposure to air should be avoided when cooking vegetables.

Modern conditions of canning fruits and vegetables have overcome these difficulties, and freshly canned products properly prepared are but little inferior in vitamin content to the natural article. Time and temperature, however, are factors which gradually bring about a diminution of the vitamin during storage of the produce—though this is retarded somewhat by acidity.

Fresh fruits and vegetables lose a certain amount of their vitamin content during storage due to respiratory oxidation, continuously in process in living tissues. Vitamin C is very sensitive to environmental conditions, and although it is readily destroyed by heat, the presence of acids or alkalis and oxygen may considerably modify the effects of heat one way or the other, and the importance of oxidation should always be borne in mind. Vitamin C is destroyed during the bacterial acid fermentation of materials originally rich in this factor, so that silage, in which any appreciable bacterial fermentation has occurred, is not an antiscorbutic ration. Vitamin C has now been prepared in tablet form from the concentrated products of tomato and lemon juice, and is sold as a commercial pharmaceutical article. One of the quickest rough and ready methods, however, of obtaining it, which was employed by the British Army during the first World War, is slightly to germinate seeds such as the pulses. The writer has found that wheat kept moist and allowed to germinate for 24 hours had its vitamin C content increased by 600 per cent.

It has been shown by workers in England and the U.S.A. that the pulp of teeth deteriorates when diets deficient in vitamin C are taken over any length of time, and that dental caries and pyorrhea respond to dietary correctives such as lemon juice, oranges, tomatoes and lettuces. There is no doubt that much dental trouble is caused by vitamin and other dietary deficiencies, both pre-natal and post-natal, and that a stricter adherence to dietary hygiene at early stages would do much to rectify the amazing amount of dental trouble which afflicts most civilised peoples.

Vitamin D

Vitamin D—the antirachitic vitamin—is a preventive of rickets, a disease which has been known for long, and is prevalent among children and young animals. Bone is chiefly composed of calcium phosphate, and there is a very close relationship between the bone forming function of calcium and phosphorus and vitamin D. In rickets normal ossification of the bones does not take place, and they become soft and cartilaginous, with the result that the general skeletal structure becomes mis-shapen. A similar disease known as osteomalacia occurs in adult animals. When normal and rachitic bones respectively are split longitudinally, and examined after treatment with silver-nitrate, they show conspicuous differences. Normal bones have a well calcified metaphysis and the epiphyseal cartilaginous line is smooth and narrow. In rachitic bones the metaphyses are ill formed and the epiphyseal line is wide and irregular owing to faulty calcium and phosphorus deposition.

Mellanby [1934] produced experimental rickets in dogs which he was able to cure by the administration of cod liver oil, and concluded that it was essentially a dietary disease and could be cured by an improved dietary regime. At the time of Mellanby's experiments the only fat soluble vitamin then recognised was vitamin A and it was not till McCollum and his co-workers [1922] demonstrated that when cod liver oil was heated in contact with oxygen, it retained its antirachitic properties after the xerophthalmia preventing properties due to vitamin A were destroyed, that the existence of a separate fat soluble vitamin related to rickets was discovered.

The origin of vitamin D

Vitamin D appears to be formed universally in all plant and animal tissues by the action of sunlight or ultra violet rays lying in that part of the ultra violet spectrum between wave lengths 260--300 million the class of bodies known as the sterols, of which ergosterol and cholesterol are examples, and the origin of vitamin D is explained by its relationship to these bodies. The possibilities of the formation of vitamin D are therefore limited only by the prevalence of the necessary rays, and the presence of the precursors—the sterols.

Until comparatively recently ergosterol was the only member of the sterols which was known to be capable of activation, and conversion through a number of intermediary products into vitamin D (calciferol), thus:—

Ergesterol—lumisterol—protachysterol—tachysterol—vitamin D, the final vitamin being isomeric with the original ergosterol. So long as ergosterol was the only known pro-vitamin D, it was assumed that all vitamin D of animal or vegetable origin was activated ergosterol. Many other sterols, however, are now known to act as pro-vitamins and are capable on activation by the same rays which activate ergosterol, of being transformed into antirachitic substances similar in action to vitamin D (calciferol). The most important of these, and the only one of biological interest is 7-dehydrocholesterol, which, when activated is, in modern nomenclature, known as vitamin D_3 which is isomeric with the pro-vitamin, 7-dehydrocholesterol. According to the same nomenclature irradiated ergosterol is now known as vitamin D_2, while vitamin D_1 is a molecular combination of vitamin D_2 and lumisterol.

Vitamin D is therefore a complex, and the most important part of this complex for vertebrate animals is not irradiated ergosterol (vitamin D_2) as formerly thought, but irradiated 7-dehydrocholesterol (vitamin D_3).

This fact has an important bearing on farming practice, because while vitamins D_2 and D_3 are equally efficacious for human beings, vitamin D_2 or irradiated ergosterol has little effect on bone formation in poultry or dogs.

It is therefore better to give fish liver oils to poultry as a source of vitamin D rather than other material which would supply chiefly irradiated ergostrol, as fish liver oils are richer in that part of the vitamin D complex known as vitamin D_3 (irradiated 7-dehydrocholesterol).

All plant and animal tissues contain elements of the provitamin D complex, as is well known by the fact that plant tissues and foodstuffs can be irradiated, when they assume antirachitic properties, and the exposure of patients suffering from rickets to ultra violet light induces curative effects due to the conversion of the sterol pro-vitamins in the epidermal cells into their corresponding vitamins, which are then absorbed into the blood stream and the lymphatics. Vitamin D is found in most fish oils, particularly halibut and cod liver oils, in animal fats and in yeast and the yolk of eggs, whereas fruit and vegetables are deficient in it. Vitamin D is characterised by resistance to heat and aeration.

Function of Vitamin D

Rickets is not entirely a vitamin deficiency disease, and vitamin D does not have the same specific relationship to rickets that the other vitamins have with the diseases with which they are associated.

The relative proportions of calcium and phosphorus in a ration are also important factors in connection with the etiology of rickets, and one of the functions of vitamin D appears to be to maintain a physiological balance in the blood between calcium and phosphorus quite independently of what the food may contain. The vitamin therefore is most essential when the food is unbalanced in its calcium-phosphorus content, and plays an even greater part in regulating the metabolism of phosphorus than that of calcium.

Precisely how vitamin D acts in regulating mineral metabolism is not clear—the ultimate explanation of biological operations can never be attained—but it appears to control the normal intestinal and tissue pH (*i.e.*, acid base equilibrium) and regulates absorption and excretion along the intestinal tract. It facilitates the absorption of calcium and phosphorus from the intestines into the blood stream so that these minerals may become accessible for the developing bones, and it also enables the bones to utilise the calcium and phosphorus present in the blood. In the absence of vitamin D the bones cannot utilise the minerals of the blood even if calcium and phosphorus are injected in excess.

In India, if animals are fed reasonable quantities of fresh green fodders, including legumes, and get plenty of sunlight, they should not suffer from rickets or osteomalacia. The disease, however, is far from unknown among animals in this country and when it occurs should be treated on a dietary basis in relationship to general mineral metabolism.

Vitamin E

The existence of vitamin E was first discovered in 1922 by Evans and Bishop [1922] who announced that they had been able to induce sterility in rats by feeding them with diets which, at the time, were considered to be satisfactory with respect to all the then known vitamins. They found that a very characteristic form of reproductive failure occurred, although in all other respects the rats appeared to be in excellent health. Although ovulation took place and the foeti were implanted and appeared to start normal development, yet, sooner or later, the functions of the placenta failed before parturition, with the resulting death and absorption of the implanted foeti. When certain natural foods were added to the experimental diets it was found that sterility did not result, and correspondingly sterility could be cured by the addition of natural foods to afflicted animals.

These workers, therefore, concluded that as the diet appeared to satisfy all the other physiological requirements of the body it must be lacking in some special factor regulating reproduction.

Evans [1925] has also recorded a special type of sterility, which may be permanent, in the male rat due to destruction of the germ cells. This develops much more slowly in the male than in the female, although permanent sterility is much more rare in the female.

Our knowledge of vitamin E deficiency is still very scanty, and it has been contended by some workers that normal reproduction is not dependent on any one vitamin such as E, but that other dietary factors are also needed, and that diets deficient in vitamin A have also produced characteristic forms of sterility. It appears that vitamin A plays an important part in bringing about normal ovulation, whereas in the absence of vitamin E normal ovulation occurs, but the placenta fails to perform its normal reproductive functions.

The work of Evans and co-workers has resulted in a technique by which it is possible to determine whether oestrus is occurring or not, whether fertilisation has occurred, and whether breeding is likely to result. That is to say it can foretell whether young will be born, and if sterility is present, the type of sterility responsible for failure of the reproductive process.

Vitamin E occurs widely distributed in natural foods, and although it is soluble in fats it is not exclusively confined to the fat containing parts of plants but is found in abundance in green leaves and particularly in lettuces. Its most concentrated sources are the embryos of wheat, oats and maize and the most concentrated preparations have been obtained from wheat germ oil. In these preparations nevertheless the vitamin exists merely as an impurity and it has not been extracted in the pure state.

Cotton seed oil which has been hydrogenated also contains the vitamin in moderate quantities. Many animal tissues contain vitamin E, although not in concentrated form, especially muscle tissue, liver and milk, particularly milk fat, though eggs are lacking in it. It is also found in certain fruits such as oranges, tomatoes and bananas. In fact, nature seems to have provided for a very extensive distribution of this factor for the performance of one of her most vital requirements—reproduction—and provided animals are fed well-distributed natural rations, nutritional sterility due to lack of vitamin E should not be of great importance, particularly with domestic animals, which usually obtain a considerable proportion of cereals and green fodders in their rations.

Properties

Vitamin E is remarkably resistant to heat, light, aeration and ordinary chemical reactions, and can be distilled in steam. It is not affected by ordinary oxidising agents, but the chemical oxidation of fat or its oxidation products exerts a destructive effect on the vitamin and prevent it performing its normal functions. [Evans & Burr 1927.] On the other hand, it is unaffected by the hydrogenation of fat as is shown by the fact that hydrogenated cotton seed oil still contains it.

Little is known about its chemical nature and although concentrated preparations have some of the properties of the sterols, it does not appear to belong to any of the known members of that group of bodies. It appears to be associated in some at present unknown way with iron, and the death of the embryo in cases of vitamin E deficiency is supposed to be connected with disturbances of iron assimilation. In spite of the association of vitamin E with sterility, whatever may be the facts as regards farm animals, it does not seem to be associated with poor dietaries among human beings as those classes who stand nearest the bottom of the economic and dietary scales are usually most prolific, and whatever other defects their diets may exhibit, an anti-sterility vitamin appears not to be among them.

Speaking generally, although vitamins are of such profound scientific interest and play such an important part in dietary and rationing regimes, the average farmer need not worry unduly about them if he manages his animals on rational lines and allows them plenty of sunlight, and a well-balanced mixed ration with plenty of green stuffs. Special attention should be paid to stall fed animals in localities and seasons when green stuffs may be scarce, and this also applies to poultry, pigs and calves reared on artificial foods. Provided the rations include green fodders, particularly legumes (containing vitamins A, C and D), bran (containing B and E), and reasonable quantities of yeast (containing B) and fish liver oil (A and E) and similar rationing materials, when required, vitamin deficiency among farm animals should be the exception rather than the rule.

Natural Versus Artificial Vitamins

Ordinarily farm animals obtain all the vitamins they need from their rations, provided these contain sufficient quantities of green fodder and good quality hay, and the question of supplying artificial vitamins does not arise.

It may have occurred to some people to ask, in connection with human nutrition, whether the artifically prepared vitamins which are available to-day have the same nutritive value as those occurring naturally in food. The answer is definitely yes. Some vitamins such as vitamin B_1, vitamin B_2, vitamin C and vitamin E are now prepared synthetically in the laboratory. Vitamin D can be prepared by irradiating the pro-vitamin ergosterol, and vitamin A by concentrating it from certain fish oils, which are very rich in this vitamin.

The different vitamins, whether occurring naturally or whether prepared artificially in the laboratory, have definite chemical constitution and their properties depend on the constitution and not on the source from which they are derived. Any difference between a well balanced natural diet rich in vitamins and one lacking in vitamins, but which has been fortified by the addition of synthetic or otherwise artificially prepared vitamins, may be explained as follows:—

Artificially prepared vitamins may be employed to remedy specific vitamin deficiencies, as when vitamin B_1 is used to fortify white wheat flour. A diet which is badly balanced in vitamin content may be deficient in more than one vitamin, and naturally it would require a range of more than one vitamin to remedy this defect. On the other hand well balanced natural food usually contains all the vitamins required. It is, therefore the range and the quantity of vitamins present in a natural well balanced diet which makes this superior to a vitamin deficient diet fortified by added vitamins, and not any inherent differences in the value of natural and artificial vitamins. Furthermore, it is highly probable that natural food stuffs contain vitamins which have not yet been discovered. Diets which are deficient in one or more of the already known vitamins may also be deficient in the as yet undiscovered vitamins. It is therefore obvious that in this latter respect the addition of certain artificial vitamins will only partially remedy the defect, and the diet will still not be as suitable from a vitamin point of view as a well balanced natural one. This in no way detracts from the very great value of synthetic or otherwise artificially prepared vitamins, which may be employed to remedy specific vitamin deficiencies in the human or animal dietary.

THE VITAMINS

FAT SOLUBLE A.

Animal sources.	Oils and fats, butter, halibut liver oil, cod liver oil, cellular organs of animals, milk and its products, eggs.
Vegetable sources.	Oils and fats (traces); cereals, vegetables, green leaves, fruits, (chiefly as the precursor carotene).
Vitamin A deficiency causes:—	Retardation of growth, ophthalmia metabolic and specific tissue changes, intestinal lesions, changes in the pH of the blood, gall stones, lowered resistance to infection, with special susceptibility to infections of the lungs.
Chemical nature.	A sterol allied to cholesterol or ergosterol. Is destroyed by aeration. Resistant to heat.
WATER SOLUBLE B_1 & B_2	B. Antineuritic. Not destroyed by aeration. Destroyed by heat.

	B_2. Antipellagric, resistant to heat, growth promoting.
Sources of B_1	Cereals and yeast. The yolks of eggs, fresh vegetables and fruit. Pulses and nuts, peas and lentils, milk.
Sources of B_2.	Milk (richer in B_2 than in B_1). Yeast. Cereals (Wheat and maize poor in B_2). Animal tissues (rich), vegetables and fruits (Spinach, Tomatoes and bananas particularly rich).
Vitamin B deficiency causes:—	Beri beri, structural and metabolic derangement, especially of the nervous system and the heart.
Vitamin B_1 deficiency causes:—	Pellagra, retardation of growth, gastro-intestinal disturbances, skin lesions, and degenerative changes in the central nervous system.
WATER SOLUBLE C.	Anti-scorbutic.
Sources.	Produced during the germination of seeds. It is also present in fruits and vegetables (high), water cress and green leafy vegetables, animal tissues (low), milk (low); boiling destroys it.
Vitamin C deficiency causes:—	Scurvy. Readily destroyed by oxidation and alkalis.
FAT SOLUBLE D.	Antirachitic (Prevents rickets).
Sources.	Cod liver oil and other fish oils, and animal fats. Yeast and the yolks of eggs. Fruits and cereals (deficient). Produced in the surface cells of the body as a result of the action of sunlight or ultra violet rays.
Chemical nature.	Irradiated ergosterol.

Vitamin D deficiency causes:—	Rickets, disturbances in cellular metabolism, disturbance of normal testinal and tissue pH. It is not easily destroyed by heat, and resists oxidation.
FAT SOLUBLE E.	Controls reproductive efficiency.
Sources.	Wheat germ oil (potent), green leaves, lettuce (potent). Animal products (negligible).
Vitamin E deficiency causes:—	Disturbances in the reproductive processes which are considered to be associated with defective assimilation of iron. It is unaffected by heat, light or aeration.

BIBLIOGRAPHY

1. Evans, H. M. 1925. *Invariable occurrence of male sterility with dietaries lacking fat soluble vitamin E. Proc. Bott.* Acad. Sci. 11, 373-77.
2. Evans, H. M., & Bishop, K. S. 1922. *On the existence of hitherto unrecognized dietary factor essential for reproduction. Science,* 56, 650-651.
3. Evans, H. M., & Burr. G. O. 1927. *Vitamin E. The destructive effect of certain fats and fractions thereof on the anti-sterility vitamin in wheat germ and in wheat germ oil.* J. Amer. Med., Assoc., 89, 1587-90.
4. Gautiar and Clausmann, Compt. Rendu. 1919. 168, 976; 169, 115.
5. Majumdar, B. N., Ray. S. N. and Sen K. C., 1943. Ind J. Vet. Sci. & Animal Husbandry. XIII. 95-107.
6. McCollum, E. V., Simmonds, N., Becker, J. E., and Shipley, F. G., 1922. *An experimental demonstration of the existence of a vitamin which promotes calcium deposition.* Johns Hopk. Hosp. Bull. 33, 229-230.
7. Mellanby, E. 1934. *Nutrition & Disease.* pp. 1-10.
8. Osborne, T. B. 1909. *The Vegetable Proteins.* 78.
9. Reed, O. E., & Huffman, C. F. 1930. Mich. Agric. Exp. Sta. Tech. Bull. 105, 3-63.
10. Susuki and Aso (1903). Bull. Coll. Agric. Tokyo, 5.473.

CHAPTER III

DIGESTION, ABSORPTION AND UTILIZATION OF FOOD

Having reviewed the chief classes of material which enter into the composition of foodstuffs, the changes which they undergo during digestion, their methods of absorption and the functions which they subserve in the animal's body may now be discussed.

The animal organism may, in some respects, be likened to an engine. Fuel is supplied in the form of food, just as coal or petrol is supplied to the engine, and this provides the animal with energy or heat, and the efficiency of the fuel in both cases depends on the degree to which it can be converted into energy, and in both cases waste products result. There the analogy ends, however, as the animal is able to utilize part of its own structure in cases of necessity to provide heat and energy (interchangeable terms), and it also uses the fuel or food to build up its body, perform work, and produce various products, such as milk etc., for the use of its young, other animals and man.

The object of digestion is to convert the—for the most part—insoluble constituents of the food into soluble ones which can be absorbed into the blood stream and used. A moment's reflection on the nature of the foodstuffs fed to animals, and the products obtained from them will indicate the complexity of the chemical changes involved. For example, the complicated proteins of the food have to be converted into simple soluble nitrogenous compounds, which are in turn built up into tissue proteins and animal products; the insoluble starches and other carbohydrates must be converted into glucose, and fats changed into soluble soaps and glycerine, before being rebuilt into tissue fat. The process of digestion therefore results in the various chemical substances discussed in Chapter II, present in the foodstuffs, being broken down and simplified in the alimentary canal. A selective activity then comes into play on the ultimate products of digestion; some are absorbed into the blood stream and used by the living tissues whilst others are excreted in the faeces. In this connection, the organs of digestion, the secretions they produce, the effects of these on the foodstuffs, and the ultimate fate of the digestion products of the food can now be reviewed.

The organs of digestion

The simplest digestive operation may be seen in the unicellular amoeba. When this lowly organism comes into contact with a particle of matter which can serve it as food, it forms a depression into which the particle sinks and the rest of the cell then closes round it, and, acting as a primitive stomach, proceeds to digest it, the undigestible residue being then discarded by reverse movements of the amoeba. In the course of evolution the animal body has developed very specialised organs to perform the digestive functions, which are linked together to form the complicated digestive apparatus of the higher animals.

The mouth

The mouth is the organ of mastication into which the juice of the salivary glands which perform the preliminary stage in the digestive of that carbohydrates is poured. This juice is secreted by three pairs of glands, the parotid, the submaxillary and the sublingual, and contains a special amylytic ferment, ptyalin, the active constituent, which acts best in a neutral medium, *i.e.*, neither acid nor alkaline, and converts starch into maltose. The salivary juice has, in addition to its chemical action, a special function to perform in moistening and lubricating the food, for which purpose the amount required is often very large, depending on the degree of dryness of the food, and amounting in the case of the horse or cow to as much as 100 lb. per day. Hence the necessity for a liberal water supply to farm animals.

The stomach

The formation of the stomach differs in the case of ruminants (cattle, sheep and goats,) which have a four chambered stomach and chew the cud, and non-ruminants (horses, pigs, cats, dogs, etc.,) which have, like human beings, but a single stomach, and do not chew the cud. Ruminants, which are generally mild animals, seem to have been specially endowed by nature to eat quickly large quantities of roughage, often of a fibrous nature, after which they can retire to a spot safe from the depredations of wild animals to remasticate and digest at leisure. Such animals as horses and pigs, however, need a ration comparatively low in fibre content, as the latter contains but little nutriment.

The compound stomach of a ruminant (Fig. 4) consists of four chambers:—

1. The Rumen or Paunch where the saliva starts to act.
2. The Reticulum or water reservoir.
3. The Omasum—directly communicating with the oesophagus *via* the oesophageal canal.
4. The Abomasum.

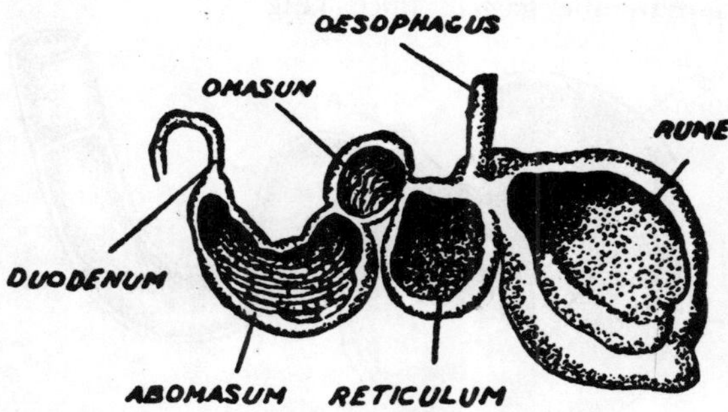

FIG. 4.—STOMACH OF A RUMINANT

The first three chambers constitute in reality an anatomical enlargement of the oesophagus, analogous to the crop of the chicken, whilst the fourth chamber is the stomach proper. In the first stage of rumination although liquid feeds may pass directly into all four chambers, the more solid feeds enter the rumen direct, which, owing to its capacity, can hold a considerable quantity of material, and also into the reticulum, though the latter serves more as a source of water for moistening the food in the rumen.

The more liquid portion of the food may now pass directly on to the omasum, but when the animal has finished feeding the major and more solid part is resurged into the mouth by contraction of the walls of the first and second stomachs for thorough chewing, and is then again swallowed.

This time the thoroughly masticated food by-passes the first and second chambers *via* the oesophageal canal and enters the third stomach where it is squeezed and ground between the corrugated folds of the stomach walls, and finally passed into the abomasum where it is acted on by the gastric juice. The four stomach chambers of the ruminants may, therefore, be said to subserve respectively the functions of a preliminary reservoir of food, a reserve chamber for moisture, a grinding or mixing chamber, and a final digestion chamber.

In non-ruminants the food passes direct to the one stomach for digestion by the gastric juice (Fig. 5).

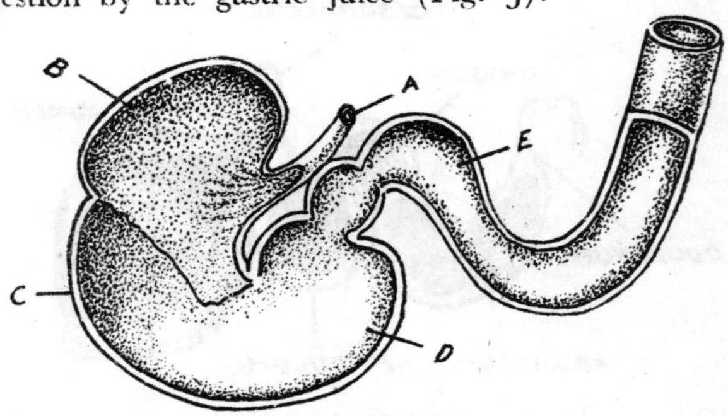

FIG. 5.—STOMACH AND DUODENUM OF A HORSE
A—Œsophagus. *B*—Cardiac portion. *C*—Fundus. *D*—Pyloric Region. *E*—Duodenum.

THE GASTRIC JUICE

The stomach may be considered as a receptacle which churns up the food and prevents too rapid an entrance into the intestines where the chief actions take place. The walls of the stomach are lined by simple tubular glands which differ in appearance and function in the fundus and pyloric ends. The ducts in the fundus end secrete a digestive enzyme pepsin, and also hydrochloric acid from specially differentiated cells, while those of the pyloric end secrete pepsin only. The pepsin is formed from a precursor—pepsinogen—and the active gastric juice consists of this proteolytic enzyme acting in conjunction with the hydrochloric acid, *i.e.*, pepsin-hydrochloric acid. This active enzyme brings about the first stage in the breakdown or digestion of proteins and produces proteoses and peptones and possibly some polypeptides, though the final digestion of proteins is reserved for the proteolytic juices of the intestines. The ducts of the stomach also secrete a ferment, rennet, which induces the coagulation of the caseinogen of milk and converts it into casein. Rennet is specially plentiful in the stomach of the calf—the chief commercial source of rennet—and constitutes a provision of nature to enable the calf to digest its natural food, milk.

THE INTESTINAL JUICES

While the digestive glands of the stomach lie in the stomach walls, most of those secreting the intestinal juices lie in organs outside the intestines into which the juices are conveyed through ducts. The gastric juice is acid, but the intestinal juice is alkaline. When the gastric contents reach a certain degree of acidity the pyloric muscle closing the entrance from the stomach to the duodenum relaxes and permits the entrance of the stomach contents into the small intestine. A remarkable chain of reactions is now set in action, initiated by the intestinal cells at this stage secreting an enzyme into the blood stream which is conveyed to the pancreas, which in turn is stimulated thereby to secrete the pancreatic juice, and at the same time the liver is stimulated to secrete bile. Also the ducts lining the intestinal canal secrete a specialised enzyme. We have therefore to consider:—

(a) The pancreatic secretion.
(b) The bile.
(c) The intestinal juice proper.

(a) The pancreatic juice

In the case of the horse and sheep the pancreatic duct and the bile duct open into the duodenum by a common exit (or entrance), while in cattle and swine the pancreatic duct opens lower down than the bile duct. The pancreatic juice which is secreted contains three digestive enzymes :—

1. The proteolytic enzyme, trypsinogen, which is the precursor of trypsin, the active agent, and acts on proteins.
2. An amylytic enzyme, amylase, which acts on starch and other carbohydrates.
3. A lipase, steapsin, which acts on fats.

It is possible that the latter two also enter the duodenum as precursors of the actual active enzymes. Concurrently with these activities the intestinal glands secrete an activising agent known as entero-kinase which converts the trypsinogen into the active trypsin, so the stage is now set for pancreatic digestion in the intestines.

(b) The bile

Bile is secreted by the liver, a large gland lying immediately below the diaphragm on the right side of the body. The juice secreted by the liver cells passes into small capilliaries which unite into small ducts and these finally unite into the bile duct which, as noted above, joins later on with the pancreatic duct in horses and sheep and conducts the bile into the duodenum a short distance from the stomach. In cattle and swine the bile is stored in the gall bladder a reservoir from which a duct leads directly into the duodenum. Sheep also possess a gall bladder, but the horse does not.

The chief functions of the bile are, firstly, to bring the fats into a finely emulsified condition so that they may be more readily acted on by the lipolytic enzymes, and secondly, to effect some degree of saponification. It also assists in precipitating pepsin which is present, thus preventing this enzyme from digesting the pancreatic enzymes and the bile, which would otherwise occur, since these substances are protein in nature.

(c) The intestinal juice proper

The walls of the small intestine contain a very large number of two specialised types of gland known as the glands of Lieberkkuhn and Brunner's glands respectively; these secrete a number of specific digestive enzymes which act chiefly on carbohydrates and derived proteins. These enzymes are:—

1. Maltase, which acts on maltose converting it into dextrose.
2. Sucrase, which converts cane sugar and sucrose into dextrose and laevulose.
3. Lactase, converting lactose into dextrose and laevulose.

and 4. A proteolytic enzyme, erepsin, which does not act on native or unchanged proteins, but completely breaks down the derived proteins, such as proteoses, peptones, etc., into their constituent amino acids.

The contents of the stomach are thus neutralised by the intestinal juices and, while digestion and the absorption of digestion products are proceeding, are gradually moved along by the peristaltic action of the intestine walls towards the large intestine.

In the large intestine the consistency of the alimentary canal contents becomes more solid due to absorption. There are also present in the large intestine—more so than in the small—very large numbers of bacteria, and a considerable part of the crude fibre of the food is digested by them. This process is often designated bacterial fermentation in the intestines. The fibre is broken down into various organic acids which are absorbed, and various gases, such as methane, hydrogen, etc., which are excreted.

Bacterial activity is far greater in ruminants than in nonruminants, so that a ruminant with its complex stomach and high intestinal bacterial fauna can be fed a far greater proportion of roughages in its ration than a non-ruminant.

That part of the food which has escaped digestion by the various digestive enzymes and by bacteria is finally voided as excreta.

We may now examine in a little more detail the chemical processes involved in converting the raw material of the food into the material comprising the tissues of the animal body, and animal products.

THE DIGESTION OF PROTEIN AND NON-PROTEIN NITROGENOUS BODIES

Digestion in the stomach

It was seen in Chapter II, in the discussion on the composition of proteins, that they were ultimately built up of a considerable number of amino acids of varying degrees of complexity, and that intermediate between the native, or whole proteins at one end, and amino acids at the other, a considerable number of intermediate products was formed during digestion, such as the metaproteins (acid or alkali proteins), proteoses, peptones and polypeptides of diminishing complexity. The gastric juice takes the digestion of proteins only as far as the peptones (and possibly some polypeptides), but does not produce amino acids. The final stage of protein breakdown is reserved for the trypsin of the intestinal juice. The gastric juice also splits up the conjugated proteins into their protein and non-protein parts, and the protein is then acted on in the normal manner.

Digestion in the intestines by trypsin

The chief active proteolytic agent operating on proteins in the intestines is trypsin formed from its precursor trypsinogen, secreted by the pancreas. The activation of trypsinogen is performed by an activating agent, enterokinase, formed by the cells of the intestinal walls following the entry of the pancreatic juice into the intestines. It acts in much the same manner as, but more energetically and thoroughly than pepsin, and produces a considerable quantity of pure amino acids although its action in this respect needs to be completed by erepsin.

Erepsin is a proteolytic enzyme formed in the intestinal cell walls, and acts somewhat differently from pepsin and trypsin in so far that it cannot act on native proteins but assists in breaking down that part of the protein molecule which is unaffected by trypsin or pepsin.

The final products of the digestion of proteins are thus the relatively simple amino acids which are then absorbed through the intestinal cell walls into the blood stream.

The nucleo proteins are similarly digested by the tryptic and peptic enzymes, and the nucleic acids which result are acted upon by a series of nucleases (*i.e.*, nuclei splitting, corresponding to lipases, or fat splitting), the final products being phosphoric acid, pentoses and various bases belonging to the purin group.

It has already been seen that the intestines contain an enormous number of bacteria, and many of these, which act in an alkaline medium, exert a putrefactive action on proteins which increases as the lower part of the intestines is reached, and diminishes again as the faeces become more solid. The products of bacterial decomposition are ammonia and various evil smelling substances such as indole and skatole, in the case of the carnivora, which may be absorbed to a considerable extent through the intestine but have no nutritive value, but rather a toxic action. The body protects itself against this poisonous action, however, by immobilising them in combination with other substances such as sulphates, in which form they are excreted in the urine as etherial sulphates.

The non-protein nitrogenous substances are much simpler in structure than the pure proteins, and resemble in many cases the ultimate digestion products of the latter. It has been assumed, therefore, that their digestion is a simple matter and amounts to little more than absorption into the blood stream for the ultimate use of the tissues. They have been shown to be more particularly abundant in green fresh fodders and other roughages, and appear to fulfil wider functions than mere absorption.

For example, alimentary bacteria are particularly numerous in the rumen of ruminants, and the non-protein nitrogenous substances being more readily attacked by bacteria than the true proteins, appear to serve the double function of acting as a readily available food supply for the bacteria, and, in consequence, protect the more valuable true proteins from bacterial decomposition before the digestive juices can perform their work.

Again, bacteria being present throughout the alimentary canal, the supply of readily available sources of food in the non-protein substances enables them to multiply and perform their specialised duty of fermentation of the carbohydrates of the food, particularly the less digestible parts.

The non-protein substances may also serve as an indirect supply of protein to the animal by being built up into protein in the bacterial cells, which are themselves in due course digested by the proteolytic enzymes, and so serve as protein amino acids from an originally non-protein source. Evidence is still vague as to the precise nutritive value of the non-proteins of food-stuffs, and an ultimate estimate could perhaps only be found by studying the amino acid yield of all the nitrogenous matter, protein and non-protein (crude protein), of feeding stuffs, or comparing the relative efficiency of the latter for maintenance and growth. Hence, while the non-protein nitrogen has undoubtedly a certain food value, the true protein is taken by many workers as the basis of protein value.

The digestion of fat

The digestion of fat is accomplished by the combined physical process of emulsification and the chemical process of saponification. The former is aided by the presence of free fatty acids and is brought about by the action of the bile. This favours the development of a greatly increased fat surface, thus facilitating the chemical process of saponifaction brought about by the steapsin of the pancreatic juice in the intestines. In this process free fatty acids and glycerol are formed, and the former further aid in the emulsification process. The two processes thus mutually assist each other. It has been maintained that the steapsin is first activated by the lecithin contained in the bile, but the bile certainly assists in fat digestion by its capacity to absorb considerable quantities of fatty acids and their soaps.

The digestion of carbohydrates

Most of the carbohydrates present in the feeding stuffs of farm animals, chiefly in the fibrous parts, consist of the group known as polysaccharides, including cellulose, starches, and the various pentosans. The disaccharides and monosaccharides are comparatively scarce except in certain crops such as beet, sugar-cane and the sorghums.

The cellulose part of the plant was for long assumed to be indigestible, but later investigations have shown that the crude fibre of feeding stuffs is partially digested in ruminants such as cattle and sheep, and in some non-ruminants. This is not true digestion, however, and takes place chiefly in those parts of the digestive tract where food stagnates, *i.e.,* in the rumen, and the caecum and colon. It is in precisely these portions of the alimentary canal that the greatest number of bacteria are found, and it is these facts which have led to the view that cellulose digestion is in reality a bacterial fermentation, particularly as heavy ingestion of crude fibre by cattle leads to a great increase in intestinal gas production (carbon dioxide, methane and hydrogen).

Cellulose does not appear to be attacked by any of the digestive enzymes of the higher animals, and the salts which are formed from the organic acids produced during fermentation appear to be the only possible nutritive factors which are derived from cellulose.

The pentosans, which are universally present in vegetable matter, and, as already stated in the previous chapter, occur in both the crude fibre and the nitrogen free extract, appear to be digested to a considerable extent by most farm animals. Their digestion, however, appears to be in the nature of a bacterial fermentation like that of cellulose, as no known pentosan enzyme has yet been discovered.

Pentosans constitute more than one third of the digestible organic matter in wheat straw pulp, and the methane produced during its digestion has been found to be proportionate to the total digestible crude fibre and nitrogen free extract which naturally includes the pentosans. This would indicate a bacterial fermentation rather than a true digestive process, and the products of the digestion of pentosans are very much the same as those from cellulose.

The hemicelluloses, or reserve carbohydrates of the cell wall, also appear to be fermented in much the same way as the pentosans and yield much the same products. The bacteria which attack the pentosans, cellulose and the hemicelluloses may also ferment any starch which escapes the action of the ptyalin of the salivary juice and the amylopsin of the pancreatic juice, and finds its way to the caecum or colon, the chief seat of fermentation, for example, in the horse.

Digestion of starch

(a) In the mouth and stomach

The first stage in the digestion of starch occurs during the brief stay of the food in the mouth. Here, at a temperature of about 40°C, the starch of the food comes into contact with the ptyalin of the saliva, which acts best in a neutral medium and is inhibited by acids and alkalis. The first stage in digestion is the conversion of the starch into various dextrins which are ultimately changed into maltose, at which stage the action stops. If starchy foods are cooked or partially dextrinised, digestion is more rapid as the heating has already performed the first stage.

When the masticated food arrives in the stomach, the gastric juice inhibits further ptyalin activity as soon as a certain degree of free acidity (HCL) has been reached. In the horse and pig, however, owing to the peculiar configuration of the stomach in the horse, and the fact that the left hand end of the pig's stomach contains no peptic glands, the hydrochloric acid of the gastric juice takes a considerable time before it reaches the mass of the food, and ptyalin digestion of starch may proceed for several hours. This is even more pronounced in the case of ruminants which secrete only a little ptyalin, since the food remains in the rumen for a long time in contact with the large amount of saliva secreted.

(b) In the intestines

Any starch which has not been acted on by ptyalin comes into contact with the pancreatic enzyme amylopsin, when the food reaches the duodenum, and is there converted into maltose.

Digestion of the disaccharides and monosaccharides

The animal organism cannot directly utilise any of the disaccharides, comparatively simple though they are, and if they are introduced directly into the blood stream they are excreted unchanged in the urine. It has already been seen in Chapter II, that they all have the same general formula, $C_{12}H_{22}O_{11}$, and that when hydrolysed, *i.e.,* heated with dilute acids or alkalis, they take up one molecule of water, and are converted into two simple monosaccharides. Very much the same thing happens in the alimentary canal. The hydrochloric acid of the gastric juice can, and does to some extent, hydrolyse or invert the disaccharides. The chief inversion, however, takes place in the intestines by a series of enzymes known as the invertases, each of which appears to be specific for its own particular sugar. Thus, cane sugar or invert sugar is acted upon by the enzyme sucrase, yielding dextrose and laevulose; maltose, resulting from the ptyalin digestion of starch is acted on and split up by maltase, yielding dextrose only, while lactose, or milk sugar is split up by lactase, and yields dextrose and galactose. These inversions appear to take place chiefly in the epithelial cells lining the intestinal lumen.

It will readily be seen that the monosaccharides need no digestion. Like the amino acids in the case of the proteins, they are the ultimate products of carbohydrate digestion, and when assimilated into the blood stream are at once available for whatever demands the body has for them.

Digestion of the minerals of the food

In determining the minerals, or ash as they are sometimes called, in a feeding stuff in the laboratory, all the minerals which existed both in the free, or electrolytic form, and also those which existed in organic combination, such as the sulphur in the cystein molecule of the proteins, are included. In natural digestion each of these groups has to be taken into consideration. Little, however, is known about the digestion of these substances, but the free salts appear to be more or less freely absorbed direct into the blood stream. It also appears extremely probable, as far as present knowledge goes, that the sulphur of the proteins and the phosphorus of the nucleo-proteins are also absorbed into the blood stream, still in organic combination with the amino acids into whose composition they enter.

We have thus seen that all feeding stuffs are made up of certain definite groups of materials, and that in the course of digestion they are broken down into a number of simpler substances which are capable of being absorbed through the walls of the intestines and passed into the blood for the direct use of the animal. It now remains to trace briefly how this latter operation is effected, and the special uses or functions which the various products of digestion subserve.

Resorbtion

Anatomically speaking the products of digestion are still outside the body until they have been absorbed into the blood and lymph by the process known as resorbtion. The small intestine is the special part of the alimentary canal through which this process takes place, and if its epithelial lining were carefully straightened out and examined under a magnifying glass it would be seen to be covered by an enormous number of little 'hillocks', somewhat resembling, to use a rough analogy, the rough surface of the familiar circular rubber pads used on toilet tables. These small hillocks or protuberances are called Villi, and are covered by epithelial cells. Each villus contains in its interior a network of chyle vessels linked up with a larger central vessel, and all these ultimately unite to join up with the main lymphatic system. In each villus there is also an intricate ramification of small arteries and veins which finally unite into larger vessels which join up with the main blood stream. The general configuration of these vessels is seen below:—

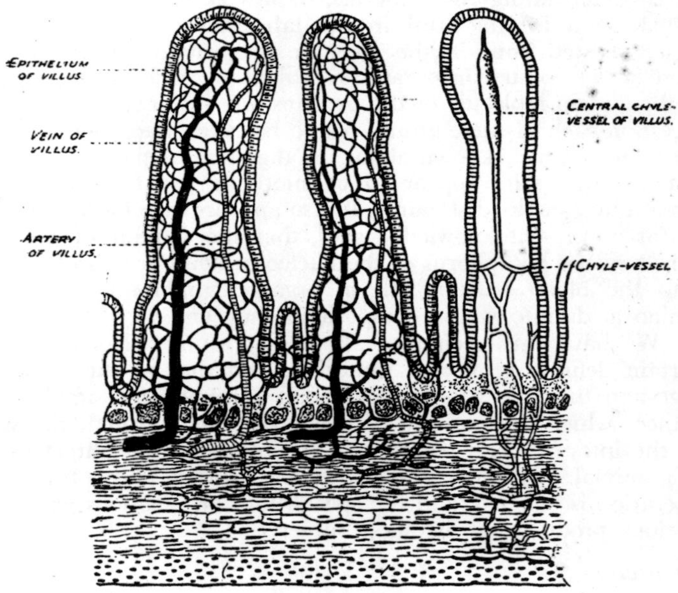

Fig. 6. Section of Villi

The products of digestion are passed across the epithelial cells of the villi into the vascular system within, and ultimately into the larger lymphatics or blood vessels, but precisely how this is effected is still a matter for debate, and chiefly a matter of physiological interest rather than of concern to the practical farmer, but it may be stated in general that the transfer takes place by virtue of the inherent function of the living epithelial cells. The products of digestion of the carbohydrates, proteins, organic acids and salts appear to pass directly through the epithelial lining into the blood capillaries in the membrane, and thence *via* the portal vein to the liver. The fats on the other hand enter the lacteal vessels of the villi, but physiological speculation is still rife as to whether the products of fat digestion are reformed into fats before the transfer across the epithelial lining, or whether they are reformed in the epithelial cells themselves. Suffice it to say that the fats are resorbed into the villi and then pass *via* the lympthatics to the thoracic duct.

A stage has now been arrived at when the products of the digestion of food are directly available for the animal and the particular uses to which each is put in the animal economy may now be considered.

THE UTILISATION OF FOOD

It is a matter of considerable importance to the stockowner to know how the different foodstuffs available for different purposes can best be utilised. The functions of the different groups of food materials considered in the previous chapter vary to a considerable degree. One of the most important functions of the food is to build up the body and keep the various tissues in repair; this is done mainly by the proteins and minerals of the food. For this purpose the proper type of proteins, minerals (and water) are all essential, and none can be neglected without detriment to the animal. Another function of the food is that performed mainly by the various groups of fats, carbohydrates and to a less extent the proteins, and is to supply the body with a source of potential energy which, in the process of metabolism, will be converted into heat and kinetic energy, these two terms being different expressions of one and the same thing—energy.

The minerals cannot be regarded as foods or suppliers of energy in the ordinary sense of the term as applicable to fats, carbohydrates and proteins, although there is an applied sense in which they may be so regarded. They do not yield energy by oxidation, but there is reason to believe that they are able to act indirectly as sources of energy by virtue of the osmotic properties which they possess. In this way the ingestion of minerals assists the absorption and diffusion of the body fluids, an action which may be considered to be equivalent to the supply of a certain amount of energy to the body, thus raising the minerals to the category of foods by virtue of the fact that they are not only tissue builders but are also sources of energy.

Physiological opinion has undergone considerable change as to the relative values of the organic constituents of foodstuffs since Liebig advanced the opinion last century that the proteins were the chief producers of muscular energy, while the carbohydrates and fat merely acted as fuel and maintained the body temperature. This is now recognised to have been a wrong view. From a physiological point of view it would seem to be immaterial to the actual cells of the body whether they derive their energy from the proteins or from the other food constituents, although they can probably get it more easily and quickly from the proteins and carbohydrates, than from the fats. It must also be recognised that body heat is not something apart and requiring to be specially catered for, but an inevitable concomitant of all cellular activity. Life and heat are inseparable and in performing its other functions in the body a cell must also produce heat, as a by-product so to speak of functional activity. Hence it is a matter of indifference as far as the cells of the body are concerned from what foodstuff their energy is derived so long as they are supplied with their dynamic requirements.

The story is different, however, in regard to the building up and repair processes of the body tissues, which can only be done by the proteins and minerals with the co-operation of water. The constituents of food may, therefore, be classified according to their functions in the body as follows:—

TISSUE FORMERS	ENERGY AND HEAT PRODUCERS
Proteins	Fats
Minerals	Carbohydrates and also the
Water	proteins and albuminoids

The proteins alone are able to perform both functions of the food, and it is this physiological competence which makes them such an important part of the ration. Without protein, growth or production would be impossible, and so would repair of the daily wear and tear of the body which protein alone can serve.

We may now proceed to examine the criteria by which the relative values of different foods may be assessed, *i.e.,* how to know whether a particular food is good, bad or indifferent for particular purposes in stock-feeding. Information on such matters can only be obtained by submitting the food under consideration to the following tests :—

1. *Chemical analysis*

This determines the percentage of each nutrient that the food contains, but it tells nothing about the extent to which an animal will be able to digest and utilise the food. This will be ascertained by:—

2. *Digestibility trials*

These will show the percentage digestibility of the various food constituents.

3. *The physical test*

This is a refinement which determines the potential energy capacity of the food in terms of calories of heat, or the gross energy content of the food. (See Chapter V.)

4. *The economic test*

When the value of the food has been determined in terms of digestible ingredients, the question is asked, what are the relative values of different foods based on the cost per food unit ? (See Chapter XVI.)

THE CHEMICAL ANALYSIS

The chemical analysis of a food constitutes the first stage in the series of operations necessary for the proper evaluation of the food in terms of its capacity to supply energy to an animal. It tells us how much of the different food ingredients the food contains, so that it is possible to make an approximate comparison between one food and another. For example, foods rich or poor in fats or proteins or carbohydrates can thus readily be distinguished and appraised, and the amount and types of minerals present determined. Chemical analysis, however, only tells us something about the gross potential of the food, but little about its actual value in feeding practice. The next stage, therefore, is to determine the digestibility of the foodstuffs by actual feeding trials, or the nett energy for production purposes by elaborate calorimetric trials.

Digestibility Trials

The term 'digestibility' employed in nutrition trials has a definite and specialised meaning. It tells the percentage of the total feed, and also the percentages of the separate ingredients of the feed which are acted on by the digestive juices, absorbed and capable of being made use of by the animal's tissues, or in other words, digested. For example, in such a trial at Lyallpur on heifers fed 4F American cotton seed cake, out of every hundred pounds of the cake fed, calculated on a dry basis, 72.8 were digested, so that the digestibility coefficient of the dry matter of the cake as a whole was 72.8. Similarly of each 100 lb. of protein contained in the cake, 85 lb. were digested, so the digestibility coefficient of the protein is said to be 85, and so on for all the other constituents. By means of digestibility trials, therefore, it is possible to compare different foodstuffs by determining how much of their food constituents are digested by a particular animal. It will be clear on reflection, however, that such determinations will not tell the whole story about the value of the foodstuff, as this will depend on the capacity of the animal's tissues to utilise the products of digestion, or in other words they do not give the complete picture about the biological value of the food. For example, the digestibility of the protein of maize is about 75, but if maize were fed as the sole protein in a ration, the high digestibility coefficient of the protein would be largely nullified by the biological incompleteness of the protein, zein, of maize in certain essential amino acids. All such factors have to be taken into account in using digestibility data for the compilation of rations, as will be more fully described in succeeding chapters.

FIG. 1. Experimental cows in outside stalls with harness for digestibility experiments.

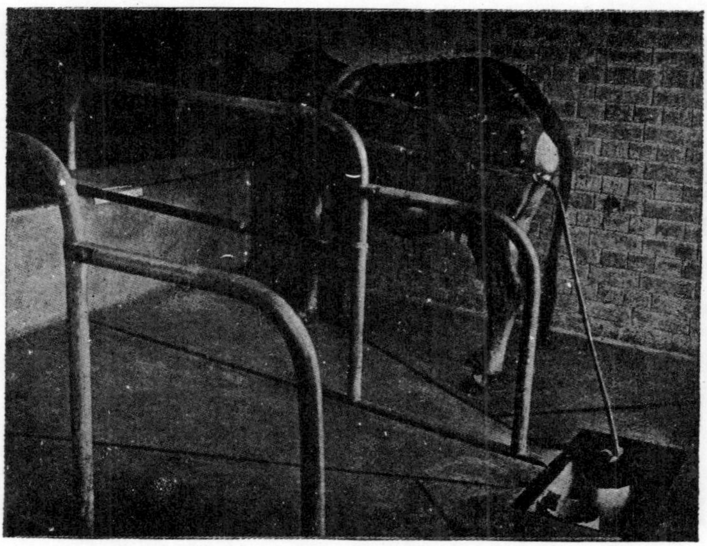

FIG. 2. A cow in inside stalls with harness.

FIG. 1. A Bullock in outside stalls with harness for collecting dung and urine.

FIG. 2. Bullocks undergoing digestibility trials with harness attached while at work on a cane crusher.

Conducting a digestibility trial

Digestibility trials are conducted with animals in specially designed stalls, and the series of operations may be compared roughly with the keeping of a banking account. It is obviously not possible to measure directly the amount of food that is digested, and the method adopted, therefore, is to measure the amount of food materials which the animal eats on the one hand, and the amount of excreta voided on the other. The balance or difference between the two tells how much of the different parts of the food have been digested and retained in the animal's body for utilisation [Plate IV, Figs. 1 and 2]

Every digestion trial is preceded by a preliminary non-experimental period varying from 10 days to a fortnight during which the animal is fed the food under trial in order that the digestive tract may be freed from all traces of previous feeds. During the ensuing trial proper the same feeding regime is continued, but the full experimental operations are followed, and the exact quantity of food eaten by the animal each day is recorded and also the amounts of urine and dung voided. Representative samples of all these are then analysed in order to compute the balance sheet previously mentioned. [Plate V, Figs. 1 & 2]

Example of a digestion trial

The writer has recorded (Lander, 1930) the results of digestion trials with four heifers fed on oat hay and American cotton seeds. Taking averages, each animal was fed 9.2 lb. of oat hay per day for 15 days during which the actual trial lasted. During these 15 days the average weight of the faeces excreted daily was 17.8 lb. per animal. Samples of the oat hay and faeces were analysed and found to contain the following percentages of dry matter:—

Oat hay	..	..	96.00 *per cent*
Faeces	..	..	20.02 *per cent*

The 9.2 lb. of hay, therefore, contained 8.83 lb. of dry matter and the 17.8 lb. of faeces contained 3.56 lb. The difference may be assumed to represent the amount of dry matter digested by the animals. This amounted to 59.68 per cent of the dry matter eaten. Therefore, we say that the digestibility of the dry matter eaten was 59.68. The digestibility coefficients of the

separate constituents of the hay were determined in exactly the same way. The average composition of the oat hay is shown in detail in the following table:—

AVERAGE COMPOSITION OF OAT HAY

Moisture.	Dry matter.	Ash.	Fat.	Fibre.	Protein.	Nitrogen free extract.
%	%	%	%	%	%	%
4.00	96.00	6.94	1.18	33.98	5.00	48.27

and the complete results from the digestibility trial on the hay were as follows:—

DIGESTIBILITY COEFFICIENTS OF OAT HAY

Dry matter.	Ash.	Fat.	Fibre.	Protein.	Nitrogen free extract.
%	%	%	%	%	%
59.68	23.44	70.59	64.54	43.47	62.61

In order to determine corresponding data for a concentrate, such as cotton seeds, it is necessary to feed the animals for a preliminary period on a maintenance ration roughage such as oat hay whose composition and digestibility data have first been determined as described above, as an animal cannot be fed on concentrate alone. The determination of the data for the concentrate is slightly more complicated. Known amounts of the concentrate are fed with the basal ration, and the complete digestibility data are obtained for the composite ration. Thus, knowing the data for the basal ration and those for the combined ration a simple calculation enables the data for the concentrate to be determined. Thus, taking the average for the four heifers used in these trials, each animal received per day in the cotton seed period 9.9 lb. of hay, and 3.0 lb. of cotton seed. The average daily excretion of faeces on the mixed ration was 22.1 lb. and the composition of the cotton seed and faeces was as follows:—

AVERAGE COMPOSITION OF COTTON SEED AND FAECES

	Moisture.	Dry matter.	Ash.	Fat.	Fibre.	Protein.	Nitrogen free extract.
	%	%	%	%	%	%	%
Cotton seed.	6.33	93.67	4.95	18.77	26.20	18.13	25.62
Faeces.	77.18	22.82	3.43	0.51	7.00	2.13	9.75

The digestible matter contained in the combined ration was computed exactly as for the oat hay. If now it is assumed that the digestibility of the oat hay was the same in the combined feed as was the case when it was fed alone, it is possible to calculate how much of each food ingredient in the total ration was derived from the hay, according to the previous trial, and by difference, the amount that came from the cotton seed, and consequently the digestibility coefficients of the latter, may be determined.

The data showing these coefficients for the cotton seed concentrate are shown below:—

DIGESTIBILITY COEFFICIENTS OF COTTON SEED

Dry matter.	Ash.	Fat.	Fibre.	Protein.	Nitrogen free extract.
%	%	%	%	%	%
56.94	-	89.27	54.33	64.82	51.95

It is clear that the indirect method which has to be used for a concentrate is not as reliable as the direct method used for a single feed. The margin of error may be considerable and will depend upon the relative proportions of concentrate to roughage, just as when estimating the data for a single foodstuff such as wheat *bhusa,* the margin of error for some of the constituents such as the protein will depend on the amount of protein in the *bhusa.* In general, it may be stated that the smaller the percentage of concentrate present in the combined ration, the greater will be the margin of error. Furthermore, as has been stated, it is assumed that the digestibility of the roughage is unaffected by the presence of the concentrate, and this may not necessarily be true. Obviously all errors arising from this cause will be assigned to the concentrate.

It sometimes happens that absurd results are obtained in a digestibility trial for ingredients which are only present in small amounts. For example, the writer has found in protein poor roughage such as wheat straw [Lander, 1932] with a 3 per cent or less protein content that negative results may be obtained. The analyses indicate that the faeces contain more nitrogen than the food, a finding which is no doubt due to the influence of excretory products, etc., from the digestive juices of the alimentary canal passing out with faeces and joining the food excretory nitrogen, and hence becoming calculated with it. Correspondingly, results showing a digestibility above 100 may be obtained, but these abnormal figures generally correspond to a very low content of the constituents to which they refer, in the feeding stuffs, such as is sometimes found in the cases of protein or fibre or fat. Nevertheless, properly conducted feeeding trials provide a method whereby most of the data have been collected for the compilation of feeding standards.

Digestible Proteins and Total Digestible Nutrients

(*See Appendix* 2)

Since proteins serve very special functions in the body the percentages of digestible protein are given separately in the feeding standards. This figure is obtained by multiplying the percentage of protein in the food by the digestibility coefficient. For example, 4F American cotton seed contains 17 per cent of protein of which 64 per cent is digestible. Thus the percentage of digestible protein is approximately 11. (For further details regarding the protein see Chapter V.)

The total digestible nutrients comprise in addition to the protein, all the other organic parts of the food, *viz*., the fats, fibre and the nitrogen free extract and their digestible percentages in the food are calculated in the same way as for the proteins, except that the fat figure is multiplied by 2.25 (or sometimes 2.3), because its value in terms of energy production for animals is approximately 2.25 times that of protein or carbohydrate. The sum of the digestible percentages of all the separate ingredients is known as the total digestible nutrients.

Digestibility trials also enable what is known as the daily nitrogen balance of the animal to be determined over the experimental period. Just as a trial shows how much food an animal is digesting, so it tells whether an animal is losing more nitrogen in the excreta than is contained in the food, or whether it is gaining in nitrogen, *i.e.*, protein. This is an important factor in connection with maintenance rations which may be briefly considered here.

Maintenance rations

In dealing with the feeding of animals it will be necessary to consider foodstuffs from two aspects, *viz.*,

(1) The food that is needed to maintain an animal in health over a long period without gaining or losing weight, and without performing any appreciable work, or producing anything. This is known as the maintenance requirement, and,

(2) The additional food that is needed above maintenance requirements for growth, work, milk production etc., etc.

A ration which meets the requirements of No. 1 above is known as a maintenance ration and will always give a positive daily nitrogen balance. Substances such as wheat straw or rice straw may not be, and for many classes of cattle are not maintenance rations, because they do not contain sufficient digestible protein of the right type to meet an animal's maintenance requirements. The ordinary functions of life in a resting non-producing animal involve a certain amount of tissue breakdown, and hence a certain level of nitrogen metabolism and excretion. If this cannot be made up from the protein of the food, then the animal has to draw on its own body tissues to restore the balance. This involves an excess of nitrogen excretion in the excreta over that contained in the food. The daily nitrogen balance thus affords a useful criterion as to whether a ration constitutes a maintenance ration or not.

Maintenance rations will be dealt with in greater detail in the next chapter, and although it is usual in using feeding standards to discriminate between maintenance requirements and production requirements, no absolute line can be drawn between them. For example, a cow giving a certain amount of milk will need a certain amount of food according to her weight merely for maintenance and a certain additional amount for the milk produced. Yet if this combined ration, so to speak, is suddenly or gradually reduced merely to a maintenance ration level, she will still continue to produce some, although a diminished amount of milk, but this will be done by the animal drawing on her own body tissues. She will ultimately dry up much quicker than if a full productive ration had been given, and her body weight will also decline in consequence.

BIBLIOGRAPHY

1. Lander, P. E. & Dharmani, L.C. 1930. Digestibility Trials, Part V. Mem. of the Dept. of Agri. in India, Vol. X, No. 6. 181.

2. Lander, P. E., & Dharmani, L.C. 1932. The Indian Jour. of Veterinary Science and Animal Husbandry, Vol. II, Part II, p. 147.

CHAPTER IV

GENERAL PRINCIPLES OF MAINTENANCE AND PRODUCTION

MAINTENANCE

An animal's body may in some respects be regarded as an engine or transformer of energy. In the case of the engine, however, the fuel supplied does not become an integral part of the body of the mechanism, but merely transforms the potential energy of the fuel into kinetic energy represented as work. In the animal body the composition of the fuel or food produces not only kinetic energy in the form of work or movement, or production in some other form, but also enters into integral relationship with, and becomes a part of, the animal's body.

Two distinct but closely related processes are in continual operation side by side. One of these is the building up of the body tissues from materials supplied by the products of digestion of the food. This process is known as Anabolism, or building up. The second process is known as Katabolism, or breaking down, and represents the destructive processes which are continually going on in all living tissues. The sum total of these two processes is known as Metabolism. If a fully grown animal is suddenly deprived of food, *i.e.*, brought into a condition of fasting, it follows that within a short time the building up process will come to a stand-still owing to the shutting off of food supplies. The Katabolic or breaking-down process, however, will continue and the products of Katabolism will be excreted in the urine faeces and sweat, as before. The amount of these excretory products if carefully estimated will give an indication of the amount of the break-down or Katabolic processes which are going on, a state referred to in physiological terms as 'fasting Katabolism'. It follows that during fasting the animal will lose weight, and if fasting is continued long enough, will die.

In the fasting animal, therefore, there is a great preponderance of the katabolic over the anabolic processes, the latter being almost negligible. If this state of affairs is to be arrested or reversed, a certain amount of food will have to be fed, and as the animal is gradually brought back to normal, a point will be arrived at when the amount of food fed, will be just sufficient to counteract the katabolic or breaking down processes. The amount of food which is necessary to achieve this end is the minimum required for the animal if it is to maintain its body weight and remain in health whilst doing no work other than taking a minimum of exercise, and producing nothing in the way of animal products. This amount of food is called the 'maintenance requirements' or 'maintenance ration' of the animal.

Another way of describing a maintenance ration, is to say that it represents that amount of food which is necessary to maintain the animal's body weight and keep it in proper running order and in health whilst nothing is being produced in the form of work, or other kinds of production such as milk, etc. For the purpose of merely maintaining an animal at constant weight and in health, therefore, the maintenance ration is the minimum needed, and any food fed over and above this minimum will be wasted if no production is required.

A maintenance ration has been referred to so far as applying to a fully grown animal. The argument as to its nature applies, however, to animals of all weights and ages, whatever the weight or the age of an animal may be. For any particular weight of an animal its body requires a certain amount of food solely for its maintenance and any food fed above that amount falls in a different category, *viz.*, the food required for production, whether that production be increase in body weight, as in the case of a growing animal, or whether it be production such as work in the case of a bullock, or production in the form of animal produce, such as milk, eggs, etc.

A very rough analogy may be taken in the case of a car the engine of which is just allowed to 'tick over' with the gear thrown out. The petrol then consumed may be considered to represent the maintenance requirements for that engine to keep it just running and doing no work. If now the gear is thrown in this represents a demand on the engine for work, but if no further petrol is supplied the engine cannot respond and will stop.

If, however, production, *i.e.*, work is demanded of the engine then extra petrol must be fed or in other words a production petrol feed must be given. So it is with the animal, with due adjustment in the argument, allowing for the fact that the animal is a living organism.

The maintenance requirements for an animal in a state of rest or non-producing condition, must suffice to provide the energy required for the continuance of normal bodily functions, such as respiration, digestion and so forth, and a limited amount of food is required for this purpose, depending on the age, weight and type of the animal concerned, and, as mentioned above, any food given in excess of these requirements would be wasted as far as maintenance is concerned, just as petrol would be wasted if the throttle of a car is open when the gear is out.

Not only would any additional food be wasted as far as maintenance requirements are concerned but (apart from the animal possibly utilising it to put on fat which may not be wanted) the animal's body will have to expend unnecessary energy in digesting surplus food and eliminating excreta. The maintenance ration must supply a certain minimum amount of energy represented in terms of digestible protein and total digestible nutrients to maintain the essential body functions, *i.e.*, its running requirements so to speak. It must in addition, supply a certain minimum of essential minerals and vitamins, as it by no means follows that because an animal is doing no work or producing nothing that any of these aspects of its ration can be neglected for any length of time.

An animal's body is capable of storing considerable quantities of some of the vitamins, such, for example, as vitamin 'A', which is stored in the liver, and of minerals which are stored in the skeletal structure, and consequently it may be able to carry on for quite a long time without any untoward results becoming apparent; but if the maintenance ration is deficient in any of these essentials this deficiency may become manifest after the lapse of time, as is well known in the case of domestic animals kept in confined spaces when their rations have been unsatisfactory. Such defects may become evident in the development of such conditions as rickets. A maintenance ration, therefore, although generally expressed in terms of food required by an animal during 24 hours, must meet its needs over a prolonged period of time.

Nevertheless, although the maintenance ration must maintain health for a long period, it is not easy to say just precisely what constitutes a maintenance ration to suit all conditions. It is necessary to take many factors into consideration in order to comply with the health factor, and the purely physiological definition of a maintenance ration may not be exactly the same as that required from a practical stand point. Thus it is essential that the animal should take a certain amount of exercise, and this will influence general metabolism, and cause a more rapid evacuation of excreta. This is an important consideration, because lack of exercise leads to an accumulation of indigestible food and waste products in the alimentary canal, which may in turn lead to entirely fallacious conclusions in regard to the weight of the animal kept under such conditions. Lack of exercise will also have harmful effects on the animal itself. It is necessary, therefore, in considering maintenance rations for ordinary farm animals, to remember that the actual requirements such as might be determined in rigidly controlled physiological experiments over a short time in which the animal is doing no work must be increased somewhat under ordinary practical conditions to allow for sufficient exercise to keep the animal in health. This fact has been taken into account when computing maintenance requirements for ordinary farm stock as given in this book, and in drawing up the feeding standards described in Chapter V and Appendix 2.

How Maintenance Requirements are Determined

The maintenance requirements of different types of animal have been largely deduced by carefully controlled experiments on cattle using an elaborate contrivance known as an Animal Calorimeter (Plate VI), and to a lesser extent on sheep and horses. Elaborate experiments have also been conducted by the late Professor T. B. Wood on pigs at Cambridge. In conducting such experiments the animal is kept in a closed chamber through which air, which can be carefully measured, is allowed to circulate. The food fed to the animal is carefully weighed and analysed, and also the excreta. The analysis of the air passing through the chamber includes the products of respiration

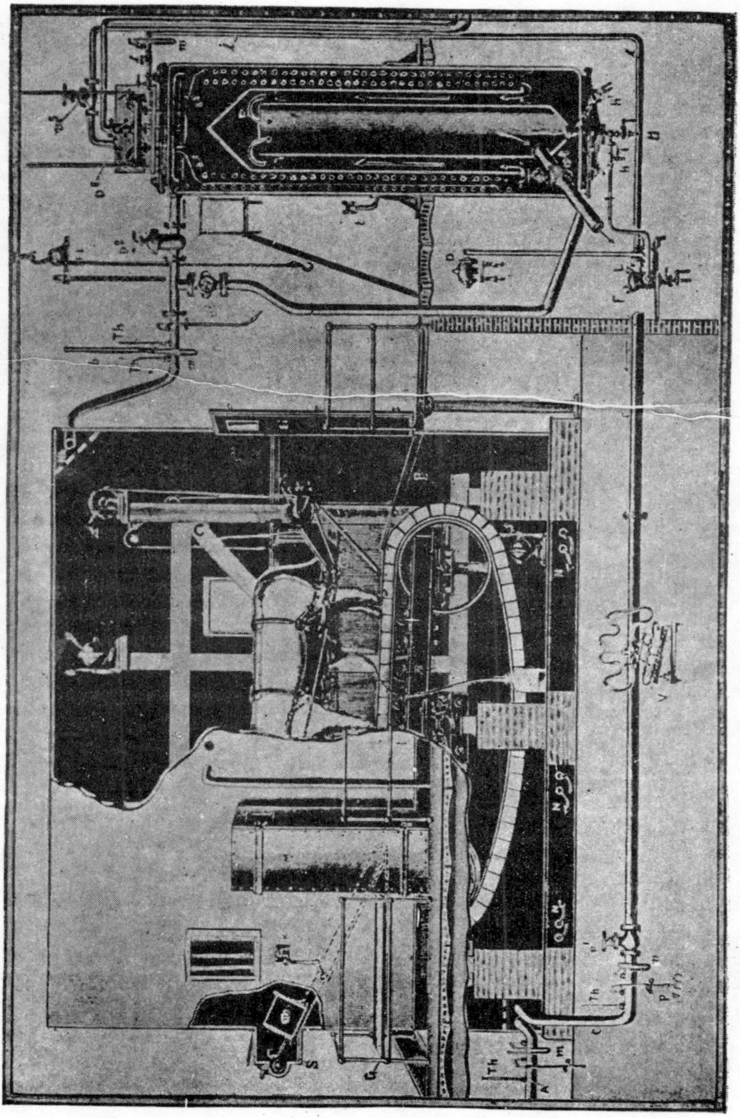

Regnault-Reiset apparatus as used by Zuntz.

and other gaseous excreta. At the same time the amount of heat which the animal produces is very accurately measured by ascertaining the rise in temperature of known quantities of water which are allowed to circulate around the inner chamber of the Carlorimeter in which the animal is kept. An elaborate series of data is thus obtained from which the amount of energy or food required by the animal for maintenance can be determined.

Such experiments involve a tremendous amount of work and are very costly, but they have formed the basis of all methods for accurately determining the maintenance requirements of animals as laid down in the various standards. It is not necessarily the case that the maintenance requirements as determined for one type of animal will be the same for another type of similar weight.

A less elaborate, but nevertheless approximate guide as to what constitutes a maintenance ration may be obtained from ordinary digestiblity trials carried out over a certain period of time. During such trials as described in Chapter III, careful daily records are kept of the weights of the animals and all the food and excreta are analysed. From the weight and condition of the animals and from the analytical data of the food and excreta it is possible to determine whether they are in a condition known as 'nitrogen equilibrium'. If an animal is found to be excreting more nitrogen (representing protein break-down) in the excreta, than is being fed in the food, it means that the protein supplied is deficient or defective, and that the animal is drawing on its own bodily reserves of protein. If, however, more nitrogen is being supplied in the food than is eliminated in the excreta, it means that no such demand is made on the body tissues, and, from the point of view of nitrogen or protein the food is both efficient and sufficient, or in other words constitutes for practical purposes a maintenance ration. Similar arguments apply to the other ingredients of the food such as the carbohydrates, minerals, etc.

In addition to the main consideration of the amount of food fed, there are certain other factors which must be taken into account in deciding whether a ration is a satisfactory maintenance ration, although many of these apply *mutatis mutandis* to production rations as well. The chief of these are as follows:—

Palatability

It is a matter of common knowledge that the mere sight or smell of food causes an increased flow of saliva, and in some cases of gastric juice, which may influence an animal in its readiness to eat. An unpalatable ration tends to check the appetite, although it does not appear that an unpalatable ration once eaten is digested to a less degree than a palatable one. The food, however, should always be both palatable and attractive for all types of animal and damaged or stale feeding stuffs should be avoided. Feeds which are suitable for one type of animal may not be suitable for others, and one type of feed may be satisfactory in certain combinations but not with others. For example, maize is not a satisfactory grain when fed alone with poor roughages to pigs or horses, but if combined with grains of higher protein content or with some legume, forms an excellent part ration.

Suitability of food

Different types of stock need different types of food, and a ration suitable for one type may be quite unsatisfactory for another, as, for example, in the cases of ruminants and non-ruminants. The digestive system of a ruminant is constructed in such a way as to enable it to deal with considerable quantities of roughage, and a roughage which in quantity and quality would prove suitable as a maintenance ration for a cow would be quite inadequate for the same purpose, weight for weight, for a horse or a pig.

Dry matter

Notwithstanding the distinction just mentioned, every ration should contain sufficient bulk to enable the stomach and intestines to be properly distended so as to promote that sense of repletion after feeding which is necessary for an animal's welfare. Depending on the type of animal, part of this bulk should always be supplied by roughages sufficient in quantity to open up the more concentrated parts of the food so as to enable the digestive juices to have free access to all parts, and to aid the peristaltic action of the intestines which assist this process, and help to move the undigested residue along the alimentary tract. The proportion of dry matter needed in a ration varies with different animals, and is described for separate types in subsequent chapters, but on an average it should constitute about 2.5 per cent of the body weight for the larger types of animals.

Protein requirements

The protein must be fed in the right quantity and be of the right type, although the amount needed for maintenance is comparatively small, and only necessary in sufficient quantity to keep the body tissues in proper repair. Even during rest the tissues of an animal are in a continuous state of 'metabolic activity', that is to say, a small part of their substance is being continually broken down. As most of the muscular tissues of an animal and a considerable part of the other tissues are composed of protein, there is thus a continual small loss from the body of the products of protein break down, and these have to be replaced by similar substances obtained from the food. Although the actual protein requirements of an animal at rest are small, it is nevertheless essential that the protein in the feed should be of the right kind, which will be broken down during digestion into the essential amino acids which the animal's tissues need for their reconstruction. If an animal is fed the wrong type of protein as its sole source of nitrogen, trouble may ensue quicker than if no protein is fed at all, because the animal has to expend energy in masticating and digesting the food, whilst the products of digestion cannot be properly utilized to rebuild body tissues and supply energy.

For example, Zein, the protein of maize is a typical example of a deficient protein, and maize should not be fed as the sole source of protein in the concentrate part of the ration, whether for maintenance or production. Again, rice straw is deficient both in protein and in minerals and experience has shown that it does not constitute a maintenance ration for dairy cows which have been accustomed to better feeds. It would, therefore, be a bad combination to attempt to make up the defects of poor quality rice straw by feeding maize as a supplement, which is in itself deficient both in protein and minerals. A suitable supplement for a non-maintenance hay or straw ration is a small amount of some suitable concentrate such as bran, or legume hay. Trials at Lyallpur have shown that legume hays constitute maintenance rations for dairy heifers of 800 lb. weight.

Good quality oat hay is also a maintenance ration for such animals.

Mineral supply

There must be a properly balanced and adequate mineral supply in the feed to provide for the body's needs in minerals, to replace those lost in the excreta and to maintain the acid-base equilibrium of the blood and lymph as described in Chapter II. Both the protein and the mineral aspects of a ration assume a much greater degree of importance in requirements for production, but both are also important for maintenance. If the calcium or the phosphorus content of a ration is deficient or ill-balanced, the animal may find it necessary, over a prolonged period of feeding with this type of ration, to draw on its own resources in the skeletal structure. Such a drain may proceed for a considerable time without any ill effects being observed, but ultimately a condition known as Osteomalacia or softening of the bones will ensue with all its subsequent ill effects. Osteomalacia is a disease brought about by continued feeding with rations deficient in calcium or phosphorus or both, often coupled with vitamin 'D' deficiency.

A number of other minerals such as iron, iodine, copper, sulphur, potassium, and magnesium, more fully described in Chapter II and under ' Production ', should also be present in small but sufficient quantity in any maintenance ration.

Critical temperature

The amount and type of ration required for maintenance will vary with the external temperature, and a further supply of energy over and above the normal amount of metabolizable energy supplied to animals in a maintenance ration, as judged from standard requirements, will be necessary where what is known as the 'critical temperature' has to be taken into consideration. The 'critical temperature' for any particular animal is that approximate temperature of the air when the heat lost by the body by physical means, such as evaporation, just balances the heat produced as a result of internal work and tissue oxidation, and below which further tissue oxidation is necessary to maintain the normal temperature of the body. It is clear, therefore, that a ration which may be a maintenance ration at or above the critical temperature will cease to be one at a temperature below it, and additional provision must be made to supply the extra energy needed in the form of additional food to maintain the body temperature. It is difficult to give precise data for the critical temperature of different farm animals as it varies considerably with different species and between different members of the same species, and it will also vary according to the nature of the animals' coats. The figure for horses, however, is about 16°C (60°F) and for cattle somewhat higher.

Proportion of Different Food Ingredients

In a maintenance ration, as for productive rations, the protein and non-protein parts of the ration should be in proper proportion to each other. This ratio is known as the 'nutritive ratio' (see Chapter V), and for maintenance purposes ma be fairly wide. For example, in the case of oat hay used in the Lyallpur trials, the ratio was 1 : 25; for a composite pasture hay obtained from Ambala (a maintenance ration for 800 lb. dairy heifers) it was 1 : 18, whilst with non-maintenance pasture hays, it varied from 1 : 30 to as wide as 1 : 70. In a maintenance ration for cattle the ratio should lie around 1 : 18 to 1 : 20, although this figure will vary for other animals.

The proportion of fat to protein should also lie within certain rough limits. Too small a percentage of fat is not satisfactory, although, admittedly, animals can do without it, and can manufacture body fat from carbohydrate sources. On the other hand, too large a proportion of fat in the ration results in deficient protein digestion and absorption. The proportion of fat to protein in good oat hay used as a maintenance ration, and also as a basal feed in the Lyallpur digestibility trials was 1 : 8.5, and may be taken as satisfactory. Our knowledge of the optimum ratio, however, is very scanty, and opinions regarding it are conflicting, and no very definite advice of real value can be offered on this point. It can, however, be stated that too high a percentage of fat is unsatisfactory both in regard to its effects on protein digestion, and the effort which the animal's digestive system has to make to cope with it.

Maintenance Rations not governed solely by the Weight of an Animal

It might be thought that the maintenance requirements of a 1,000 lb. animal would be double those of a 500 lb. animal. They are not governed solely by weight, however, but by the surface area of the body exposed to the air.

The weight of an animal bears a definite relationship to its volume, or surface area, which has been utilized by Rubner in the enunciation of what is known as Rubner's surface law.

T. B. Wood (1937) has explained this very clearly as follows:—

> "Since all animals are approximately of the same density, and since the volume multiplied by the density is equal to the weight, the weight of an animal must be proportional to its volume. Now the volume is of three dimensions, length, breadth, and thickness—and surface is of two dimensions—length and breadth only. Surface is, therefore, proportional to the square of the cube root of volume, and since volume is directly proportional to weight, surface is also directly proportional to the square of the cube root of the weight. Food requirements being proportional to surface, they must also be proportional to the square of the cube root of the weight."

Thus, if the maintenance requirements of a 1,000 lb. dairy cow in terms of total digestible nutrients is 7.5 lb. and we wish to find the corresponding requirements of a 1,200 lb. animal, the formula to be used as given by the above definition is:—

$$\frac{R_1}{R_2}=\frac{(\sqrt[3]{W_1})^2}{(\sqrt[3]{W_2})^2} \quad \text{Or,} \quad R_2=\frac{R_1\,(\sqrt[3]{W_2})^2}{(\sqrt[3]{W_1})^2}$$

Where W_1 is the live weight of an animal whose requirement R_1 is known, and W_2 the live weight of an animal whose requirement R_2 is to be found.

$$\text{i. e. } R_2=\frac{7{\cdot}5\times(\sqrt[3]{1200})^2}{(\sqrt[3]{1000})^2}=\frac{7{\cdot}5\times 112{\cdot}9}{100{\cdot}0}=8{\cdot}47 \text{ lb.}$$

Brody and associates (see list of references) have slightly modified this and conclude from their studies that the basal metabolism, and, therefore, the maintenance requirements for energy or total digestible nutrients of animals of various sizes are proportional to the 0.73 power of their live weights. According to this definition the corresponding figure for a 1,200 lb. animal will be 8.57 lb. arrived at thus:

$$R_2 = \frac{R_1 \, (1200)^{\cdot 73}}{(1000)^{\cdot 73}} = \frac{7{\cdot}5 \times 176{\cdot}9}{154{\cdot}9} = 8{\cdot}57 \text{ lb.}$$

Therefore, if we determine the maintenance requirements of an animal of any particular weight, we can calculate the corresponding needs for an animal of similar type but of different weight by means of this law. It has previously been pointed out, however, that it is not possible to feed animals strictly in accordance with a mathematical formula, and the above law should be applied as an approximate practical guide rather than as a rigid one to be followed in all circumstances. It furnishes, nevertheless, a useful basis for calculating the maintenance rations for farm stock of different weights as a prelude to computing the additional rations needed for production.

It is unfortunately the case in India that the vast majority of dairy cattle are not fed according to any scientific or practical standard. The average zamindar is inclined to neglect his animals between lactations and allow them to roam and graze whatever fodder they can find. Frequently they are fed nothing but straw or inferior grass hay or sometimes an exclusively green fodder ration, and do not get even a maintenance ration.

During the past 22 years the writer has carried out a large number of maintenance ration trials on heifers, both on dry roughages and on green fodders. In the case of green fodders these trials have shown that the following when fed alone to 800 lb.—1,000 lb. dry cows, in the amounts indicated, are effective maintenance rations able to keep the animals in health for a long time (but not indefinitely!) with a high positive nitrogen balance.

Fodder	*lb. per. day.*
Oats	40—50
Senji	70—90
Berseem	70—85
Maize	60—70
Guara	50—70
Velvet beans	60—80
Bajra	50—60

Elephant grass	60—80
Guinea grass	40—60
Sunflower (when very young) ..	50—75

The following hays have also been found to be maintenance rations for 800 lb. heifers in the amounts fed daily shown against each:

Hays.	*lb. per day.*
Oat hay	11 — 12
Berseem hay	12 — 13
Senji hay	12 — 13
Maize hay	15 — 16

PRODUCTION

When any form of production is required from an animal extra food or energy is necessary in addition to the maintenance requirements as described above, in proportion to the level of production required. It has been seen in the case of maintenance rations, that the food has both a quantitative and a qualitative aspect, and this holds good to an even greater extent in feeding for production. Chemical analyses and digestibility trials will tell us how much digestible nutrients a foodstuff contains, and how much digestible protein, but these factors only take us a part of the way in the task of ensuring that a ration shall be efficient for production purposes.

For example, the products of digestion of protein from a foodstuff may be digested and absorbed into the blood stream, but if they are biologically deficient or, in other words, do not contain the particular amino acids that the animal needs for production purposes, the products of digestion cannot be properly utilized, and consequently, the energy value of the food will be lowered proportionately.

Therefore, comparisons between different feeding stuffs for production values, when these are based on digestible nutrients alone, are relative, and cannot show the absolute value of the feeding stuffs, although such data form a very useful working basis as to the nutritive value of the food, provided the ration is properly balanced.

It may at this stage be advisable to mention the meaning of gross, metabolizable and nett energy values of a food. The gross energy value of a foodstuff may be stated in terms of the total heat, expressed in calories, which the foodstuff will yield when completely burnt in a calorimeter. The metabolizable energy, however, represents that which is available to the animal from that part of the food which is actually digested. This may be approximately computed by measuring the total heat of combustion of the food on one side, and of the solid, liquid and gaseous excreta on the other. The difference between the two represents the metabolizable energy.

The nett energy is that amount of energy left over for production from the metabolizable energy after the energy required in mastication and digestion of the food is accounted for.

The metabolizable energy as already explained is thus the energy capable of transformation in the body both for production in the form of increase of the body weight, manufacture of produce and in the evolution of heat. The metabolizable energy of a foodstuff does not measure its 'production value', since part of it is lost as described above, as far as production is concerned. The metabolizable energy as thus expressed, however, is an apparent and not a real value, because various influences effect the digestive processes on the feeding stuffs of which metabolizable energy takes no account. Thus, for example, Kellner [1908] found in one of his experiments that when beet molasses was added to a basal ration, the amount of energy bound up in the methane evolved was diminished by 138.5 calories, but at the same time, it so depressed the digestibility of the basal ration that the amounts of energy lost in the faeces and urine on this account were increased by 186.5 calories and 272.3 calories respectively. These two amounts of energy, therefore, represent losses as a result of adding the molasses and must be subtracted from the gross energy of the latter to obtain its metabolizable energy. This might be expressed in a different way by stating that certain ingredients of feeding stuffs if fed in excess may influence the availability of others, or in other words, that the metabolizable energy of the total feed is reduced. Practical illustrations of this are shown in the well known fact brought out by digestibility trials, that excess of sugar or fat in a feeding stuff will interfere with the availability of the protein.

Our knowledge of the chemistry of feeding stuffs and the behaviour of the various constituents comprising them in animal metabolism is still imperfect, and the actual nutritive values of any feeding stuff can only be determined by direct trials with animals. Accurate data can be obtained by the use of an elaborate and expensive animal calorimeter already described, but the earlier animal calorimeter experiments are being supplemented by trials designed to ascertain the actual production effects of different types and combinations of feeding stuffs. Hence, whilst some American authorities still adhere firmly to the principle of expressing production values of feeding stuffs in terms of nett energy of calories, the method of expressing production requirements in terms of digestible nutrients is becoming increasingly prevalent.

A great deal of confusion has arisen in the past even in the minds of nutrition experts in defining their terms, and stating their requirements in terms of energy, and the writer is of the opinion that the practical farmer is more likely to be at home with figures of digestible nutrients than with data expressed as calories. Admittedly, the exponents of the calorie school state, and rightly so, that the ultimate value of a feeding stuff is to be judged in terms of the nett energy it supplies, but the energy values expressed in calories or therms per 100 lb. of feeding stuff, are, as a matter of fact, of the same order of magnitude as the more familiar term—digestible nutrients. The writer prefers the use of digestible nutrients, and in the chapters dealing with different types of farm animals the requirements for different production purposes are given in terms of digestible nutrients which can readily be computed from the data in Appendices 1 and 2. Reference to Appendix 2 shows the actual amounts of digestible nutrients which are required for animals of different weights both for maintenance and production, and these have been utilized for computing the rations described in Chapters IX—XV.

Special Aspects of Feeding for Production

In considering the maintenance requirements of an animal it was pointed out that a certain minimum of digestible nutrients was required in order to make up the losses involved in the general metabolism of an animal at rest, also that it was necessary to take into account the mineral supply and vitamins. In other words the ration has to be considered from two points of view, firstly, the total energy or nutrient value of the feed, and secondly, those aspects of the feed which do not necessarily supply energy but are otherwise essential. Immediately an animal passes from a state of maintenance to a production state, both of these aspects acquire greater significance.

We will consider first of all the quantitative aspect in regard to digestible nutrients. The amount and type of food required by any animal for production over and above its maintenance ration will depend upon the nature of the production. For example, an animal which is merely being fattened, or in other words producing mainly fat, will require a different type of feed from one which is producing milk, eggs, etc., or again from one whose production is mainly work, such as a fully grown horse or working bullock. These special requirements have been dealt with in detail in the following chapters but as an illustration of the general principles involved we may consider the case of a young growing pig. In the first few months of its life the pig is required to put on body weight in the form of new tissues and new bone structure, and, therefore, the food required for these purposes, *i.e.*, for production, will need to be rich in those nutrients which are essentially body forming, *i.e.*, the proteins and minerals. Hence as described in Chapter XIV, the nutritive ratio must be a narrow one in the neighbourhood of 1 : 3.5. The feeding stuffs must, therefore, be chosen accordingly. When, however, the pig is approaching maturity there will be less demand for protein or body forming food and a greater demand for non-proteins and the nutritive ratio will accordingly be widened. When the fattening stage is reached, the main increase in body weight will be fat, built up chiefly from the carbohydrate part of the food, which must, therefore, be largely carbohydrate in nature. Again, in the case of a fully grown dry heifer the maintenance requirements will be adjusted according to her weight. When the heifer is in calf, however, the maintenance requirements will need to be progressively augmented to cater for the growth of the developing foetus, and shortly after parturition the ration must be sufficiently rich in digestible nutrients to provide for the milk produced.

As production is in this case milk which is rich in protein, the food must also be rich in protein. Similarly, for egg production, a hen will require almost twice as much of the right type of food when laying, as she will when not laying, and as eggs are rich in protein the rations of the laying hen must be sufficiently rich in protein also.

Minerals and Vitamins

The general principles as regards mineral and vitamin requirements described for maintenance rations apply also to production rations, although the latter must be much richer in both minerals and vitamins than is necessary for the former. The mineral and vitamin requirements for different types of farm stock will be described in more detail in Chapters IX—XV.

It is clear that a growing animal, or an animal carrying a foetus, or otherwise producing, will need considerably more minerals in its rations, than a fully grown one. The minerals must as a rule be several times greater in the food supplied than are actually found in the produce, and they must also be well balanced. The same remarks apply to vitamins. An animal only needing a maintenance ration may get on quite well for a considerable time on a limited vitamin intake, but serious results may be expected if the rations of the producing animal are deficient in vitamins either in quantity or quality.

Qualitative Aspects of Production Rations

The nature of the proteins fed to an animal for production is of the utmost importance for the reasons explained in Chapter II. Gliadin the protein of wheat is deficient in lysin, which although adequate for maintenance, is unable to promote growth. The proteins of the cereals in general, whilst not absolutely lacking in certain essential amino acids, are however, unbalanced for the purpose of animal nutrition as compared with animal proteins, and they are all deficient in certain amino acids, particularly lysin, which appears to be essential for growth. It is obvious, therefore, that quite apart from specific defects in certain amino acids, excessive quantities of such foodstuffs would need to be given in an endeavour to supply the body with the particular amino acids it needs. Our knowledge in regard to the specific protein values of different feeding stuffs is not so far advanced as it is in respect to nett energy values, or total digestible nutrients, because, we still do not know what many of the proteins of feeding stuffs actually are, and our knowledge of the exact amino acid requirements of different farm animals for different purposes is as yet by no means complete. Even if we were in a position to determine the production values of the single proteins which happen to be known, it would not be easy, or even possible, to estimate from them the production values of the mixed proteins of feeding stuffs, since a deficiency in any one protein might be compensated by a sufficiency or surplus in another, and hence the mixture would show a much higher production value than the sum of the separate ingredients if fed individually. For example, whilst wheat gliadin, which lacks lysin, when fed as a sole protein, has little growth value, it may have a high value when supplemented by other proteins which supply lysin. Thus wheat alone fed to a young animal would not promote growth, but the substitution of a part of it by milk containing lactalbumin and casein would. Any such mixture of proteins, therefore, would have a production value which might differ very considerably from the mean values of the individual proteins. Recognition of this fact should always be borne in mind in practical feeding and computing rations, and whenever deficient protein feeds are used they should be properly balanced by others. This explains why it is generally better to feed mixed feeds rather than feeds consisting of only one or two ingredients. If this is done, the protein deficient feeds can play an important and valuable part in the computation of rations.

Osborne and Mendel [1914] state that the protein composition of maize is approximately as follows:—

Zein	41%
Maize glutelin	31%
Globulins, albumins and proteoses ..	22%
Insoluble in alkali	6%
	100%

Although maize is deficient as a whole in protein or amino acids, the glutelin nevertheless provides, at least qualitatively, those amino acids which are lacking in zein and the other proteins in maize. Many trials have been conducted with maize alone or supplemented by other protein feeds particularly on pigs, and Hart and McCollum [1914] state that they were able to obtain normal growth with pigs on rations whose only protein was supplied by maize with the addition of salt. On the other hand Waters [1914] has produced definite evidence of the unsatisfactory nature of the proteins of maize in four trials on pigs. One lot received only maize, whereas others were given maize with the addition of salt, whilst again other groups received complete protein supplements as nearly as possible ash free, in the form of blood and milk. He showed that pigs receiving only maize, or only maize and ash or added salt, practically ceased to grow, but that the groups which received the complete proteins added to the maize maintained steady and normal growth. Other trials carried out by Hart, Humphrey and Morrison [1912] and McCollum [1914], on growing heifers and young pigs respectively, support the conclusion that maize alone cannot produce efficient growth but that maize when supplemented by other high quality proteins forms a valuable constituent of the ration. When thus supplemented, a much higher retention of protein takes place in the body than when maize is fed alone.

The plate reproduced here is taken from a book published by Palmer [1919] and constitutes an interesting commentary on the effect on pigs when maize is fed without being properly balanced by other rations. [Plate VII.]

Value of Non-Protein Nitrogen for Production

Details have been given in Chapter II, on the distinction between protein nitrogen and non-protein nitrogen. The latter consists of a heterogeneous group of substances containing nitrogen in an amido group, but not necessarily containing the amino group. A typical example of such non-protein nitrogenous bodies as described in Chapter II is asparagin, which is not found amongst the break-down products of proteins, but belongs to the class of bodies known as acid amides. It does not appear that

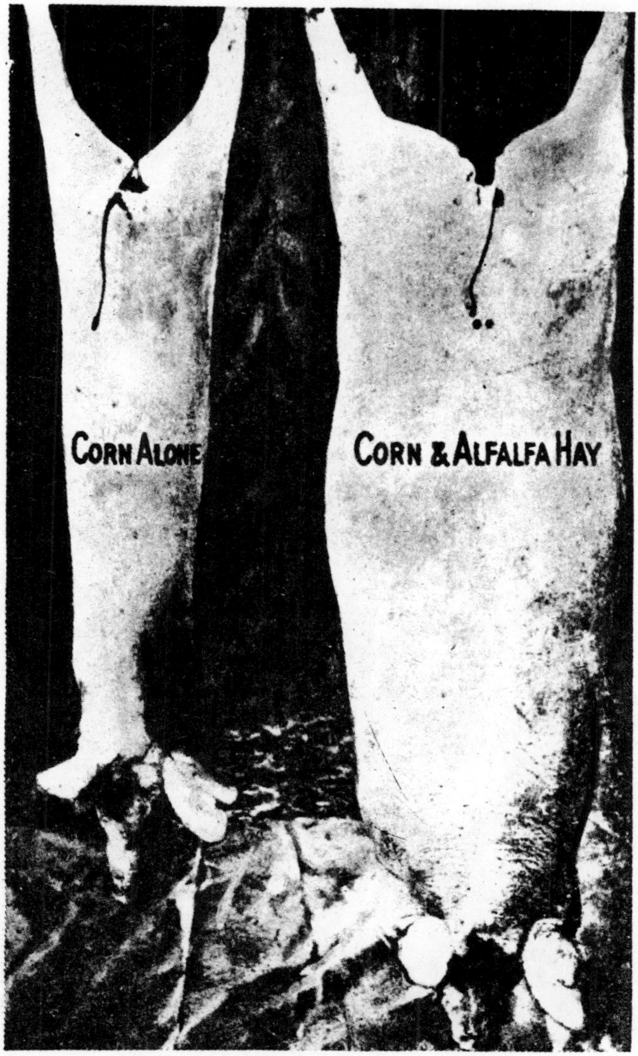

An object lesson on the value of balanced rations. The above carcasses are from hogs of the same age. The small carcass resulted from feeding an unbalanced ration of corn alone. The large carcass resulted from feeding a balanced ration.

asparagin can play any part in the building up of body protein unless it is first converted into aspartic acid, and no evidence is yet forthcoming that the body can effect this conversion.

Opinions differ considerably as to the extent to which the non-protein nitrogen of feeding stuffs can act as body protein builders, and it may be that their value in this direction lies rather in the extent to which these substances are utilized, first by the micro-organisms of the digestive tract and the resultant bodies built up by bacteria which themselves become ultimately digested and of use to the animal. As previously pointed out recognition has been taken of the possible value of the non-protein bodies in computing what is known as the 'protein equivalent', a factor comprising the combined digestible true protein and non-protein nitrogen. Various writers use different terms in designating 'digestible protein'. Some only consider the digestible crude protein, others only the digestible true protein, whilst others use the composite protein equivalent. The figures as given in Appendix 1, and designated as 'digestible protein' are computed in terms of digestible crude protein.

BIBLIOGRAPHY

1. Brody, Procter, and Ashworth. Mo. Res. Bul. 220.
2. Hart & McCollum. 1914. Journal of Biological Chemistry, 19, 373.
3. Humphrey and Morrison, 1912. Jour. Bio. Chem. 13, p. 133.
4. Kellner, O. 1908. Scientific Feeding of Animals. p. 48.
5. McCollum. 1914. Jour. Biol. Chem. 19, p. 323.
6. Osborne and Mendel 1914. Journal of Biological Chemistry, 18, 1.
7. Palmer C. C., Live Stock Feeding Association, Pleasant Hill, Ohio, U.S.A. 14.
8. Waters, 1914. Proceedings of the Society for Promoting Agricultural Science, U.S.A., p. 7.
9. Wood T. B. 1937. Animal Nutrition. (University Tutorial Press, London.) p. 151.

CHAPTER V

FEEDING STANDARDS AND BALANCED RATIONS

Feeding standards enable us to express the nutritive value of foods, and the food requirements of animals in some common terms. Rations may be changed, and the proportion of their relative constituents may vary, hence for the compilation of rations definite standards which can be expressed in terms of energy or nutrients for various purposes are necessary.

Various methods have been devised to accomplish this. Armsby [1917] in America based his methods on the principle that all foodstuffs are sources of energy, and Kellner [1908] in Germany devised a standard by reducing all the food constituents to a common denominator in terms of starch. All foodstuffs can ultimately be expressed in terms of energy, and, as we have seen, it does not matter much which food ingredients supply the animal cells with the energy they need, provided they get the necessary proteins to repair waste or for production purposes such as growth, work, milk, eggs, etc.

If, therefore, the stock owner could be given a simple formula which told him that the complex composition of his foodstuffs could all be expressed as though they were simply starch or energy, and that his animals needed so much 'starch' for various purposes, or according to Armsby's system, if the foodstuff could be expressed in terms of energy or 'calories' and so many calories were needed for maintenance and production, his task in devising rations from the tables of analyses would be easy.

The matter is not quite so simple as the above suggests but perhaps the simplest system of all is one which is now becoming increasingly adopted and expresses foods and food requirements in terms of Digestible Proteins and Total Digestible Nutrients, while the method so long in vogue in America of using calories or therms is going out of favour.

If, therefore, the basal or fundamental fact is kept in mind that the food must supply, firstly, a certain 'total amount of energy', and secondly, a certain 'minimum amount of protein' according to circumstances, the stock owner or the scientist can choose whichever standard he prefers for building up rations from the data given in Appendix I.

The various standards that may be considered, therefore are:

1. The energy standard of Armsby.
2. The starch equivalents standard of Kellner.
3. The standard based on total digestible nutrients and digestible proteins.

1. ENERGY VALUE

The energy value of a foodstuff is measured in terms of heat units. The amount of heat that will be required to raise the temperature of one gramme of water through one degree centigrade is called a calorie (small c). A large Calorie (large C) is the equivalent of 1,000 small calories, and a therm is 1,000 Calories.

If now a small quantity of any foodstuff is taken and completely burnt in a special apparatus called a bomb calorimeter, by electrical ignition in oxygen, the energy which the food contains is liberated in the form of heat which can be measured by noting the rise in temperature of the water in the calorimeter.

When combustion is complete this heat represents the total, or 'gross energy' of the food. The whole of the energy of the food eaten cannot, however, be utilised by the animal and a considerable part leaves the body in the form of waste products which are not completely oxidised or burnt. The energy which is still available is that represented in the products of digestion which are absorbed, and as these can be utilised for the various activities of metabolism it is called the 'metabolisable energy.' Hence metabolisable energy is the gross energy minus the energy still contained in the excreta. During digestion and the various other metabolic activities of the body, however, a certain amount of heat is produced. Metabolisable energy is thus primarily a measure of the heat and production forming capacity which can make good the losses of heat, etc., from the body in various ways and thus has a definite value.

A maintenance ration (Chapter IV) must contain sufficient nutrients to provide for the various metabolic activities of the animal in health and maintain its weight, but during consumption and digestion of food, metabolism is increased, there is an additional loss of energy from the body in the form of heat. Therefore, of the total metabolisable energy, that part which remains for purposes of production is the nett energy.

Therefore, nett energy = Metabolisable energy minus the energy required for digestion and the increased metabolism following food ingestion.

It has been seen that certain deductions must be made from the total energy as determined in the calorimeter to arrive at the metabolisable energy available from the proteins, carbohydrates and fats of the food.

Careful experiments have been carried out to determine these deductions, as a result of which it has been laid down that the metabolisable energy in terms of Calories of the proteins, carbohydrates and fats may be obtained by multiplying the percentages of these in a foodstuff respectively by 2,133 ; 1,707 and 4,000. Thus:—

% Proteins x 2,133 Calories (C)
% Carbohydrates x 1,707 Calories
% Fat x 4,000 Calories

= The metabolisable energy per pound of each respectively.

We can now see how the metabolisable energy in terms of Calories can be arrived at if we know the chemical composition of the feeding stuff to be examined.

For example, 100 pounds of 4F American cotton seed contains:—

(1) 11.2 pound of digestible proteins.
(2) 42.0 pounds of digestible fibre and carbohydrates.
(3) 15.3 pounds of digestible ether extract.

Multiplying 1, 2 and 3 respectively by the figures shown above we get:—

11.2 x 2,133	=	23,889	Calories
42.0 x 1,707	=	71,694	Calories
15.0 x 4,000	=	60,000	Calories
		155,583	Calories or say 155 therms.

Another way of expressing this metabolisable energy is to compute it in terms of starch. For example, one pound of starch or digestible carbohydrates yields 1,707 Calories of metabolisable energy.

The corresponding value for protein would be $\frac{2133}{1707} = 1\ \ 5$ (expressing the protein in terms of starch), and for the fat $\frac{4000}{1707} = 2.3$ (expressing the fat in terms of starch).

Thus if we wish to express the whole of the metabolisable energy of the 100 lb. of cotton seed as though it were derived from starch we must take the separate digestible nutrients and multiply them by the values as shown. Thus:—

Digestible protein	11.2 x 1.25 = 14.0	starch pounds.
Digestible fibre and carbohydrates	42.0 x 1.00 = 42.0	starch pounds.
Digestible fat	15.3 x 2.3 = 35.2	starch pounds.
	91.2	

Thus 100 pounds of cotton seed are equivalent to 91.2 lb. of starch in terms of metabolisable energy. This figure is sometimes referred to as the Maintenance Starch Equivalent of a food and must not be confused with Kellner's Starch Equivalent, to be described shortly, which is based on the nett energy value of the food. If a certain amount of food is given for productive purposes over and above the maintenance requirements, there will be a certain amount of energy available for the body, represented by the difference between the metabolisable and the nett energy of the food given. Consequently, it is the nett energy of the food which really matters for production purposes, whereas for the computation of maintenance rations it is the metabolisable energy. However, as will be mentioned later it is simpler and easier to use total digestible nutrients for the compilation of maintenance rations—or even production rations.

It has for long been a debatable point whether feeds should be compared, or rations computed in terms of nett energy values or digestible nutrients.

There is much to be said to commend the first, but there are great scientific and practical difficulties in the determination of nett energy values. The respiration apparatus which is needed (Chapter III) is very costly and comparatively few feeds have actually been investigated with cattle by this means, and still less in the case of other farm animals.

It has also been found that nett energy values as determined by elaborate trials differ in the case of the same feeds with different classes of stock, and, therefore, the nett energy values as determined in experiments with one class of animal may not necessarily furnish a correct index of the values of the same feeds for others.

The nett energy values of individual feeds, therefore, have definite limitations as a practical basis for evaluating feeds or computing rations for livestock, and the practice of using total digestible nutrients, or, if preferred, starch equivalents, is more common.

There is no doubt, however, that the best measure of the actual value of any feed for a particular class of stock can be obtained only through carefully planned feeding experiments.

This is the method employed by the writer in the Lyallpur digestion trials and he considers that the expression of results and evaluation of rations in terms of Digestible Proteins and Total Digestible Nutrients is the simplest and is to be preferred.

2. Starch Equivalent (Kellner's Production Starch Equivalents)

The principle which Kellner followed to measure the values of foodstuffs for production purposes in which one pound of digestible starch was taken as the unit, was based on their value in terms of starch instead of nett energy. These values were determined on bullocks using an elaborate and expensive respiration calorimeter. The nett energy value of a feeding stuff was then expressed as the number of pounds of starch which it was computed would have the same nett energy value for production purposes. These starch values are similar to Armsby's nett energy values which are stated in 'Therms' or 'Calories' except that the manner of expressing them is different. For example, Kellner found by his method that the starch value of maize was 81.5 lb. (which means that 100 lb. of maize has the same nett energy value as 81.5 lb. of starch). Armsby, employing his method, gave it a figure of 85.5 therms.

Remembering that the nett energy value of one pound of starch is 1,071 Calories or 1.071 Therms, it will readily be seen that the starch value of any feed expressed in terms of starch equivalents can be converted into nett energy expressed in terms of Therms by multiplying it by the above factor 1.071 *i.e.*,

Kellner's (starch values)	*Armsby (Therms)*
Starch value x 1.071 =	Therms

Kellner fed bullocks kept in his respiration chamber a basal or maintenance ration, and then added to the ration certain known quantities of the pure ingredients of which foods are composed, such as proteins, fats, carbohydrates, and recorded the increases in body weights.

Taking as his unit one pound of digestible starch, he found that one pound of digestible pure protein produced 94% as much body fat as one pound of digestible starch.

In oil bearing seeds and oilseed meals in which most of the 'Ether Extract' or fat is practically pure fat, he found that one pound of digestible fat had a starch value of 2.4 pounds, while the fat from cereals and industrial food by-products had a starch value of 2.12 lb., while the corresponding value for fat from roughages such as dry fodders was 1.9.

Kellner determined the starch values of a few typical feeding stuffs by means of the respiration calorimeter, and then, using the factors he obtained, and knowing the amounts of digestible nutrients in the respective feeds, computed a table of starch values. Thus:—

Multiplying the digestible proteins by 0.94	=	Starch value
Multiplying the digestible fat in coarse fodders by 1.9	=	—do—
Multiplying the digestible fat in cereals by 2.1	=	—do—
Multiplying the digestible fat in oilseeds by 2.4	=	—do—
Multiplying the digestible carbohydrates and fibre by 1.0	=	—do—

(The fat in foodstuffs is generally taken in practical computations, however, to have a starch value of 2.3 whatever its source of origin.)

We may take 4F American cotton seed to see how a simple calculation may be made. The respective percentages of digestible nutrients as found by chemical analysis and digestibility trials are multiplied by the appropriate factors indicated above.

Thus:—

4F AMERICAN COTTON SEED

Digestible nutrients.	Per cent		Starch factor.	Starch Equivalent.
Proteins or protein equivalent (see below) ..	11.2	x	0.94	10.5
Fat ..	18.65	x	2.3	42.9
Carbohydrates ..	10.0	x	1.0	10.0
Fibre ..	14.8	x	1.0	14.8
			Total:	78.2

That is to say, 100 pounds of 4F American cotton seed, when fed to bullocks in addition to the basic maintenance ration should produce as much fat as would be produced by 78.2 lb. of starch when similarly fed.

Correspondingly for purposes of work the energy values of the starch and cotton seed respectively may be converted into Therms by means of the factors already given, and the relative work production evaluated. In this way the production starch equivalents of any feeding stuff may be determined from the data given in Appendix 1.

Certain corrections however, are necessary because, as Kellner found, it is not possible to calculate the starch values of feeds with a high fibre content with any degree of accuracy, and from the comparatively few feeds which he studied he found that it was necessary to make very considerable deductions from the calculated starch equivalents, of from 5 to 30 per cent in the case of certain factory by-products of high fibre content, and even up to as much as from 50 to 70 per cent with straw and certain other roughages, in order to get correct starch values.

In the case of oats, for example, the number that must be deducted from the calculated starch equivalents was found to be 5. This led to the evaluation of what are called Keilner's value numbers for the different foodstuffs, and show the corrected values for starch equivalents. Thus the value number for oats is 95, and the value number for straw, say, 50. As an illustration we may calculate the approximate production starch equivalents of a typical composite hay obtained from the Military Dairy at Murree in December, 1927.

MURREE HAY

Digestible nutrients.		Per cent		Starch Factor.	Starch Equivalent.
Protein equivalent	..	0.51	x	0.94	0.48
Fat	..	0.71	x	2.3	1.63
Carbohydrate	..	14.0	x	1.0	14.0
Fibre	..	25.0	x	1.0	25.0
				Total:	41.11

This hay proved to be very poor and was not a maintenance ration for Montgomery heifers. Hence if we take the 'value number' of this hay as, say, 55, then its real production starch equivalent would be:—

$$\frac{41.11 \times 55}{100} \text{ or } 26.6$$

i.e., 100 lb. of the hay has the value of only 22.6 lb. of starch.

Thus the Production starch equivalent, or more simply, the starch equivalent (S.E.), gives us a measure of its nett energy value, and one pound is equivalent to 1,071 Calories or 1.071 Therms, because the nett energy value of one pound of digestible starch is 1,071 Calories or 1.071 Therms.

Protein Equivalent

We have already seen that the total nitrogen in a feed is made up partly of 'true protein' and partly of 'non-protein' nitrogen, and that the non-protein nitrogen, although having a definite food value is inferior to the true proteins in this respect.

If, therefore, the protein equivalent is calculated as though all the protein in the food were true protein, too high a value will be obtained, which will vary according to the particular foodstuff. Kellner and Armsby based their feeding standards on digestible protein only.

Many workers now-a-days, however, do not take the pure protein but the sum of the digestible pure protein and half the digestible non-protein nitrogen, and the figure thus obtained is called the protein equivalent and is used in calculating standards.

Other workers use a somewhat different figure, and take half the sum of the digestible crude protein, as their 'protein equivalent'.

The Nitrogenous or albuminoid ratio of a food

We have previously seen that the actual tissue cells are not specially concerned with the nature of the particular nutrients they get to provide them with energy. It is a matter of importance, however, in connection with production whether the nutrients are derived from protein, or non-protein sources, and therefore a method has been devised to show the proportion in which these occur in a foodstuff. If the fat is multiplied by the factor 2.3 to convert it into terms of starch (see above) and then the sum of the protein nutrients balanced against the non-protein ones, the ratio so obtained is called the "Nutritive ratio" (N.R.), or "Nitrogenous ratio", or "Albuminoid ratio" (A.R.) and is called narrow, medium, or wide according to the relative quantities of protein and non-protein. Obviously rich protein foods will have narrower ratios than carbonaceous ones. Different workers have employed different methods to evaluate the nutritive ratio, according to whether they use the crude protein or the true protein found in chemical analysis or without reference to the amount digested. Again, others have omitted the digestible crude fibre. It therefore follows that a somewhat wider ratio will be obtained if the 'protein equivalent' as just described is used, than if digestible crude protein is used.

For example, in the first case oats will have a nutritive ratio of about 1 : 7.6, and in the latter case 1 : 7.0.

Similar differences will be obtained according to whether the digestible fibre is included or omitted in the calculation.

The best formula for calculating the nutritive ratio, which is now generally used, is as follows:—

$$\frac{\text{Digestible carbohydrate} + \text{Digestible fibre} + (\text{Digestible fat} \times 2.3)}{\text{Digestible protein (crude)}}$$

If this formula is used for 4F American cotton seed and Murree hay, their respective nutritive ratios will therefore be:—

Cotton seed	1 : 6.0
Murree hay	1 : 55.0

The nutritive ratio is an important factor in computing rations for different types of animals and in regard to the purpose for which they are given.

Young growing animals, cows in milk, and horses and bullocks doing hard work need a narrow ratio.

Very often cows in the early stages of lactation, and fattening animals, are given rations with too wide a ratio, while animals needing little more than a maintenance ration are given one which is too narrow. If the ratio is too wide there may be interference with proper digestion and absorption, thus causing unnecessary waste, whereas if it is too narrow, there may be corresponding waste in feeding unnecessary protein which the animals must use energy to get rid of.

For the purpose of computing rations the stock owner should consult the data given in Appendix 1 and then, according to the standard he adopts, compute his rations accordingly. Absolute accuracy in feeding precisely the right quantity or mixture of food stuffs can never be attained because many factors intervene which it is impossible for him to cater for and arithmetical exactitude can never be attained.

For example, the analytical composition of feeding stuffs vary considerably according to the locality from which they come, and the time of year at which they are harvested, and these variations may depend even more on climatic factors than on the soil. Furthermore, as has been described more fully in Chapter VII, the composition of the plant depends on its stage of growth, its species and a number of other variables.

Again, different breeds of animals may digest corresponding foodstuffs to a different degree though these may not be wide. It has been found, for example, that Kangra hill cattle can exist on rice straw as their main feed for long periods but this straw was not a maintenance ration for Montgomery heifers at Lyallpur. [Lander & Dharmani 1931.]

Hence such variations may partially mask some of the other factors to be taken into account in computing rations, such as the evaluation of the digestible crude protein and protein equivalents.

[illegible]. Total Digestible Nutrients, or Food Units

It has been recorded that the old energy standard of Armsby

and the starch equivalents of Kellner are being superseded by the method of expressing food values in terms of total digestible nutrients.

These are ascertained by multiplying the sum of the digestible protein or protein equivalent, and the digestible fat by 2.5 and adding to this the sum of the digestible carbohydrates and fibre. Thus:—

Total digestible nutrients (Food Units). = (Protein Equivalent + Digestible Fat) x 2.5 + Digestible Carbohydrate and Fibre.

This method, however, takes no account of the relative amount of protein present though it gives a reasonably correct idea of the relative values of foods.

It is usual therefore to specify both the 'digestible protein' or 'protein equivalent' and the 'total digestible nutrients' and use these as a simple, and perhaps the easiest, standard for computing rations.

The writer [Lander 1942] has recently adopted this method of computing rations for determining the feeding value of cotton seed cake and cotton seed for dairy cows, to be described in Chapter IX.

The feeding standards reviewed above have been worked out chiefly in the U.S.A. and European countries on cattle, and the data thus obtained have also been largely used in computing feeding standards or feeding regimes for other classes of stock. For example, if it is found that the digestibility coefficient of the protein of cotton seed cake by a cow is 80%, the same figure has been used in compiling standards for other classes of animals on the assumption that they would digest the protein neither more nor less efficiently than the cow.

It does not necessarily follow, however, that because a food conforms to a certain standard or has a certain value for cows, that it will have a corresponding value for other types of animal. These may be physiologically and anatomically different from the cow, and their different types of alimentary tract may, and no doubt do, digest food materials differently.

It would be an advantage if standards could be worked out for other types of farm animals but at present this knowledge is not available. Rations which have been computed for these as they stand to-day have not the same scientific authority behind them as have rations for cattle, and they have been evolved partly in accordance with generally accepted principles, partly along empirical lines and partly according to local customs and the feeding stuffs available. Our knowledge of Indian feeding stuffs is still in its infancy, although a considerable amount of work has been done in recent years, and we are largely dependent for feeding standards on those evolved in other countries.

The required information will no doubt gradually be filled in, and in Appendix 1 will be found most of the chemical and digestibility data which have so far been obtained in India.

BALANCED RATIONS AND THE COMPUTATION OF RATIONS

A balanced ration is one which provides an animal with all the nutritive ingredients it needs for any particular purpose. In the previous Chapter the various feeding standards have been reviewed by means of which the first step may be taken in the computation of balanced rations.

In the first instance a balanced ration whether for maintenance or production must conform to the general principles outlined in Chapter IV, both qualitatively and quantitatively, and in computing a balanced ration for any particular purpose both these aspects must be kept in view. Suppose, for example, it is required to compute a ration for a cow of a certain weight giving 20 lb. of milk daily. The first step will be to ascertain from the data given in Appendix 2 what quantities of digestible protein and total digestible nutrients the cow needs for maintenance and the production of that amount of milk. Correspondingly for any other level of milk yield, or other type of stock. After this first step, the tables in Appendix 1 may be consulted, which will show how much of the various nutrients required will be furnished by the various foodstuffs from which choice can be made. Armed with these two sets of data it is an easy matter to select and compute a ration which should provide for an animal's requirements, due regard being paid to the fact that different types of animals require different types and proportions of foodstuffs as described in later chapters.

It will not be possible to compute satisfactory rations from the data given in Appendices 1 and 2, unless other factors are also considered, and the specific requirements of different types of animals are met. There are other matters just as important in determining the efficiency of a ration as are the amounts of digestible protein and total digestible nutrients. Some of these are the quality of the protein, the minerals and vitamins and the general suitability of the ration for a particular class of animal and for a particular purpose. Again, the proportion of concentrates to roughages is an important consideration in preparing rations for different types of animal. For example, heifers or idle horses may be satisfactorily maintained on good quality roughage alone, but a horse doing hard work or a cow yielding milk need a considerable amount of concentrates of certain kinds which depend on the amount of work done or milk produced, and as production increases it will be necessary for an increased proportion of the total digestible nutrients of the ration to be provided by the concentrates.

The analyses, including the digestiblity data of all the more important foodstuffs likely to be used in India are given in Appendix 1, and Appendix 2 shows the various amounts of nutritive material required for different classes of animals according to Morrison's standards, and corresponding data worked out at Lyallpur. The Morrison standards are widely recognised today among all workers in animal nutrition, and as similar standards, apart from those evaluated by the author for cattle and working bullocks, are not yet available for Indian conditions, the Morrison standards may be used as a satisfactory guide for computing rations.

They are shown in Appendix 2 marked 'M', whilst corresponding data determined at Lyallpur using heifers, milch cows and working bullocks are shown marked 'L'. Morrison has no standards for working bullocks, as bullocks are not employed for work purposes in the U.S.A. The evidence so far available from the Lyallpur trials show that, taken as a whole, the requirements of Indian cattle for maintenance, milk production and work are somewhat lower than western standards.

Bearing the above facts in mind and the general principles outlined above and in the previous chapter, it is now possible to show how representative balanced rations may be computed. It should be borne in mind that in computing all rations care should be taken to see that they are properly balanced not only in regard to digestible protein and total digestible nutrients but also in regard to all the other factors mentioned, such as the relative proportions of roughage to concentrates and in their mineral and vitamin content.

Computing a Balanced Ration for a 600 lb. Dairy Heifer

Before proceeding to discuss in detail how to estimate maintenance and production rations a simple example will now be given showing how to compute a properly balanced ration for a growing dairy heifer of 600 lb. body weight.

We will assume that the feeding stuffs available are:—

1. *Dry roughages*: Wheat *bhusa* and berseem hay.
2. *Green fodders*: Berseem and maize.
3. *Concentrates*: Cotton seed.

The table shown below may then be computed from the data relating to these foodstuffs as given in Appendix 1:—

	Dry matter lb. per lb. of the feed	Total digestible nutrients lb. per. lb. of the feed	Digestible protein lb. per lb. of the feed
Dry roughages			
Wheat *bhusa*	0.92	0.45	nil
Berseem hay	0.87	0.50	0.097
Green fodders			
Berseem	0.15	0.10	0.022
Maize	0.21	0.15	0.010
Concentrates			
Cotton seed	0.94	0.70	0.120

If now we refer to Morrison's standards as shown in Appendix 2, the heifer would require the following nutrients:—

	Dry matter lb.	Total digestible nutrients lb.	Digestible protein lb.
Morrison ..	12.0—13.6	7.7—8.7	0.94—1.06

The corresponding figures derived from the Lyallpur trials are somewhat lower than the American figures and hence more economical, but it might be advisable for stock owners in India to select figures intermediate between the two according to the breed and condition of their animals and general climatic environment.

	Dry matter lb.	Total digestible nutrients lb.	Digestible protein lb.
Lyallpur	10.0	6.2	0.74

Assuming for the sake of illustration that we wish to feed the heifer on *bhusa* alone, in order to provide the requisite amount of dry matter it would obviously be necessary to feed $\frac{10}{0.92}$ lb. of *bhusa* as *bhusa* contains 92 per cent dry matter; this works out to 10.9 or say 11.0 lb. of *bhusa*. But although this 'all-bhusa' ration would furnish the necessary amount of dry matter it would provide only 0.45 x 11 = 4.95 lb. of total digestible nutrients and practically no digestible protein, because *bhusa* for all practical purposes may be considered to contain little or no digestible protein. It is clear therefore that *bhusa* alone will not meet the protein requirements of the animal.

Let us see how far berseem hay alone will meet these. In order to obtain 10 lb. of dry matter the animal would need $\frac{10}{0.87}$ or 11.5 lb. of berseem hay which would contain 0.50 x 11.5 or 5.75 lb. of total digestible nutrients; and 0.097 x 11.5 or 1.116 lb. of digestible protein. Thus 11.5 lb. of berseem hay will provide the requisite 10.0 lb. of dry matter, considerably more digestible protein than is needed, but slightly less total digestible nutrients. Therefore approximately 12.0 lb. of berseem hay will constitute a maintenance ration as regards dry matter and total digestible nutrients but will supply far more digestible protein than actually needed.

In order to make the calculations so far clear the data for the two types of hay in regard to the nutrients they provide and the nutrients required by the animal are shown in tabular form below:—

According to Lyallpur data the 600 lb. heifer requires:—

Dry matter lb.	Total digestible nutrients lb.	Digestible protein lb.
10.0	6.20	0.74

and the dry roughages mentioned above provide the following nutrients:—

	Dry matter lb.	Total digestible nutrients lb.	Digestible protein lb.
Wheat *bhusa* 11 lb.	10.0	5.75	—
Berseem hay 11.5 lb.	10.0	4.95	1.116

Neither of these in the above amounts constitutes a balanced ration for the animal in question and to obtain one it is necessary either to increase the amounts of the single fodders or to mix two of the dry roughages available, or one dry roughage and one green fodder. So we may proceed to balance the ration by mixing two of the dry roughages in certain proportions, and one green fodder.

If we feed 35 lb. of green berseem to the heifer, which contains, on the basis of the digestibility data 0.15 lb. of dry matter, 0.10 lb. of total digestible nutrients, and 0.022 lb. of digestible protein per pound of green berseem, the amounts of these respective nutrients contained in the berseem are shown below:—

Nutrients in 35 lb. of Green Berseem

Dry matter lb.	Total digestible nutrients lb.	Digestible protein lb.
0.15 x 35 = 5.25	0.10 x 35 = 3.50	0.022 x 35 = 0.77

But the requirements of the heifer are:—

10.0	6.20	0.74

therefore, if the nutrients supplied by 35 lb. of green berseem are subtracted from the total requirements of the heifer we will know how much must still be provided by some supplementary ration. The amount is as follows:—

	Dry matter lb.	Total digestible nutrients lb.	Digestible protein lb.
Required by the heifer	10.00	6.20	0.74
Supplied by 35 lb. berseem green	5.25	3.50	0.77
Still needed	4.75	2.70	the D.P. is in excess by 0.03 lb.

We can now calculate how many pounds of wheat *bhusa* are required to furnish the additional 4.75 lb. of dry matter, and 2.70 lb. of total digestible nutrients. The amount of *bhusa* needed will be $\frac{4.75}{0.92}$ or 5.16 lb., or, call it for the sake of convenience in weighing, 5.5 lb. If corresponding calculations are made for the total digestible nutrients and digestible protein the total amount of nutrients supplied by 5.5 lb. of *bhusa* and 35 lb. green berseem are as follows:—

		Dry matter	Total digestible nutrients	Digestible protein
		lb.	lb.	lb.
Bhusa	5.5 lb.	5.06	2.48	—
Berseem green	35 lb.	5.25	3.50	0.77
		10.31	5.98	0.77

This table shows that a mixture of 35 lb. of green berseem and 5.5 lb. of *bhusa* would provide the heifer with the necessary amount of the various types of nutrients for maintaining it in health.

The ration thus decided on provides about 0.31 lb. of dry matter and 0.03 lb. of digestible protein more and slightly less (0.22 lb) total digestible nutrients than the standard requirements but it may be considered suitable as the slight excess of digestible protein will be an advantage rather than otherwise.

In India it is the exception rather than the rule to feed both berseem and *bhusa* to animals of this type which are generally turned out into the fields to pick up whatever they can find. At the other end of the scale however, on scientifically conducted stock farms, a small amount of concentrate is usually fed in addition to the roughage.

Examination of the Mineral Balance of the Above Ration

Rations for heifers up to the time of calving should contain at least 10 grs. (0.36 of an ounce) of phosphorus (expressed as P_2O_5) and approximately 20 grs. of calcium (CaO) per day during the first year, and somewhat more afterwards.

It will be seen from Appendix 1, that the phosphorus requirements of the heifer are well met, and the calcium requirements much more so, cheifly because green berseem is one of the richest calcium containing feeds available.

Feeding stuffs vary very considerably in their mineral and

other nutrient contents and their values can never be computed on a strictly arithmetical basis, but the above figures are reasonably accurate.

Thus the amount of calcium and phosphorus, based on the analytical data, contained in 5.5 lb. of wheat *bhusa* and 35 lb. of berseem green are as follows:—

		Calcium grs.	Phosphorus grs.
Wheat *bhusa*	5.5 lb.	5.65	2.72
Berseem green	35 lb.	63.15	13.38
		68.80	16.10

Example 2

The same roughage as above but with some added cotton seed as concentrate.

We may now ascertain how the nutrient requirements of a 600 lb. heifer can be met if some of the *bhusa* and berseem are replaced by a certain amount of cotton seed.

If the *bhusa* is reduced to 5.0 lb. and the green berseem to 25 lb. this combination will provide the heifer with the following nutrients:—

		Dry matter lb.	Total digestible nutrients lb.	Digestible protein lb.
Wheat *bhusa*	5 lb.	4.60	2.25	—
Berseem green	25 lb.	3.75	2.50	0.550
		8.35	4.75	0.550

Therefore, if these figures are subtracted from the requirements necessary according to the standard, it will be seen how much additional nutrients must be supplied by the cotton seed, and consequently, from the digestibility data of the cotton seed, to compute how much seed will actually be needed:—

	Dry matter	Total digestible nutrients	Digestible protein
	lb.	lb.	lb.
Required per standard	10.00	6.20	0.74
Supplied in *bhusa* 5 lb. and Berseem green 25 lb. ..	8.35	4.75	0.55
Still required	1.65	1.45	0.19

One pound of cotton seed contains 0.94 lb. of dry matter therefore $\frac{1.65}{0.94}$ = 1.75 lb. or say 2 lb. of cotton seed would be needed.

The addition of 2 lb. of cotton seed to the above ration consisting of 5 lb. of *bhusa* and 25 lb. of green berseem would add the following nutrients: —

		Dry matter	Total digestible nutrients	Digestible protein
		lb.	lb.	lb.
Bhusa	5 lb. ..	4.60	2.25	—
Berseem green	25 lb. ..	3.75	2.50	0.550
Cotton seed	2 lb. ..	1.88	1.40	0.240
		10.23	6.15	0.790

We thus see that the ration is well balanced, and although it is slightly richer in all nutrients compared with the standard, it is generally safer to feed an animal slightly more rather than less of the calculated requirements.

The Mineral Balance of the Second Ration

The *bhusa*, berseem and cotton seed of the ration supply calcium and phosphorus as shown below:—

		Calcium	Phosphorus
		grs.	grs.
Bhusa	5 lb. ..	5.13	2.48
Berseem green	25 lb. ..	45.10	9.56
Cotton seed	2 lb. ..	2.54	11.06
		52.77	23.10

Both calcium and phosphorus are more than adequately provided for.

The above two balanced rations represent winter feeds, and for comparison a summer ration for the same heifer will now be shown.

Summer Ration

If the available fodders are green maize and berseem hay, and 30 lb. of green maize are fed the amount of nutrients contained in the maize can be determined as before. Thus 30 lb. of green maize will contain:—

Dry matter	Total digestible nutrients	Digestible protein
lb.	lb.	lb.
6.30	4.50	0.300

If these respective nutrients are subtracted from the corresponding figures needed according to the standard it will be clear that the balance will have to be supplied by the berseem hay:—

	Dry matter	Total digestible nutrients	Digestible protein.
	lb.	lb.	lb.
Requirements	10.00	6.20	0.74
Nutrients supplied in 30 lb. of green maize	6.30	4.50	0.30
Still required	3.70	1.70	0.44

One pound of berseem hay contains 0.87 lb. of dry matter, 0.50 lb. of total digestible nutrients, and 0.097 lb. of digestible protein, therefore, the 3.70 lb. of dry matter still needed may be supplied by $\frac{3.70}{0.87} = 4.25$ lb. of berseem hay, correspondingly the 1.70 lb. of total digestible nutrients still needed may be provided by $\frac{1.70}{0.50} = 3.4$ lb. of berseem hay, and the 0.44 lb. of digestible protein still needed will require $\frac{0.44}{0.097}$ or 4.54 lb. of berseem hay. If, therefore, the nutrients thus provided by the berseem hay are added to those supplied by the 30 lb. of green maize the following figures will be obtained:—

	Dry matter	Total digestible nutrients	Digestible protein	Calcium CaO	Phosphorus P_2O_5
	lb.	lb.	lb.	gms.	gms.
Maize green 30 lb.	6.30	4.50	0.300	14.99	30.41
Berseem hay 4.5 lb.	3.92	2.25	0.437	58.97	7.26
	19.99	6.75	0.737	73.96	37.67

The figures for dry matter and total digestible nutrients are slightly above the standard but show nevertheless that the ration is well balanced.

Thus if one wishes to know how to compute a properly balanced ration for an 800 lb. dairy cow giving 20 lb. of milk a day, the information can be found in the same way by reference to Appendices 1 and 2 and the observance of the aforementioned general principles. (See Chapter IX.)

BIBLIOGRAPHY

1. Armsby. 1917. The Nutrition of Farm Animals. The Macmillan Company, New York.
2. Kellner, O. 1908. Scientific Feeding of Animals. p. 92.
3. Lander, P. E. 1942. Preliminary report to the Imperial Council of Agricultural Research on a scheme for studying the nutritional requirements of milch cows.
4. Lander, P. E. & Dharmani, L. C. 1931. The feeding value of Kangra rice straw. Ind. Jour. Vet. Sci. & Animal Husbandry, 6, p. 177-191.

CHAPTER VI

FEEDING STUFFS

In this chapter a detailed account will be given of the various feeding stuffs met with in India, the chemical analyses of which are used in computation of rations according to the feeding standards. In a country the size of India, with its variations in climate and soil features, uniformity in distribution of feeding stuffs cannot be expected. Furthermore, it is inevitable that the analysis of any particular material will vary according to the locality from which it is obtained, the soil on which it is grown and the stage of growth at which it is sampled (this applies to growing plant material). Such variations may be considerable. Nevertheless, whilst bearing these facts in mind, the computations of rations in this book are based on what may be considered to be average standard analyses.

Feeding stuffs may be divided into two main types:—

(1) Roughages.

Succulent feeds such as green fodders and roots.
Dry feeds such as wheat *bhusa* and hays (see Chapter VII).

(2) Concentrates.

Carbonaceous—such as the cereals.
Protein rich—such as legume seeds and oilseed cakes, etc.

SUCCULENT FEEDS

(a) NATURAL GRAZING

During the monsoon in India, grass both in the hills and plains makes prolific growth, and furnishes grazing to millions of cattle, goats and sheep. With the onset of the cold weather, however, the grass available decreases in amount and in quality, until in due course, the majority of grazing grounds in the more arid areas provide very little natural grazing of any great value. In the provinces certain forest lands are available for grazing, but only about one in twenty of the animal population of the country can obtain access to them, the rest subsisting on whatever pasturage they can find on land adjacent to the villages.

In some of the Government forests, controlled grazing is practised over limited areas which are periodically closed. The available grazing on lands in charge of the Forest Department can exert only a slight influence upon the larger problem of finding grazing facilities for India's animal population, and it can truly be said that there is little or no attempt either to cultivate grazing grasses or to control grazing on natural grazing areas over the major part of the waste land of India. In the hilly tracts the picture is somewhat different owing to the higher and more evenly distributed rainfall, but even there, over grazing is leading to the destruction of vegetation, and incidentally the soil. [Lander 1942.]

On certain large estates and Remount Depots, systematic efforts have been made to set apart areas of land for permanent grazing, or for the production of grass for hay. This development was advocated as long ago as 1886 by Symonds who urged that pasture grasses should be systematically cultivated on permanent pasture lands and not solely as one of the crops in a rotation. The development of good quality pasture both for grazing and hay making is well worth more attention than it at present receives, and the general farming community appears still to be unaware that with careful attention, and provided the necessary water is available, it is possible to obtain from grass a feeding stuff for farm stock as rich as many of the commonly used concentrates.

Factors Affecting Nutritive Value of Pasture Grasses

Data are still somewhat meagre on the nutritive values of Indian grasses.

Young and actively growing plants are richer in protein calculated on a dry basis than when they are usually cut for hay. They are also softer, contain less fibre and are more digestible than when old. As a rule they contain more calcium and phosphorus on a dry basis, and are decidedly richer in vitamin A.

Grasses from well grazed and controlled pastures will invariably show a higher protein and general nutritive content than so called mixed grasses cut at the usual stage of maturity. The latter may sometimes show a protein content no better than wheat *bhusa,* or even less.

Young actively growing grass of good type and species is an invaluable source of roughage on any stock farm, and in many ways comparable in feeding value with some of the protein rich concentrates such as bran, linseed cake, etc.

Grass, however, generally contains a greater proportion of fibre than concentrates and therefore cannot, as a rule, fully take their place in farm rations.

Pasture Improvement

Good management is essential in any system of pasture improvement in order that the grass may be kept actively growing over as long a period as possible, for encouraging the growth of desirable grasses and legumes, and at the same time keeping down weeds and inferior grasses to a minimum.

In dry districts where water is not abundant it is important that the pasture should not be grazed so intensely that seeding cannot develop, although in wetter areas it is better if matters are so arranged that grass is not allowed to go to seed.

Both under-grazing and over-grazing should be avoided. In the former case grass tends to become fibrous and unpalatable and of low nutritive value and young growth is inhibited. On the other hand over-grazing will reduce the yield, and the grass will never get the chance of arriving at the optimum nutritive stage of maturity, nor will the roots be able to replenish their stores of plant food nutrients. Over-grazing also brings in its train attendant evils such as impoverishment of the soil and erosion.

When animals are kept on one pasture throughout the year the system known as rotational grazing should be adopted. This involves fencing the pasture area into fields or paddocks which are then grazed in succession, so that the animals can be periodically removed from one field to another; the grass will thus have a chance to develop new growth unhindered. This system will give greater yields of total digestible nutrients and increase the milk yield of dairy cows. It is particularly useful for high yielding cows and is also commonly employed in rearing sheep.

A system of grazing long practised in Germany where pasture production is brought to a high level of efficiency, is the well-known Hohenheim system, which involves rotational grazing combined with liberal manuring. The pasturage is divided into a number of paddocks and milch cows are first allowed to graze on each plot in succession for some days so as to obtain the most nutritive part of the grass, and then dry cows, heifers or other stocks are turned in to graze the remaining less nutritive parts. This system naturally requires a considerable area, sufficiently large to permit a part always to be free so as to enable new growth to develop.

Pasture land should receive periodic applications of dung and should then be harrowed so as to spread it evenly over the surface. There is undoubtedly a very strong case to be made out for a more systematic cultivation of pasture grasses, both as natural pastures, and also for hay to serve as reserves in times of scarcity.

The following are the more important common grasses of India in an approximate order of merit which might be used for such purposes:—

1. Dhub (*Cynodon dactylon*).
2. Anjan (*Pennisetum cenchroides*).
3. Palwan (*Andropogon pertusus*).
4. Chhimbar (*Eleusine flagellifera*).
5. Spear grass (*Andropogon contortus*).

Dhub grass. (CYNODON DACTYLON)

Dhub or *hariali* grass is one of the best and most ubiquitous of Indian grasses and is superior to every other herbage in the country for sweetness, heaviness of yield and nutritive value. It needs a deep, rich and moderately light soil to obtain the best results, but judgment is needed in selecting the ground, though *dhub* will flourish even in poor paddy soils if they are carefully prepared. For getting the best results from *dhub,* well rotten farmyard manure should be dug in to a depth of at least one foot. An excellent way to plant is to make a long furrow in which fresh *dhub* roots are systematically laid and then covered up. Once the grass is well up, watering twice a month should suffice. The yield from *dhub* grass land varies, but when the crop

Cynodon dactylon, *dhub*.

Pennisetum cenchroides, *anjan* or *dhaman*.

is well established and a pasturage has developed, it should be possible to obtain a total yield of 10 maunds per acre per annum from successive cuttings. (Plate VIII.)

The question may be asked—Will it pay to incur the expense of manure and cultivation to grow such pasturage ? The answer will naturally depend on the nature of the soil, the locality, the availability of natural and compost manures and the yields and nutritive value of the grass produced. In southern India even dry lands will yield two crops of *dhub* during the monsoon with an average protein content on a dry basis varying from 8 to 14 per cent, depending on the stage of ripeness at cutting time.

The writer has found 20 per cent of protein, calculated on the dry material, in *dhub* from Hissar in the Punjab.

The table on p. 150 shows the analysis of *dhub* grass from the Government Cattle Farm, Hissar, in comparison with those of certain other representative fodders.

Anjan or kolukattai. (PENNISETUM CENCHROIDES)

Anjan or *kolukattai* grass is widely distributed in India and has a higher feeding value than most other grasses except *dhub*, and contains about 4 per cent of digestible protein and 55 per cent of total digestible nutrients based on the dry matter. It thrives best on a good loam soil with an average rainfall of 25 inches a year. It will produce excellent natural pasture if periodically manured and harrowed. It may also be used for silaging. It is palatable and nutritious and highly relished by all kinds of stock. (Plate IX.)

Palwan grass. (ANDROPOGON PERTUSUS)

Palwan grass is found all over India in the plains, and in the lower hills. It grows quickly, will stand cutting very well, and makes an excellent fodder grass which is very much relished by cattle. It compares favourably in protein content with most other pasture grasses, and investigations at Lyallpur have shown that from June to November the protein content of *palwan* grass grown at Sirsa in the Punjab may contain as much as 6 per cent protein or as little as 3, depending on the time of year it is cut and the climatic and other conditions prevailing. (Plate X.)

TABLE I

SHOWING CHEMICAL COMPOSITION OF VARIOUS FODDERS ON DRY MATTER BASIS

	Ash percentage.	Fat percentage.	Crude fibre percentage.	Protein percentage.	Soluble Carbo-hydrate percentage.	Calcium as Ca O percentage.	Phosphorus as P_2O_5 percentage.
Dhub grass (Hissar).	12.58	2.71	18.63	21.94	44.14	0.81	0.82
Anjan grass (Lyallpur).	18.23	1.64	29.51	10.08	40.54	0.60	0.67
Palwan grass (Lyallpur).	11.66	1.40	33.20	6.06	47.68	0.62	0.40
Oats (Lyallpur).	10.19	1.97	34.89	7.56	45.39	0.51	0.50
Juar (Lyallpur).	9.10	1.46	29.65	4.29	58.50	0.66	0.17
Bajra (Lyallpur).	9.21	2.12	27.96	10.56	50.15	0.73	0.50

Andropogon pertusus, *palwan*.

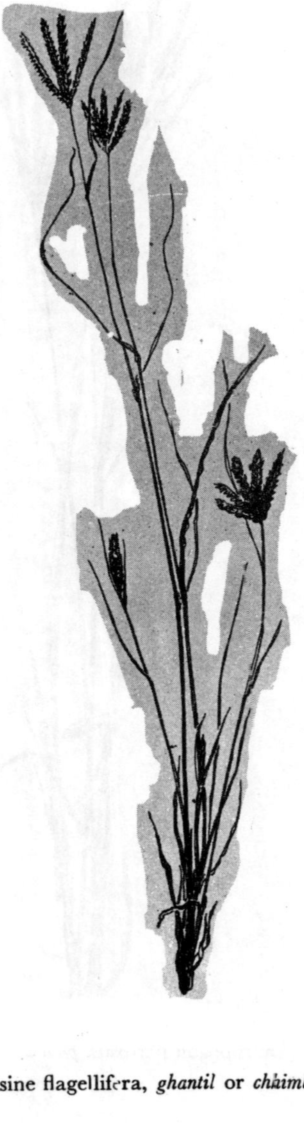

Eleusine flagellifera, *ghantil* or *chhimbar*.

Chhimbar grass. (ELEUSINE FLAGELLIFERA)

Chhimbar or *ghantil* is a sinuous grass which sends out runners, and is probably the next most abundant grass in northern India after *dhub,* and forms a useful fodder both when grazed or when made into hay. It does particularly well after the rains and takes the place of *dhub* on sandy soils in which the latter does not thrive well. In the young stage *chhimbar* grass forms a nutritious fodder containing from 5 to 9 per cent protein and compares favourably with *dhub* and *palwan.* (Plate XI.)

Spear grass. (ANDROPOGON CONTORTUS)

There are many species of grass of the genus *Andropogon,* of which *Andropogon contortus* is probably the most common. This is commonly known as spear grass and is found all over India.

Andropogon contortus is a good fodder crop when young, although it is coarser and contains more fibre than many pasture grasses. Good hay can be made from it if it is cut before it flowers. The only objection to the grass is the presence of sharp spike-like awns on the ripe seeds which become twisted together like the strands of a rope and cause considerable annoyance to animals which eat it.

This grass is one of the chief pasture grasses in the hills of northern India and many of the hays made from grasses collected in the hills contain a very large proportion of spear grass. It is also one of the most commonly found grasses in Madras and southern India.

A machine for removing these awns is shown in Fig. 7.

EXPLANATORY NOTE TO THE PARTS AND WORKING OF THE COMB.

1. Seat.
2. Two iron wheels of 1 foot diameter.
3. Adjustable fork. The fork can be made adjustable by having holes in it. A bolt can be passed through the frame and fixed in the required hole.
4. Framework. This can be made of iron or wooden reapers to stand the strain. If wooden, the reapers should be 2 feet thick and 2 feet broad. The framework is 6 feet by 2 feet with intermediary reapers for strength.

5. Teeth of the comb. The teeth are $\frac{1}{10}$ inch by $\frac{3}{4}$ inch by 9 inches. They are riveted to a flat iron by twisting at the base to form the shape of an ordinary comb. The interspace between tooth and tooth is $\frac{1}{8}$ inch. The front portion of each tooth is sharpened like that of an ordinary comb.

6. Shaft. This is an ordinary wooden pole.

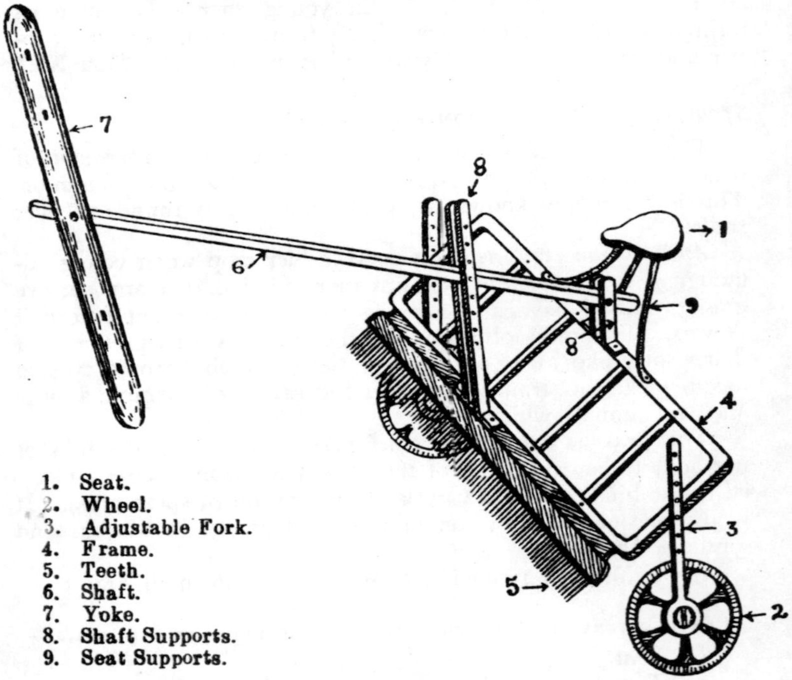

FIG. 7. COMB. IMPLEMENT USED FOR COMBING SPEAR GRASS.

7. Yoke.

8. Shaft Supports. These are iron flats fixed to the frame. The shaft is fixed to these supports in turn. Two supports are necessary to give sufficient strength. One is raised from the front and the other from the back reaper of the frame.

9. Seat Supports. These supports are raised from the framework simply to bear the weight of the driver.

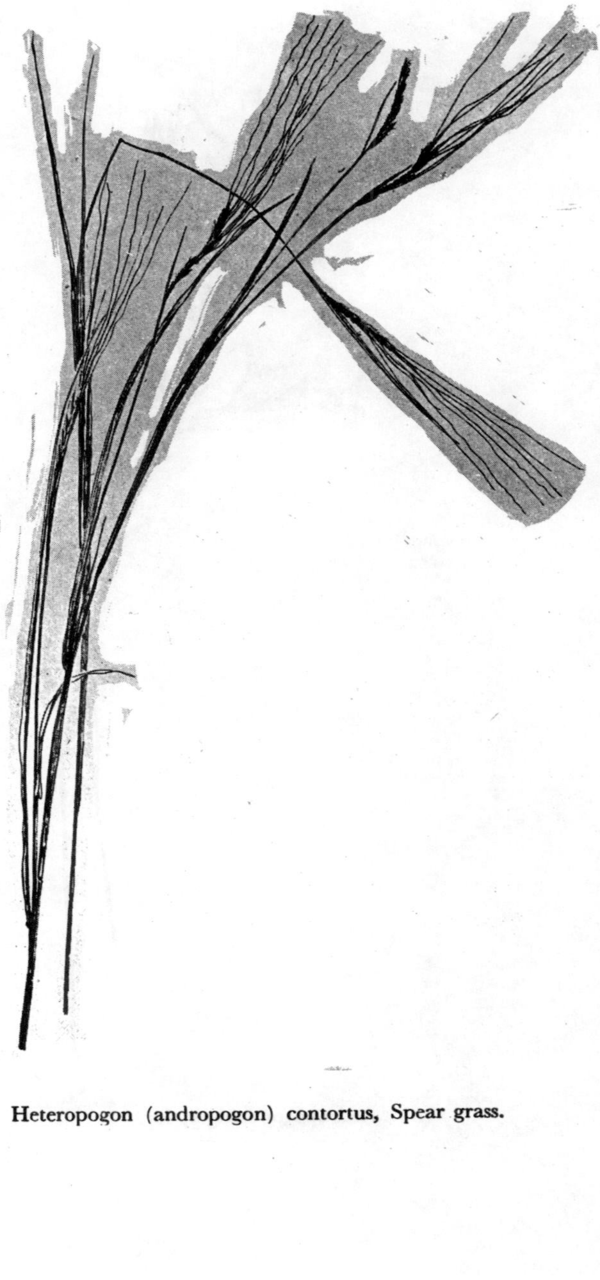

Heteropogon (andropogon) contortus, Spear grass.

Andropogon laniger, *bur*.

Fitting up the parts

No. 5. The flat iron to which the teeth are riveted is fixed to the framework No. 4.

Wheel 2 is fixed to the fork 3.

The fork is fixed to the frame 4 by means of a nut and bolt so that the fork will be at right angles to the frame, *i.e.* the comb will be parallel to the ground. The shaft supports are fixed to the frame and shaft fixed to the supports in turn. The shaft and teeth should face the same direction. A yoke is attached to the shaft and the seat fixed to the frame by means of seat supports.

Animals are tied to the yoke and driven. The machine moves on the wheels and the teeth which are below the level of the flower heads begin to comb the grass. The awns which are easily detachable when well ripe are caught in the teeth. When a sufficient quantity of awns is deposited on the teeth, the animals are stopped and the awns pushed down by means of a forked stick. The awns thus collected are removed by coolies in baskets.

Some other common species of Andropogon found widely distributed throughout India and generally regarded as good cattle fodders are:--

1. *Andropogon foveolatus,* Del.
2. *Andropogon moticola,* Schult.
3. *Andropogon annulatus,* Forsk.
4. *Andropogon laniger,* Dest. (Plate XIII.)

Perennial rye grass. (LOLIUM PERENNE)

Efforts are continually being made to improve fodder grasses and more attention might be directed to the cultivation of wild perennial rye grass, specially in the hills.

This grass can be seen growing luxuriously in many places in hill stations in northern India along road sides, where it has spread apparently from lawns where it had been planted for ornamental purposes in the past. Perennial rye is an excellent fodder grass and resists frost, and if it could once be introduced it should continue growing in village lands throughout the winter when the ordinary indigenous grasses are damaged by frost.

Such cultivation would reduce the amount of hand fed grasses necessary, as cattle would be able to graze all the year round on the rye grass. To cultivate this grass one should use only the best seed, which should be scattered over the land and raked or harrowed in. If the ground could be covered with manure before this operation so much the better. Rye grass does well in Australia even in places where dry conditions prevail and the heat may be as great as in India. It should flourish in all the lower foothills and valleys and in the highlands of southern India, and should prove invaluable if properly cultivated for supplementing the all too scanty grazing grasses available.

It might prove particularly useful if cultivated extensively in such places as the Kangra valley in the Punjab, where grazing is very deficient, specially if combined with leguminous crops such as berseem. (Plate XIV.)

(b) CULTIVATED GRASSES

Some particularly useful grasses for soiling or silage crops are elephant grass, guinea grass, Sudan grass, Rhodes grass and teosinte.

Elephant grass (*Napier*). (PENNISETUM PURPUREUM)

This grass does well on well drained sandy loam soils, but is likely to impoverish the soil unless it is periodically manured. It gives very high yields, but is not as palatable as some other grasses and green crops such as juar and maize. It compares well with green oats as regards its total digestible nutrient content and in this respect there is little to choose between elephant, guinea and Sudan grass, except that Sudan grass is somewhat low in digestible protein. With proper care six cuttings of green fodder, and, under irrigated conditions up to 600 maunds per acre, can be obtained from elephant grass.

Elephant grass needs to be repeatedly ploughed, and if possible should be irrigated immediately after each cutting. After a few years the clumps become so large that they hinder cultivation and should be thinned by digging out and planting elsewhere. When the plant is about three feet high, the grass becomes hard and fibrous and unsuitable as fodder. It should therefore be harvested before it becomes too fibrous.

Lolium perenne, perennial ryegrass.

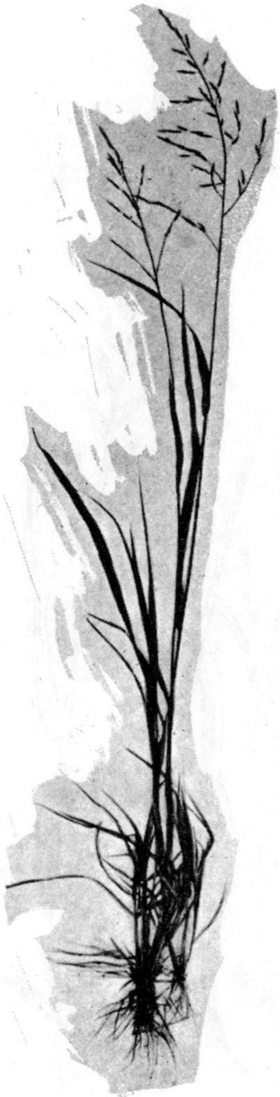

Andropogon sorghum var. Sundanensis, Sudan grass.

Guinea grass. (PANICUM MAXIMUM)

Guinea grass is a perennial plant with finer leaves and is more nutritious than elephant grass (Napier). Once planted it forms a source of green fodder for many years. It is palatable and relished by all kinds of livestock. It also has the advantage, unlike elephant grass, of not becoming coarse even when fully grown.

The grass can be grown along water channels on any land which cannot be used for other crops. It thrives best under warm moist climatic conditions and on almost any type of soil provided the soil is well drained. The best crops are obtained on fertile loams but it is advisable to apply farmyard manure at the rate of 15 to 20 cart loads per acre before planting.

The first cutting can usually be taken about 2½ months after planting, and then at intervals of 1½ to 2 months throughout the growing season, varying with the type of soil and the water available. It should be possible under favourable conditions to obtain a total yield of from 400 to 600 maunds per acre *per annum*.

Guinea grass continues to give good crops for a number of years with regular manuring and proper interculture, but after about six years, when the clumps have become very big and ploughing between the rows is difficult and the tussocks are dead in the centre, the grass should be transplanted in another field.

Sudan grass. (ANDROPOGON SORGHUM VAR. SUDANENSIS)

Sudan grass (Plate XV) is somewhat inferior to guinea and elephant grass and when cut at the milk stage has not proved to be a maintenance ration for dry heifers at Lyallpur. It yields only about 0.4 per cent or half the quantity of digestible protein present in guinea grass.

Rhodes grass. (CHLORIS GAYANA)

Rhodes grass is a perennial cultivated plant which thrives best on good loam soils which retain moisture well. It also does well on light loams if these are manured. This grass is capable of giving a considerable number of cuttings of palatable and nutritious green fodder during the year. Cuttings may be made every month on good soils if these are well manured and watered, but growth generally stops in the winter in northern India.

Rhodes grass produces runners which take root at the nodes thus giving rise to new plants. It has fine leaves which do not become coarse and is very palatable and relished by livestock, especially horses. It is usually ready for cutting two months after planting, and besides being used as pasture can be made into excellent hay. The protein content of Rhodes grass varies from 4 to 8 per cent.

Teosinte. (EUCHLAENA MEXICANA). Vern. MAKCHARI

Teosinte is believed by some people to be the wild form of the cultivated varieties of maize and does well under conditions similar to those which suit maize. It thrives best under hot humid conditions but also does well under irrigation and requires a rich soil and periodic manuring. The plants tiller profusely and give rise to a number of shoots forming thick clumps, and under favourable conditions they reach a height of 12 feet. The crop is usually ready for harvesting in 3 to 3½ months after sowing and may yield as much as 400 maunds of green stuff per acre. In northern India if teosinte is sown late at the end of July it is usually ready for fodder in October and November when there is usually a scarcity of green fodder.

Teosinte can also be cultivated mixed with leguminous crops such as velvet beans and soybeans, both of which take about the same time as teosinte to mature. When grown for seed the dry stalks are usually fed to bullocks after removing the seeds.

If the crop is sown early in March and April in northern India two or three cuttings may be obtained; the first being available in May or June depending upon the time of sowing. In this case, however, the soil should be very rich and receive liberal irrigation.

Teosinte also grows well in southern India.

(c) CULTIVATED FODDERS

Juar. (ANDROPOGON SORGHUM)

There are several varieties of *juar* (*cholam* in south India), which are widely cultivated, but the sweet ones are best, and

Andropogon sorghum, *jowar*, *chari* or *cholam*.

Pennisetum typhoideum, *bajra* or *cumbu*.

are extensively used as fodder crops in the Punjab and also for food grains in the Central Provinces and southern India. Some of the poorer types of *juar* are not as good as the fodder grasses, and may contain only about 0.4 per cent of protein and are not maintenance rations for dry heifers. Some of the improved types, however, have higher nutritive values, and the yields will vary according to the type, locality and rainfall, and may be as high as 500 to 800 maunds per acre. The sorghums, described more fully in Chapter VIII, in drought conditions are liable to cause prussic acid poisoning and death among animals which eat them. (Plate XVI.)

Bajra. (PENNISETUM TYPHOIDEUM)

Bajra, known as *cumbu* in south India, is an important millet grown both in the *barani* and irrigated tracts of India. It is the second most important crop of southern India and covers an area of nearly five million acres in the Madras Presidency. It makes an excellent green fodder and, if well cured, a good hay. (Plate XVII.)

The straw is not as good as cereal straws but is much better than paddy or rice straw and animals prefer it to the latter.

Some of the less important millets common to south India are *tenai* (*Setaria italica,* Beauv) and *varagu* (*Paspalum scrobiculatum.* L) which are poor in quality and are not frequently used.

Ragi or mandal. (ELEUSINE CORACANA, GAERTIN)

Ragi, known as *mandal* in northern India, is an important millet grown wherever there is a well distributed rainfall of 30 inches or more and it will tolerate salinity in soil better than most crops. *Ragi* is usually fed green after the ear heads have been harvested as animals much prefer it in this condition to the dry straw. *Ragi,* however, is not extensively used as a fodder crop in northern India where it is mostly grown for seed. In some parts including south India and Bengal it is much preferred, however, to paddy straw.

Oats. (AVENA SATIVA)

The cultivation of oats, of which there are many varieties, is largely restricted to places where horse breeding is practised.

Oats are of high nutritive value when cut green soon after the flowering stage, when they may contain 10 per cent of total digestible nutrients and 1.2 per cent of digestible protein on a dry basis. The broad leaved varieties give higher yields as a rule than the fine leaved ones. Oats are moderately rich in calcium, which varies from 0.73 to 1 per cent and phosphorus which may be somewhat higher; they thus have a better balanced calcium-phosphorus ratio than lucerne. Oats require the same type of climate and soil as wheat, but as they mature later they require more water.

French oats have a thick straw and broad and coarse leaves, while the Algerian variety and Fo.S.-1/29 both have very fine straw and leaves which make them more desirable both for green fodder and for hay. Green oats, especially the fine varieties, are particularly suitable for horses, dairy cattle and farm stock generally, and should be grown wherever possible.

Cheena. (PANICUM MILIACEUM)

Cheena is one of the inferior millets and is cultivated in some parts of the Punjab plains for grain and fodder, and is common in many parts of the Himalayas up to the Indus and also in other parts of India. Its grain is digestible and nutritious and may be used for poultry. It is also used in human dietaries especially by the poorer classes.

Velvet beans. (STIZOLOBIUM DEERINGIANUM). Vern. MAKHMALI SEM

The velvet bean is a leguminous plant with broad tender leaves and sends out many sinuous vines which trail on the ground giving a luxuriant growth which completely covers the soil, thus reducing evaporation to a minimum and enabling the plant to resist drought to a considerable extent.

It is cultivated in many parts of India and thrives best under moist warm conditions where rainfall or irrigation is sufficient.

The velvet bean is a new-comer to northern India where it can be sown from April to August, the best results being obtained from July sowings during the monsoon. The seed germinates after about a week, and the crop should be ready for fodder after 3 to 4 months and may yield up to 250 maunds per acre.

Zea mays, maize.

Medicago Sativa, lucerne.

The green fodder is very nutritious and its protein is highly digestible. It is also rich in calcium, phosphorus, iron and iodine, and on account of these qualities is an excellent fodder for farm animals, particularly young growing stock.

Maize. (ZEA MAYS)

Maize thrives in warm damp regions and requires a rich, well manured soil. It provides a continuous supply of nutritive fodder of approximately the same value as green oats, but with a slightly lower protein content. It is valuable for all types of farm stock. An average crop of 250 maunds or even 300 maunds or more per acre may be obtained if grown with cow-peas which enrich the soil with nitrogen. Maize has a digestible protein content of about 1 per cent and 13 per cent of total digestible nutrients. (Plate XVIII.)

Lucerne. (MEDICAGO SATIVA)

Lucerne (Plate XIX) is a valuable perennial which may yield 6 or 8 cuts a year continuously for several years under irrigated conditions. It has a deep tap root and can resist drought but not water-logging. It thrives on well drained medium calcareous loams, and if well manured should yield up to 600 maunds or more per acre annually. The best time to cut lucerne is when it is still young, as later on it becomes fibrous and indigestible. It is best to feed lucerne when in a slightly wilted condition. It is an excellent forage for horses, cattle, sheep and pigs but if given in too great quantities may cause bloating. It is also valuable for poultry and may be sown in poultry runs, but it should not as a rule be grazed as it is very easily trampled down and damaged. Lucerne is a valuable supplement to pasture as a soiling crop, and it may also be silaged, preferably with an equal quantity of green oats.

If cut just before the flowering stage lucerne has 3.2 per cent of digestible protein and 14.7 per cent of total digestible nutrients. It is moderately rich in calcium, distinctly poor in phosphorus, but has a high carotene content.

Lucerne should find a place in the rationing schemes for all stock as it is one of the best all round green fodders.

Berseem or Egyptian Clover. (TRIFOLIUM ALEXANDRINUM)

Berseem is an exotic crop having been imported into India from Egypt and is specially suited to irrigated tracts and areas of high rainfall. It cannot be grown under drought conditions. It is essentially a cold weather crop, and, belonging to the legumes, enriches the soil by virtue of the nitrogen fixing bacteria present in the root nodules. If these bacteria are absent from the soil the plant remains sickly in growth and gives scanty yields. In such cases heavy and careful manuring must be done before sowing, or better still, the seed should be inoculated with the proper nodule forming organism. Once the seed has been inoculated and a crop produced, further inoculation is unnecessary. (Plate XX.)

It is an extremely prolific crop, and from 4 to 7 cuttings weighing up to 1,000 maunds per acre can be obtaining during the season. Berseem contains up to 2.3 per cent digestible protein and 10 per cent total digestible nutrients, and is one of the most valuable fodder crops for stock, including horses. Working bullocks can be maintained on berseem even when doing hard work and receiving no added concentrates. A small quantity of dry roughage, however, is advisable.

Persian clover or Shaftal. (TRIFOLIUM RESUPINATUM)

Indian Clover or Senji. (MELILOTUS PARVIFLORA)

Shaftal and *senji* are closely related to berseem and contain a somewhat higher content of digestible protein and total digestible nutrients, but they are more liable to cause bloating in cattle. Partly on this account, and partly because they give considerably lower yields, they are now being largely replaced by berseem.

The cultivation of *shaftal* is similar to that for berseem but unlike the latter, *shaftal* does not need to be inoculated. On the other hand berseem need not be inoculated if *shaftal* has been grown on the land in the previous year. It yields only about half to two-thirds as much as berseem, an average being 400 maunds per year.

Senji has been cultivated from time immemorial and is still extensively grown, but owing to its tendency to cause bloating it it, like *shaftal,* being replaced by berseem.

Trifolium Alexandrinum, berseem. Showing greater returns after inoculation (left).

Sunflower. (HELIANTHUS ANNUUS). Vern. SURAJ MUKHI

The sunflower is an excellent green fodder and has the advantage over some other fodder crops of being cultivatable throughout the year by successive sowings, and does well both in north and south India. It does best on well drained medium loam soils which should be well manured, but it will not tolerate water-logging. Under careful cultivation sunflower has yielded in the Punjab as much as 450 maunds of green fodder per acre which forms a palatable and useful feeding stuff for all classes of livestock. Cattle show a particular liking for it. The best time to feed the crop is when the seed is in the soft milky stage, when it contains about 2.5 per cent protein and constitutes a maintenance ration which compares very favourably with many other green fodders and pasture grasses. Later on the stems become hard and fibrous.

It also makes excellent silage, the best stage for which is when the seeds are well formed but have not become hardened. If cut at an earlier stage than this the crop is too succulent for making good silage unless it is partially dried before being ensiled. It is advisable in any case to leave the crop on the ground for 24 hours after cutting before being ensiled.

The sunflower is not a good crop to use for preparing hay, as after the seeds have ripened the rest of the plant is generally too hard and fibrous to be attractive to cattle.

It is very exhausting to the soil and should be followed by green leguminous crops or by fallowing and manuring the land.

DRY FODDERS

Lucerne hay. (MEDICAGO SATIVA)

Lucerne hay perhaps varies more in nutritive value than other kinds, the extent of variation depending in a large degree on how it is prepared and handled when ready for use. This is on account of the fact that the stems are very brittle and liable to break off and be lost. The hay should be prepared during dry weather, and when mature should be of a rich green colour. Its value on the farm lies in the high protein content of the leaves which contain nearly 80 per cent of the entire protein content of the plant. Great care, therefore, should be taken in hay making to guard against loss in leaf material and to avoid exposure to rain in the late stage of curing as this may cause great diminution of food value.

The crop should be cut at an early stage when the percentage of protein is high, the fibre lowest, and the digestibility greatest. This will produce suitable hay for dairy cows, sheep, pigs and poultry. On the other hand, hay made from a more mature crop may be better for horses as it is less laxative. The curing process, however, is even more important than the precise stage at which the crop is cut, and bad curing may overmask the difference in feeding value between hays cut at different times.

The protein content of lucerne hay is considerably higher than that of most hays, averaging from 12 to 15 per cent and it is much more digestible, but it may show wide variations, and hence there may be greater differences in the amount of concentrates needed to balance rations for cows on lucerne, than on a corresponding good grass hay.

One of the great merits of lucerne hay for dairy cows and all young stock is its richness in calcium which may amount to 1.5 per cent on a dry matter basis. The phosphorus content is, however, only about 0.2 per cent on an average, and hence it is not well balanced in this respect. Carefully cured lucerne hay is also extremely rich in carotene (vitamin A), and in vitamin D.

Lucerne hay, therefore, furnishes one of the most valuable roughages for almost all classes of stock, particularly growing animals and milch cows.

Berseem hay. (TRIFOLIUM ALEXANDRINUM)

Berseem hay is approximately as rich in protein as lucerne, averaging about 14 per cent and is an invaluable fodder for almost all classes of stock. It is very rich in calcium, exceeding lucerne in this respect, with an average figure of nearly 3 per cent on the dry material, and a high phosphorus content around 0.355 per cent. One hundred pounds of berseem hay contain 9.7 pounds of digestible protein and 43.5 pounds of total digestible nutrients.

Oat hay. (AVENA SATIVA)

All the small grain cereal plants are suitable for making into hay, and the one most commonly used for dairy stock in northern India is oats. It is the basic ration for all the experimental work on cattle done at Lyallpur and constitutes a good maintenance ration for 800 lb. dry cows.

Well cured oat hay resembles in composition some of the better pasture hays, but it is low in protein when cut at the usual stage for hay making. If cut for hay at an earlier stage it may contain up to 12 per cent protein on a dry basis. An average oat hay contains about 5 per cent protein on a dry basis, and is low in both calcium and phosphorus. It has a digestible protein content of 0.20 pound and a total digestible nutrient content of 4.62 pounds. Oat hay constitutes an excellent feed for most types of stock and is a suitable roughage for mature horses. When oat hay is used for dairy cattle it is better to give it as a part of the roughage with a legume hay.

Straw and Chaff

As fodder plants approach maturity most of the nutrients which have been built up in the green growing leaves are transferred to the seeds, and consequently the straw which ultimately remains after threshing, consists of the mature and more or less dead leaves and stems and contains but little fat, protein and carbohydrates, but is chiefly composed of the more indigestible fibre. It is also lacking in minerals although silica may be high. The straws of the different types of cereals vary considerably according to the stage at which the plants are cut, the climate and the soil, but they are invariably lower in nutritive value than good hays. Their chief value is to serve as bulk roughages for ruminants, and they should not under normal conditions form any appreciable part of the roughages for dairy cattle and sheep, or for horses and mules at work. It is nevertheless undoubtedly a fact that a very large percentage of Indian cattle have to depend largely on straw for their rations.

Oat straw

Of the various straws, oat straw is by far the best due to the fact that oats are usually cut before they are fully ripe. It contains less silica than wheat straw and is softer and more easily digested. It is the only straw which should be fed to horses, and then only in small amounts. It may, however, be used for fattening cattle and dairy cows either long, or cut up into a moderate length as chaff. It has no special merits, and need not be used if a more economical hay is available.

It is invariably lower in nutritive value than most grass hays except, of course, those over-mature hays which are little better than dead leaves and stems. A reasonably good oat straw contains about 1.0 per cent of digestible protein and from 35 per cent to 40 per cent of total digestible nutrients.

Barley straw. (HORDEUM VULGARE)

Barley straw comes next in order of nutritive efficiency; it is generally threshed at a late stage and contains approximately the same percentage of digestible protein and total digestible nutrients as oat straw.

When cut into chaff and fed with roots, barley straw is better than wheat straw and likely to give better results than poor oat straw.

Wheat straw. (TRITICUM SATIVUM)

Wheat straw is one of the poorest feeds, though for a large proportion of animals in northern India as *bhusa* it forms a considerable part of the roughage which they get. It should be used chiefly for bedding.

If an ordinary analysis of wheat straw be compared with oat straw, not much difference will be found, but the former as a rule contains a higher percentage of crude fibre and silica, which lower its digestibility, hence the total digestible nutrients will invariably be lower than in the case of oat straw.

It must be borne in mind that the digestible nutrients contained in straws have been determined chiefly on ruminants and they would be considerably lower in the case of non-ruminants.

Thus, while oat straw has some small value for horses, wheat straw is of little or no value, as the energy expended in digesting it is greater than that which it supplies. It should, therefore, never be fed to horses.

Wheat straw becomes softened in the paunch of ruminants, and is slowly and partially digested during its stay there; hence they derive some benefit, chiefly from the carbohydrates. It may be fed to cattle over seven months old, either in the long condition or *bhusa,* in which form it is usually seen in India.

The author has found negative digestiblity values for the

protein of wheat straw at Lyallpur in samples containing 3 per cent or less of protein, hence when used it should always be combined with some other food rich in protein.

Lucerne straw

This straw if well prepared is of greater value than the straw from the coarser legumes such as soybeans, and may be used to replace part of the hay used for fattening lambs, although this is not to be recommended if plenty of grass hay and pasture is available.

Legume straw

Legume straw is the material left after legumes are threshed for seed, and it is often fed to stock. It is less rich than the corresponding hay, and its nutrient value will depend largely on the amount of leaves it contains. It is always high in fibre, but is a richer feed than cereal and millet straws, and contains more protein, especially the upper leaves and culms, but it is somewhat more indigestible.

Legume straw may safely be fed to dairy cattle and sheep, but gives much better results when combined with other roughages, or with a well matured silage. It is not as a rule satisfactory for horses as it contains too much dust.

Legume straws generally contain more calcium and magnesium than other straws.

Soybean straw. (GLYCINE HISPIDA)

Soybean straw has a low feeding value as it consists mostly of coarse stems with but few leaves, and should therefore not be fed as a sole roughage, but combined with a good legume hay. It is not satisfactory for high yielding dairy cows, but may be given to dry cows and heifers. It may also be fed to sheep, but they thrive better on the whole crop rather than on a mixture of soybean straw plus the beans. [See list of references.]

Bean straws. (VICIA FABA)

Bean straws are somewhat coarse, but they form a useful fodder when the crop has been cut early and not allowed to overmature. The less fibrous parts such as the leaves and pods contain more nutrients than most of the cereal straws. Bean straws should be chaffed and may then be fed to dairy cattle, sheep and horses in moderate amounts with other coarse fodders, but they should not exceed one third of the total roughage.

Pea straw. (PISUM ARVENSE)

Pea straw is usually of higher nutritive value than that from other legumes, and can be used for dairy cattle and sheep as a part roughage. It varies greatly in quality according to the state of maturity when cut, and when obtained in good condition from peas that have been picked green, can be used for almost any kind of stock except horses.

Rice straw. (ORYZA SATIVA)

Rice straw is the staple fodder for cattle in southern and eastern India, the Kangra district of the Punjab, and in many other rice growing tracts. There is a number of varieties of rice, but the straw from all shows certain notable defects. Trials carried out by the writer [Lander and Dharmani 1931] at Lyallpur on Sahiwal heifers showed that rice straw from Kangra was deficient in protein, containing only about 2.25 per cent. It was also low in calcium, 0.40 per cent, and phosphoric acid 0.25 per cent. With all the experimental animals the daily nitrogen balances and the digestibility coefficients of the protein were negative, thus showing that if fed alone to Sahiwal cattle it was not a maintenance ration, although the small hill cattle in Kangra apparently exist on it, with very little concentrates. It should not be fed to cattle if any other more suitable roughage is available, but its deficiencies can be made good by supplementing it with concentrates such as maize grain and toria cake, the latter being more economical from the point of view of protein supply, and to some extent by a green fodder such as green oats when added in the ratio of 1 : 3, or some leguminous fodder. fodder.

Work in Bengal [1942] has also shown that rice straw is deficient in protein and minerals and is usually supplemented in that province by a small amount of concentrate such as linseed cake and a green fodder when available.

ROOT CROPS

Roots and tubers belong more particularly to what are called the succulent feeding stuffs, which contain from 80 to 90 per cent water, but in spite of this high water content their nutritive value is considerable.

Roots accumulate large reserves of food nutrients in that part where the root merges into the stem during the first year's growth.

All root crops are very watery and may contain as little as 9—10 per cent dry matter, which is less than half the percentage of dry matter in a good maize silage. This dry matter, however, is very low in fibre and is highly digestible, and consists largely of carbohydrates, such as sugar in beets, carrots and mangolds, and starch in potatoes. The nitrogen or protein content is low and consists largely of non-protein nitrogenous bodies. Calcium and phosphorus are also low as a rule, vitamin D is practically absent, and, apart from carrots which are rich in carotene, there is very little vitamin A. Roots are, therefore, the precise opposite from a nutritive point of view to legume fodders and hays which are highly efficient in all these respects. On account of the high digestibility of the dry matter and their high total digestible nutrients, roots can be substituted for a part of the grain usually fed to dairy cows or sheep, and may be considered as a sort of concentrate solution.

It is an economic question of some importance as to how far roots could be fed instead of cereal concentrates, as it has been estimated that if the cost of production of 20 maunds of roots exceeds the cost at which one maund of cereal food can be produced or purchased, then, from the economic point of view, it would not pay to feed roots. As in the case of the cereals the nutritive ratio is a wide one, and it should be an easy matter when computing rations (Chapter V) to decide how much carbonaceous concentrate may replace roots and *vice versa*. Average maximum economic quantities which may be fed are about one maund to a hardworking bullock, about half this amount to a dairy cow, and about 12—15 pounds to an ordinary sheep. Some stock owners prefer to give no roots to cows in milk. Roots and tubers, however, may be considered as appetisers or bulk feeds and should not be fed in unduly large quantities, as a heavy proportion of roots may depress the digestibility of the other parts of the ration.

In addition to the nutrients they furnish, roots and other succulent feeding stuffs have a tonic effect on animals and are regarded favourably by stockmen. Horses in particular are fond of roots, but the amount fed should be limited to about 10 pounds per day. All roots have a decided laxative effect, and are therefore useful at times when the rest of the ration is mainly dry matter. Roots are valuable for colts and may be given chopped or sliced, or mixed with chopped hay, or grain,

Roots may replace a part of the concentrate given to pigs but the amount should not ordinarily exceed about 5 pounds a day.

Roots are also greatly relished by poultry and may be given as a part substitute or in conjunction with green fodder.

Root crops in general should be regarded as nutrient adjuncts and appetisers rather than the main part of rations.

Turnips. (BRASSICA RAPA)

There are several types of turnip, including the white and yellow varieties, and also various hybrid crosses between the turnip and swede. The white variety usually contains only about 7—8 per cent dry matter, while the yellow has slightly more. Both types contain only 1 per cent protein and little or no fat. Both are excellent for sheep and cattle, the tops generally being consumed with the roots.

It is advisable to store both turnips and swedes before feeding, as a mild fermentation takes place and the sugar content then rises, particularly in swedes, whereas if eaten when freshly gathered they may cause scouring.

Both turnips and swedes have a tendency to cause a disagreeable taste in milk if fed before milking, or if kept near the milking shed. This may be reduced to a minimum by feeding them sometime after milking, and not just before. They should not be fed in too large quantities.

Should a taint occur in the milk this can be completely removed by pasteurising.

Swedes. (BRASSICA CAMPESTRIS)

Swedes are in many ways very similar to turnips but have a

greater feeding value and contain more sugar. They are also considerably harder, have a greater percentage of dry matter (9 per cent), and keep better. The composition of swedes, however, varies considerably according to climate and soil, and they thrive best on a light loam with plenty of calcium, otherwise the disease known as 'Finger and Toe' is likely to occur. Climate is of greater importance than soil in influencing yield, although an adequate amount of phosphorus, nitrogen and potash in the soil are essential. Swedes will not thrive if any of these, particularly phosphorus, are deficient.

Swedes are particularly relished by sheep and pigs, but precautions similar to those for turnips should be taken to avoid tainting the milk if they are fed to cows. The average yield for swedes and turnips should be about 260—300 maunds per acre, but may vary widely beyond these limits according to the soil and climate.

Mangold wurzels. (BETA VULGARIS)

Mangolds grow higher out of the ground than swedes or turnips, and may be grown on a wide range of soils but they do best on a moderately heavy deep loam. They respond well to potassium, nitrogenous and to a lesser extent phosphatic manures, and on suitable soil may yield up to 800—900 maunds per acre. The amount of dry matter they contain is small, being only about 10 per cent, but it is very digestible, and varies with the degree of ripeness. Mangolds should not be given to animals until they have been stored for sometime, as the freshly harvested crops contain both nitrates and amides, which are likely to cause scouring, whereas if left till maturity these substances will become largely converted into proteins and help to increase the very small amount of protein previously present. The quality of mangolds varies with the variety and is generally better in the smaller varieties, which contain about 1 per cent of digestible protein and 7 per cent total digestible nutrients. Mangolds are specially suitable for dairy cows, as they do not taint the milk. They are also relished by sheep, but there is a danger that rams and wethers may be affected by urinary calculi if fed on mangolds for any length of time. They are very little used in India except on certain big farms.

Beet. (BETA VULGARIS)

Beet is not commonly used for feeding to cattle except in some parts of northern India and is similar in composition to mangolds but has a higher percentage of sugar, 15—20 per cent for the extraction of which it is usually grown; the crowns and pulp from the factories are then fed to stock.

Beet may be fed to dairy cattle and sheep, but not to the males, as it is reputed to affect adversely their capacity for breeding. It is an excellent substitute for barley meal for pigs, 3½ pounds of beet being equivalent to 1 pound of meal.

Beet tops

The crowns which contain very little sugar, and the tops are usually lopped off, and cattle and sheep are allowed to graze them in the field. They may, however, be matured in heaps and then fed as such or ensiled.

The tops form a valuable substitute for roots in rations for cattle, about 25 pounds of tops being equal to about 40 pounds of mangolds. More than this should not be given, and a sufficient quantity of roughage such as hay should be fed as an antidote against the laxative effect of the tops. The silage is less laxative than the fresh tops, but not more than 30 pounds per day should be fed to cattle, or more than 3 pounds to sheep.

Horses and pigs should not be given as much, either of the tops or the silage as is usually given to ruminants, as their stomachs do not provide the necessary fermentation processes required to destroy the somewhat high percentage of oxalic acid present. This may cause actual poisoning if beet tops are given in too large amounts. In any case the tops should be allowed to wilt for a time, and a small amount of powdered chalk spread over them will obviate any likelihood of oxalic acid poisoning.

Sugar beet pulp

The pulp from the beet after the sugar has been extracted may be used either in the wet or dry condition on farms near the factory. The wet pulp contains about 90 per cent. water, and the same amount of carbohydrates as turnips, but less than mangolds. About 60 pounds may be fed to bullocks and not more than 30 pounds to cows per day.

The bulk of the pulp produced at the factory is usually dried and sold as dried pulp. This contains only about 10 per cent water and can be used as a substitute for roots. One pound of the dried pulp is equal in nutritive value to about 8 pounds of mangolds or swedes or 10 pounds of turnips. The pulp may also replace part of the cereals in the ration; 1 pound being equal to 1 pound of oats. The dried pulp is most suitable for dairy cows and working bullocks and is also useful for horses and pigs. Kellner [1908] recommends 8 pounds for cows, 12 pounds for bullocks, 6 pounds for horses, and up to 4 pounds per day for fattening pigs. Beet and beet residues are all deficient in protein, calcium and phosphorus and these must be properly provided for in the rest of the ration.

Potatoes. (SOLANUM TUBEROSUM)

Potatoes do not as a rule form any considerable part of rations in India, but they may be used for this purpose when very small, or if a ready market cannot be found. Potatoes contain about 24 per cent dry matter, including 2 per cent protein, most of which is in the layer immediately under the skin, and 20 per cent carbohydrates as starch. Potatoes are essentially a carbohydrate food comparable to mangolds and swedes, two pounds of the latter being equal to one pound of potatoes. The nutritive ratio is 1 : 15 and therefore sufficient protein rich supplements should always be included in the ration when potatoes are used to any extent. About 450 pounds of potatoes are required to furnish as much total digestible nutrients as 100 pounds of grain; this furnishes a measure of their nutritive value, though, when cooked, they have a somewhat greater value than this for pigs.

Raw potatoes are laxative and large amounts should not be fed to start with, but ultimately cattle may be given up to 40 pounds per head per day and sheep 4 pounds per 100 pounds of live weight. Potatoes should be boiled for young or pregnant stock, and for pigs, although this entails extra expense.

Raw potatoes should be fed more sparingly for horses than for cattle, and more than about 5 pounds a day should not be given.

To get the best results from pigs, potatoes should not be fed in larger amounts than 4 pounds per pound of concentrate, and should always be boiled thoroughly with salt added to increase palatability.

Potatoes contain no vitamin A or D.

Unripe potatoes and the sprouts of stored potatoes contain small amounts of a poisonous substance—solanin—and these should be removed before feeding if the potatoes are badly sprouted.

Carrots. (DOUCUS CAROTA)

Carrots contain about 88 per cent water, 1 per cent protein, traces of fat, 9 per cent carbohydrate and 1 per cent of ash and fibre respectively and are very rich in carotene. They are superior to turnips, swedes and mangolds for stock, and are especially relished by horses and milch cows, though some farmers advise against feeding carrots to brood mares for a few weeks before and after foaling.

The fattening value of carrots is not as great as is often thought, and is actually only half that of potatoes. Seven pounds of carrots are equal to 1 pound of oats in feeding value; they are nevertheless more valuable than raw potatoes, chiefly on account of their taste, and carotene content. Hard worked horses should not be given large quantities but it frequently happens that sick horses will eat carrots when they will eat nothing else.

THE PROTEIN RICH CONCENTRATES

Oilseed cakes and meals

Oilseed cakes and meals are the products left over from the various processes devised to extract oil from oil seeds. Formerly, these cakes were regarded more as by-products of the oil industry, and used as manure, and only to a small extent for feeding purposes, but with the increased use and demand for them as feeding stuffs they have become as important a branch of the oil industry as the oil itself. Before modern extraction methods had attained their present degree of efficiency, practically all oil cakes on the market contained considerably more oil than is found in present day products, which are frequently very hard and dry owing to a high degree of extraction, as a result of which their value as feeding stuffs is somewhat diminished as far as the oil content is concerned.

There are three main methods of extraction:—

(a) The hydraulic pressure method by which the seeds are pressed under high pressure between cloths through which the oil runs into collecting channels.

(b) The expeller process, which is continuous, and in which the seeds are fed into a revolving 'screw' of diminishing circumference, the oil being collected and carried off in small channels, while the extracted cake is gradually moved on and expelled through a small opening at the narrow end of the screw.

(c) The extraction process.

In this method the seeds are crushed and placed loosely in a large container, and a solvent such as ether, benzene or petrol is allowed to percolate in a continuous stream, and by an automatic process, repeats the operation until all the oil is extracted. The extracted material is then heated with superheated steam to remove any remaining traces of the solvent. The hydraulic method usually leaves about 8 per cent oil in the cake, whereas the extraction process may not leave more than 1 per cent, and has the disadvantage that some traces of the solvent are usually left unremoved and a very hard, dry cake is produced almost devoid of fat. One of the disadvantages of the expeller process is that, owing to the friction set up, the heat produced may tend to char the cake giving it a burnt odour. The amount of oil left behind by this process is intermediate between the other two processes, being usually about 6 per cent.

In the case of cottonseed and groundnuts there are two types of cake in the market, *viz.*, the decorticated and the undecorticated. In the process of decortication the husks of the seeds are removed by a special process before the oil is extracted. The husks are fibrous, and hence the resultant product obtained is richer both in digestible protein and total digestible nutrients than the undecorticated variety; it is also more digestible.

The undecorticated cake naturally contains much more fibre, and is of lower nutritive value.

Oilseed cakes and meals are all high protein concentrates, and are especially valuable for balancing rations which would otherwise be deficient in protein. They are not only rich in protein but also in total digestible nutrients, and therefore a wise choice must be made in the amount to be added to rations which need only a protein addition.

If a ration needs to be balanced by adding both extra protein and also non-protein nutrients, an oilseed cake of moderate protein content should be selected.

Oilseed cakes should never be allowed to get wet in transit or storage as deleterious effects are often produced by feeding mouldy or fermented cake.

The farmer has a considerable range of cakes and meals to choose from, varying in degree of richness in protein. For convenience in making a selection, therefore, these have been set out in an approximate order as regards their protein value, together with their analyses and digestible protein and total digestible nutrient content. (Table II).

The data for those marked L have been obtained from digestibility trials conducted at Lyallpur.

Ground nut cake. (ARACHIS HYPOGAEA)

The ground nut, or earth nut, is chiefly a product of south India, and grows below ground. The kernel is enclosed in an outer fibrous husk, and the cake may be made by the usual methods of extracting oil with the husk still remaining, when it is known as undecorticated cake, or with the husk partly or entirely removed, when it is known as semi-decorticated or decorticated. Ground nuts contains about 45 per cent oil and 27 per cent protein. The extraction process, however, leaves about 7—8 per cent oil in the decorticated cake, and 48 per cent protein. The undecorticated cake naturally contains considerably more fibre than the decorticated, about 24 per cent compared with 6, and hence a lower digestible protein and total digestible nutrients per cent. The semi-decorticated cakes on the market vary considerably in composition according to the amount of husk left in, and some knowledge of their composition, in comparison with the decorticated cakes is desirable.

The decorticated products compare very favourably with other concentrated cakes or meals, and are among the richest protein feeding stuffs and should be used only sparingly, and in all cases the full amounts needed should not be fed at once but led up to gradually. Ground nut cake is particularly suit-

TABLE II

SHOWING CHEMICAL COMPOSITION AND FEEDING VALUES OF SOME OILSEED CAKES AND MEALS

Name of the cake.	Moisture %	Dry matter %	Ash %	Fat %	Crude fibre %	Protein %	Nitrogen free extract %	Total digestible nutrients per 100 lb.	Digestible proteins per 100 lb.	Albuminoid ratio.	Remarks.
1. Groundnut cake	6.26	93.74	5.39	6.06	15.24	37.57	29.48	79.09	31.11	1 : 1.6	L.
2. Linseed meal ..	8.70	91.30	5.50	6.30	8.00	35.20	36.30	78.20	30.60	1 : 1.6	
3. Toria cake ..	3.79	96.21	7.53	8.58	9.75	35.00	35.35	73.95	30.47	1 : 1.5	L.
4. Decorticated Cottonseed cake.	5.81	94.19	7.67	8.68	5.94	36.25	35.65	71.96	29.08	1 : 1.1	L.
5. Taramira cake ..	6.42	93.58	7.60	11.63	8.76	33.25	32.34	85.30	28.92	1 : 2.2	L.
6. Rapeseed meal ..	10.00	90.00	8.00	5.10	11.70	34.80	30.40	61.30	28.20	1 : 1.2	
7. Sarson cake ..	6.56	93.44	8.81	14.22	10.02	28.19	32.20	81.60	25.70	1 : 2.3	
8. Linseed cake ..	5.61	94.39	9.44	4.20	9.05	28.94	42.76	82.57	23.38	1 : 2.6	L.
9. Cottonseed meal	5.30	94.70	4.40	6.30	23.80	28.50	31.70	70.90	23.00	1 : 2.1	L.
10. Cottonseed cake (Undecorticated)	7.48	92.52	6.01	8.47	22.31	21.13	34.60	75.70	17.97	1 : 3.1	L.
11. Rape cake ..	6.00	94.00	9.33	13.41	7.70	36.37	33.19	92.88	27.83	1 : 1.8	Pusa.
12. So; bean meal ..	8.30	91.70	5.70	5.70	5.60	44.30	30.30	82.20	37.70	1 : 1.2	Morrison.

able as a rich protein concentrate for dairy cows, and up to three or four pounds per head per day may be given, depending on the size of the animal, the milk produced and the composition of the rest of the ration.

The decorticated cake may be used mixed with calf meals, milk substitutes and pig meals, and for fattening cattle, sheep and pigs, and up to half a pound per day may safely be fed to sheep. A certain amount of the carbonaceous concentrates fed to horses at hard work may be partly replaced by decorticated or undecorticated ground nut cake if no other protein rich food is given. The meal is very useful in poultry feeding.

The decorticated cake is especially suitable for young stock when mixed with soft foods. Too large a quantity should not be given to pigs, or there will be a danger of the fat of the animals becoming too soft. A suitable proportion of ground nut cake when fed with maize for pigs is 7 parts of the former to 10 parts of the latter, but when gram is used the amount of cake should be very considerably reduced.

Ground nut cake is especially liable to develop moulds if allowed to become damp, with the result that the protein will become decomposed, with bad effects on animals which eat it.

Soybean cakes and meals

Soybeans are not grown extensively in India, and the cake and meals are not used to anything like the extent they are in foreign countries. They differ in composition from ordinary peas and beans in that they contain nearly twice as much protein, more than ten times as much fat or oil, and contain but little carbohydrates, chiefly in the form of starch.

Soybean cakes are consequently very rich in protein which may be as high as 45 per cent, and are highly digestible. The oil content will vary according to the method of extraction used. The cake contains less fibre than cottonseed or linseed cake, the amount usually being about 5—6 per cent. Taken as a whole the cake is extremely digestible if not given in too large quantities, and supplies over 80 pounds of total digestible nutrients per 100 pounds. This is somewhat more than linseed or cottonseed cakes contain. These cakes are not rich in vitamins A and D, and both calcium (average 0.3 per cent) and phosphorus (average 0.7 per cent) are rather low, hence the question of mineral balance and vitamin supply should not be overlooked. The whole beans are sometimes fed to stock but the cake is preferable on account of the high oil content of the former.

Soybean meal is an excellent protein supplement for all classes of stock if its highly concentrated nature is kept in mind. It is of equal value to linseed or cottonseed cake for dairy cows, and fattening cattle, but the amount given should not exceed from 2—4 pounds per head per day. Horses may be given small quantities, not exceeding from half to one pound per day when doing hard work, and the ration is otherwise poor in protein.

The cake when cooked is better for pigs and poultry than when raw, and hence a cake which has been partly cooked in the extraction process is of greater value, though for cattle, sheep and horses there is little or no difference in nutritive value between the cooked and uncooked cake or meal. It can generally be seen at a glance, whether the cake has been cooked as the uncooked cake has a raw beany taste and is lighter in colour than a well cooked hydraulic process or expeller process cake. The latter is light brown in colour and has a dextrinous or nutty taste.

Soybeans are inclined to be laxative, and may be used with undecorticated cottonseed cake for cattle, as the latter has an astringent property which helps to counteract the laxative properties of the former.

Soybean meals contain less oil than the cakes owing to a higher degree of extraction, and more protein.

Linseed cake. (LINUM USITATISSIMUM)

Linseed is the seed of the flax plant, and the cake is produced by heating and pressing the seed in hydraulic presses in much the same way as for preparing other types of cake. Linseed meal is the finely ground cake, and both are sold primarily on the basis of their protein content. Good quality cakes are brown in colour and contain about 3 per cent oil and 36 per cent protein.

Although linseed cake is not used as extensively in India as it might be, it is very digestible and has a slight laxative effect. These qualities make it especially valuable for young stock, which require an easily digestible ration of high protein content. It is also one of the most valuable protein supplements for dairy cattle, horses, sheep and pigs, and there is no other feeding stuff that will so readily bring cattle into good condition. It is especially valuable when the ration otherwise has a tendency to be constipating. Cows may be given from 2 to 3 pounds a day, but if larger amounts are given the butter produced has a tendency to be soft and greasy. Linseed cake produces excellent results when fed as the only protein supplement to sheep and horses, and half to one pound a day may advantageously be fed to hard worked or sick horses, although for the latter the meal is to be preferred. Linseed meal should preferably be used in conjunction with another protein supplement of high biological value, such as fish meal, dairy by-products, or meat scraps, for swine and poultry.

Linseed cake is often considered to have a higher value for fattening purposes than cottonseed cake when fed as the only protein supplement, but a combination of equal proportions of each will produce just as good results as linseed cake alone, and is as a rule less costly.

There may be some confusion between what is generally called linseed meal and linseed cake meal. Linseed cake meal is the ground up linseed cake, while linseed meal is made by grinding the whole seed to a meal, and as it contains all the oil originally present in the seed, is naturally very different in composition from linseed cake meal proper. It may be used for calves and sick animals but should otherwise be used very sparingly owing to its very high oil content.

The term 'linseed meal' is often used by American writers to designate the ground cake, while they refer to linseed meal containing all the oil as 'oil meal' or 'oil cake.'

Palm kernel cake or meal

Palm kernel cake is not used to any great extent in India except perhaps in the south. The cake is the product of the seed kernels of certain oil palms, obtained after removal by extraction of a part of the oil, from the ground kernels. The cake or meal varies very much in composition, particularly in its fibre and oil content. The better grade meals contain somewhat more protein than wheat bran, and also more oil, if made by the hydraulic process (8 per cent oil), and supply a higher percentage of total digestible nutrients than bran.

The cakes and meals are rather unattractive, and animals take some time as a rule to get used to them, but, if used sparingly to begin with, they become accustomed to them and will eat them readily. It is as well to add some tasty material such as molasses or locust bean meal to the ration if palm kernel cake is used, and to keep the latter to not more than one third of the total ration. These cakes should always be fed cold and not soaked in warm water.

The cakes keep well if stored under dry conditions.

Palm kernel cakes are suitable for dairy cows when fed 3 to 4 pounds per day, and have a stimulating effect on the production of butter fat, and the butter produced is firm and of good quality.

The cakes should be fed to horses in only small amounts occasionally, not exceeding 1½ pounds per head per day, as they have a tendency to cause excessive salivation. They are useful in small quantities in pig rations as they help to give the fat a firm and even consistency.

Cocoanut cake. (COCOS NUCIFERA)

Cocoanut cake is produced from the dried kernel of the cocoanut, known as *copra*, and is the residue left after extraction of the oil by expression. The cake has about the same nutritive value as palm kernel cake, but varies very considerably in composition, and usually contains 8 per cent oil. A good cake has a slight brownish colour and a pleasant and characteristic smell. It is rather tasteless when dry, but absorbs a large quantity of water and increases in bulk, and should be soaked before being fed to stock.

Cocoanut cake is extremely susceptible to damp, and unless stored in a very dry place will turn rancid and cause diarrhoea, hence it should be in good condition when purchased.

Cocoanut meal contains somewhat more protein than wheat bran. The protein is better than that of maize, but not so good biologically as the protein of soybeans or ground nuts.

The cake or meal is easily digested and dairy cows may be given from 2 to 3 pounds per head per day, and it may constitute up to 50 per cent of the total concentrates in the ration for fattening animals. It is considered by some people to increase the production of milk and the amount of fat contained in it, and like palm kernel cake, it helps to give the butter a firm consistency. The meal may be given to pigs as a substitute for wheat middlings or linseed meal, and it helps to produce a firm clear-textured fat.

Sesame or gingelly cake. (SESAMUM INDICUM)

Sesame cake contains about 40 per cent high grade protein, nearly as much as is contained in cotton seed cake, and is produced from the seeds of the sesame plant sometimes known as *til* (N. O. Pedalineae), which is grown extensively in India. It is rich in calcium and phosphorus and forms a satisfactory feed for dairy cows, and is said to exert a stimulating influence on the flow of milk, but not more than 2 to 3 pounds per head per day should be fed or there may be a tendency for the butter to be soft. It is excellent for fattening cattle and sheep, and also for pigs in small quantities only, as it is inclined to produce a soft body fat.

Rape seed cakes. (BRASSICA CAMPESTRIS, var. SARSON AND BRASSICA NAPUS, var. DICHOTOMA)

Confusion sometimes exists as to the difference between the terms sarson, toria, rape and mustard. Botanically, *toria* is classified as a rape, but in popular parlance all are referred to as rapes, and there is no specific rape which is botanically or otherwise independent of the others. All belong to the Natural Order Cruciferae, which includes also cabbages and turnips.

Sarson, the Indian Colza (*Brassica campestris* var. *Sarson*) and *toria*, sometimes called the Indian rape, are commonly known as rapes and are grown extensively as cold weather crops. The seeds of both *sarson* and *toria* are very similar in appearance and are crushed chiefly for their oil (the Colza oil of Europe) which is used for lighting and cooking purposes.

Sarson and *toria* cakes are fed to cattle, although *toria* cake has a slightly bitter taste.

Mustard or *rai* (*Brassica juncea*) is sometimes confused with the Indian rape (*Brassica napus* var. *dichotoma*) but it is a distinct and separate species. The seed is used as a condiment and in medicine, but the oil is very pungent and is comparatively little used as food. The cake emits a pungent smell on being moistened with water and is not as a rule given to stock as it is unpalatable to them.

Cotton seed cake

There are two chief kinds of cotton produced in India—the *desi,* and the American varieties represented by 285F, 289F, 4F and others.

In the extraction of oil from the seed, various processes are employed. If the whole seed is crushed under great pressure without any preliminary heating, the cake produced is known as 'cold pressed' cake. If, on the other hand, the whole seed with some remaining lint attached in the case of the American varieties, is heated before crushing, the residual cake is known as undecorticated cake.

When, prior to crushing and heating the adhering lint and hard seed coat is first removed, decorticated cotton seed cake is produced. This is rich in protein and contains considerably less fibre than the undecorticated varieties.

There are thus important differences between the two types, and the protein content while averaging 22.74 per cent in the undecorticated cake may be as high as 40 to 44 per cent in the decorticated varieties, the corresponding fibre contents being 23 per cent and 9 per cent respectively. Hence undecorticated cake should not be fed as a rule to young stock on account of its high fibre content, and the high protein content of the decorticated variety must govern the amount fed to other classes of stock.

Both types are extensively used abroad but they have only been employed to any extent in India comparatively recently. There is a widespread prejudice against the use of certain American cottonseeds and cakes such as 285F and 289F on account of the adhering lint or 'fuzziness', which is supposed to be bad for cattle. Until recent years the only variety usually fed was the *desi.* Work carried out at Lyallpur, however, has shown that there is no foundation for this prejudice and that from 3 to 6 pounds of cottonseed cake made from fuzzy varieties may safely be fed per day as part of the concentrate ration to 800 lb., milch cows giving an average of 20 pounds of milk per day for long periods without any deleterious effects on the health of the cows or on the milk yield, provided sufficient high quality hay or green fodder is given to guard against vitamin A deficiency.

The protein of cottonseed cake is well suited for supplementing the cereal grains for sheep and horses, though other cakes are probably more suitable for horses. Better results are obtained for swine and poultry if cottonseed cake is used as a part of the protein supplement only, the rest being made up of milk products, fish meal or tankage.

Cottonseed cake is very rich in phosphorus, the higher grade cakes containing an average of 1.2 per cent. It is only moderately rich in calcium—0.2 per cent and is deficient in vitamins A and D.

Cottonseed cakes tend to produce a hard body fat and a butter of firm consistency, hence they are useful complements to concentrates such as soybeans and ground nut cake which have opposite tendencies. Cottonseed, however, contains an alkaloid, gossypol, which is poisonous to certain types of animal if consumed in large amounts. Cattle do not appear to be affected by it, while swine may suffer serious harm. The cakes should not be given to calves till they are about 6 months old. Horses and sheep are also affected to some extent. Swine should not receive more than 9 per cent cottonseed cake in their rations, while horses and mules may be given from one to two pounds daily, with good quality roughages. Sheep should only be allowed small amounts sufficient to balance the ration, and the use of cottonseed cake should not be long continued.

THE OILSEEDS

Apart from cottonseed, the oilseeds have not been used to any great extent as such in India as feeding stuffs.

Desi and American Cotton Seeds

Desi cotton seed is a naked seed and has always been popular with the average zamindar, but with the development of American varieties and oilseed presses for producing cake it is not as plentiful as formerly.

The writer [Lander and Dharmani, 1929] has recorded, in presenting the results of feeding trials conducted at Lyallpur with *desi* and American varieties fed at a rate of 3 pounds per head per day to heifers of 800 lb. weight, the fact that, while the American seeds 4F, 289F and 285F are richer in fat and protein than *desi* cotton seeds, yet the digestibility of the various seeds do not present sufficiently marked differences to warrant any definite distinction being made in their nutritive values. The digestible protein and total digestible nutrients per 100 pounds of seed, and also the albuminoid ratio however reveal the superior value of 285F, and to a lesser degree that of 289F and 4F over the *desi* seeds. Hitherto American seeds have been sold, and are still sold at a cheaper rate than the *dest* for feeding purposes. It would pay therefore to use them rather than the *desi* seeds because besides being cheaper they are also more nutritious.

The results of the trials mentioned above are summarised in the following table:—

TABLE III

FEEDING VALUE OF THE COTTON SEEDS PER 100 LB.

NAME OF THE SEED	Per 100 lb. of the seed. TOTAL DIGESTIBLE NUTRIENTS.	DIGESTIBLE PROTEIN.	ALBUMINOID RATIO
	lb.	lb.	1:
285F	97.44	12.23	6.0
289F	85.31	11.50	5.6
4F	78.56	10.67	5.5
Desi	86.96	8.00	8.6

Cotton seed may also replace part of the concentrate ration for working bullocks, and trials at present in progress (1943) at Lyallpur show that 3 pounds per day may be fed to 1,000-1,200 pound bullocks working 8 hours a day, with 1 pound of bran and 10-15 pounds of *bhusa*.

Cotton seed may also be fed to sheep in moderate amounts, and Jones [1933] has recorded that lambs made satisfactory gains on a ration of 0.8 pound grain, 0.4 pound of whole cotton seed and 1.4 pounds of lucerne hay.

Linseed

Whole linseed is not usually fed to farm stock and they do not much relish it. It is, however, given boiled to horses, and may be used in place of linseed cake, or given crushed, boiled or steamed to sick animals and to calves with skim milk, as a part substitute for the removed butter fat. The seed only contains about two thirds as much protein as linseed meal, but is one of the richest feeds in total digestible nutrients. For fattening cattle, camels and pigs, linseed has been found equal or superior to linseed meal, but too much must not be fed to pigs or soft pork will be produced.

In trials at Lyallpur it has been found that 800 lb. heifers would not take more than 2-2½ pounds per day, and that they consumed less of the basal oat hay ration with linseed than with other concentrates. The appetite was adversely affected due to the mucilaginous character of the seed.

Soybeans

Soybeans provide proteins of better quality than most other seeds or their by-products, and are excelled in the quality of their proteins only by feeding stuffs of animal origin. Besides being rich in protein, they contain over 17 per cent fat, and are low in fibre, and hence furnish more total digestible nutrients (78 per cent) than the grains and other common concentrates. They are also rich in both calcium, 0.53 per cent and phosphorus 1.64 per cent. They are lacking in vitamins A and D.

Trials at Lyallpur showed that dairy cattle relished soybeans which caused no digestive disturbances.

Soybeans are as satisfactory a protein supplement as cotton seed meal for dairy cattle, sheep and horses, but if fed to pigs for any length of time they cause the lard to become unduly soft and damage the quality of the pork. It is best to cook the beans before feeding to pigs. They should be ground for dairy cattle, the grinding being done with the beans mixed with the grain, and they should not be prepared too long before feeding in damp or warm weather as they are liable to become rancid.

Sunflower seed. ((HELIANTHUS ANNUS)

The seed of the sunflower is sometimes fed to animals and is highly nutritious and digestible. An average sample of seed

Cyamopsis psoraloides, *guara*.

contains 18 per cent total, and over 17 per cent digestible protein, and shows a narrow nutritive ratio of about 1 : 4. It is thus a highly concentrated protein feeding stuff of high digestibility and particularly suitable for young growing stock, and in moderate quantities for milch cows.

THE CARBONACEOUS CONCENTRATES

Grains and Seeds

(a) *Leguminous grains*

Gram. (Cicer arietinum)

Gram is the ripe seed of a leguminous plant, sometimes known as the chick pea, which thrives in light sandy soil and is one of the most universally used concentrates in India for almost all classes of stock. Strictly speaking, it should be classed as a nitrogenous food rather than a carbonaceous one, as it contains nearly 20 per cent protein and consequently should always be employed in rations with due regard to this fact. It contains 0.7 per cent calcium and 0.6 per cent phosphorus, and, though only moderately rich in these minerals, they are well balanced.

Gram used in the Lyallpur trials was found to contain 14 pounds digestible protein and 77 pounds total digestible nutrients per 100 pounds.

In India, gram is usually fed to horses up to 3-4 pounds per day, and other equine stock in somewhat smaller quantities. Mules should get about 1½ pounds per day, and bullocks and buffaloes 3-4 pounds per day. From 2.5 to 3 pounds of gram can be incorporated in the rations of 800 lb. milch cows giving 20 pounds of milk a day, to make up the total digestible protein needed.

Arhar. (Cajanus indicus)

Guara. (Cyamopsis psoraloides)

Arhar and *guara* (Plate XXI) or clusterbean are leguminous plants whose seeds somewhat resemble horse beans and are largely used in India as a cheap substitute for gram; they can be fed to practically all kinds of stock, except horses. *Guara* contains nearly 40 per cent protein, nearly double that contained in *arhar,* and it is more digestible than the protein of *arhar,* but the grain as a whole is not.

Arhar contains 13 per cent digestible protein, and 68 per cent total digestible nutrients, and compares very favourably with gram, except that the latter contains slightly more total digestible nutrients.

Guara is not as palatable as *arhar*, and is consequently fed in smaller amounts, usually not exceeding 1-2 pounds per day. It is far richer in protein than either gram or *arhar*, and this fact must be taken into account in using it as a substitute for the latter.

Guara is rich in phosphorus and is almost as good as bran in this respect, as it contains over 1.3 per cent against bran with 1.9 per cent. Its calcium content, however, though higher than that of the cereal grains is low, 0.4 per cent. Hence it is not well balanced as regards calcium and phosphorus.

If due regard is paid to these differences both *arhar* and *guara* can be used as cheap substitutes for gram.

Matri. (Pisum sativum)

Matri, the common Indian field pea, is a legume, and like most legumes, is rich in protein. In the Lyallpur trials the *matri* used contained 23 per cent total protein and 18 per cent digestible protein. The total digestible nutrients, however, are lower in *matri* than in gram, on account of the higher carbohydrate content of the latter,—62 per cent as compared with 67 per cent. *Matri* can be fed to all kinds of farm stock including horses and may be used in slightly larger quantities as a substitute for gram.

Rawan (Dolichos lablab) and *Moth* (Phaseolus conitifolies)

Both *rawan* and *moth* are common kharif beans and are allied to Morrison's adzuki beans. Their protein contents are practically the same, *viz.*, 24 per cent and 23 per cent respectively, of which 18 per cent and 17 per cent respectively are digestible. The total digestible nutrients in *moth* are slightly greater than in *rawan*, owing to the higher carbohydrate content of the former. These beans are in consequence somewhat richer than gram and *arhar*, and with others of the same type, may be fed to cattle. They may also be given to horses in amounts depending on the amount of work being done. Up to

3 pounds per day may be given to horses in the absence of any other protein rich concentrate such as cakes or peas.

The seed coat of the beans is tough and fibrous, and it is best to feed them coarsely crushed, or in the form of meals, when they may be included in pig meals for fattening and in milk substitutes for calves.

Bean meals are susceptible to damp and soon decompose. They should therefore always be kept in a dry place and in the minimum amounts needed. The *chuni* of peas and beans, especially *arhar* and *moth*, are also useful foods.

(b) *Cereal grains*

Oats. (AVENA SATIA)

When oats are threshed the seed or grain comes away encircled within the inner and outer *paleae* called the husk, and the glumes or chaff are left with the straw. The composition of oats will vary somewhat according to the amount of the husk that is present, some large varieties having as much husk as 30 per cent, while smaller or thin varieties may have as little as 20 per cent.

Oats form a better balanced concentrate than any other cereal, particularly for horses, though the Indian oat contains somewhat less protein than foreign grown oats. Samples used in the digestion trials at Lyallpur contained slightly over 8 per cent protein with a total digestible protein content of 4.5 pounds per 100 pounds of oats, and 67 pounds total digestible nutrients, and a nutritive ratio of 1 : 14. Many English and American oats contain considerably more protein and have a nutritive ratio of 1 : 8. Such variations may make a considerable difference to the stockowner if he is compiling his rations with Indian grown oats from data which refer to foreign samples whose composition may not be quite the same. This naturally applies to other feeding stuffs also.

Oats contain about twice as much fat as maize, averaging over 6 per cent, but due to the presence of the husk they have a considerable percentage of fibre, and hence yield somewhat less total digestible nutrients than wheat or maize (total digestible nutrients = 70).

Oats, although their proteins are of high quality and contain moderate amounts of certain of the essential amino acids, nevertheless share some of the nutritional deficiencies of the other cereals. They are rich in calcium compared with most other cereals, containing 0.3 per cent, while phosphorus is the highest for all cereals—0.8 per cent. They are lacking in vitamin D and contain little or no vitamin A.

Oats may be fed to all classes of stock, and are one of the safest feeding stuffs for horses, as, owing to the husk, the endosperm is prevented from caking in the stomach, which sometimes happens with other grains when fed without sufficient roughage. Oats should be stored for some time before use, because if they are fed while still new they are liable to cause indigestion and colic. Oats are not essential for horses, for though their special value is due to their well balanced composition, yet a mixture of other grains and concentrates will serve equally well if properly balanced. Young colts may be given up to 3 pounds per day, this being increased to 5 pounds when two years old, and progressively up to 10 pounds when four years old, though horses which are being hard worked may be given considerably more.

They are also one of the best feeds for dairy cattle, but are often more expensive than other grains which will supply equally well the necessary digestible nutrients.

Oats are also useful for fattening sheep and cattle in the early stages, and as part of the concentrate ration for ewes, young lambs, brood sows and young pigs, but are too bulky to form any considerable part of the ration for fattening pigs.

It is largely owing to their comparatively high content of calcium and phosphorus that they are so valuable for young growing stock. Oats should be ground or run through a hulling machine before being fed to swine or cattle as this removes some of the husk. There is considerable difference of opinion as to whether it is advisable to crush or bruise oats for horses, but it is generally accepted that bruised grain is less likely to be swallowed unmasticated than whole grain. Crushed oats are very suitable for calves from three to four weeks old, and may be fed with separated, skim or butter-milk. There are certain disadvantages in using bruised oats, as the process considerably increases the bulk and they tend to absorb moisture and become musty if kept for any length of time. The stockman should therefore bruise his oats himself as and when required. They need only be lightly crushed for horses, though when fed as meal for young stock and pigs they should be ground more finely.

Germinated oats

If oats are damped and allowed to sprout they have been found to have very beneficial effects on horses, most likely due to the development of vitamin C, to partial hydrolysis of the proteins to peptones and amino acids, and of the starch to dextrin and sugar. Germinated oats are also valuable for poultry where there may be difficulty otherwise in getting sufficient vitamin C in the rations. Sprouted oats have also been considered useful in cases of sterility in dairy cows and heifers, although evidence on this point is inconclusive.

Wheat. (TRITICUM SATIVUM)

Wheat differs from barley and oats in that the seed comes away naked on threshing, and consequently has a much lower fibre content. It resembles the other cereals however in general nutritive characteristics, but is seldom fed to farm stock, except poultry, although unsaleable grain may be ground and mixed with other concentrates for cattle and pigs, or it may be fed when prices are low. Wheat contains slightly more protein than oats or barley, but the amount is markedly affected by climate. It should not be fed in large quantities, or, as a rule, more than would constitute above 25 per cent of the total concentrates or it may cause digestive troubles.

Wheat has a fairly wide nutritive ratio and needs to be balanced by protein rich foods, such as bran or cakes.

It is advisable to crush wheat before feeding to all classes of animals, except sheep and pigs, as the grain is hard and small, but it should not be too fine as it is then less palatable and forms a sticky mass in the animal's mouth.

Ground wheat is equal in nutritive value to ground maize when forming about one third of the concentrate mixture for dairy cows, but for pigs it is somewhat superior to maize as regards rate of growth and economy of gain produced, although for fattening cattle maize is perhaps superior. If wheat is used instead of maize it should be remembered that it contains more protein, and hence less protein rich concentrate will be needed to balance a ration with wheat than with maize.

Wheat by-products

During milling, wheat is cleansed of immature grains and other seeds which constitute the screenings, samples of which have been found in Lyallpur digestibility trials to have a protein, digestible protein, and total digestible nutrient content very similar to bran. The calcium content however was 15 times that of bran, being over 3 per cent, and the phosphorus 1 per cent, or half that of bran.

In the process of milling, the outer bran cover and the underlying layer which contains much of the protein and vitamins, and the endosperm which contains the valuable vitamin E, are torn off and constitute the bran.

The bran may be separated into various parts which are sometimes confused, and may be defined according to the size of the particles concerned.

Broad bran

Broad bran consists of the coarsest part of the wheat grain removed in the milling process, and contains much fibre and is comparatively poor in food nutrients.

Medium bran

The particles of medium bran are finer than those of the coarse variety, although they have much the same composition, and may contain more of the endosperm layer and flour.

Bran

Bran is one of the commonest ingredients of rations for horses and dairy cows. It is bulky and palatable and has a laxative effect when fed as a mash although it has the opposite effect when fed dry. It is, however, not indispensable for dairy stock and may be replaced by other suitable feeding stuffs such as rice *kura*. Owing to its high fibre content and light flaky nature it is best when fed with more concentrated food so as to ensure thorough mastication. In the Lyallpur trials the bran used contained 11.5 per cent protein, although much higher figures are commonly found, the amount of digestible protein was 7 pounds per 100 pounds and the total digestible nutrients 71 pounds. The proteins of bran are superior in quality to those of wheat flour but are not as well balanced in their amino acid content as are the proteins of animal products such as milk or fish meal. The phosphorus content of bran is higher than that of any of the cereals; it was 2 per cent in the bran used in the Lyallpur trials but the calcium was only 0.23 per cent. Much lower figures for calcium are frequently found. Bran is thus ill balanced as regards calcium and phosphorus, an important matter with high yielding dairy cows which may have a tendency to draw on their own skeletal stores of calcium if the balance needed is not made up in the rest of the ration, and a condition known as osteoporosis may result. Bran contains a high percentage of magnesium, and the combined effects of this and the phosphorus are sometimes considered to be responsible for the deposition of magnesium phosphate compounds as intestinal calculi in horses. Bran is particularly useful for horses as a laxative, and mixed with heavy concentrates which might otherwise cause digestive troubles. Not more than from 2 to 3 pounds a day should usually be given to horses and cattle. The best way to feed it to sick horses is in the form of a mash made by adding a small quantity of salt to 2 to 3 pounds of bran in a bucket and then pouring boiling water over it and stirring into a consistency of thick porridge.

Bran may also be fed to sheep, pigs and poultry, although the finer wheat offals are better for pigs of all ages as bran is too bulky. It may also be used for fattening lambs in the earlier stages, but should be replaced by other concentrates as fattening proceeds. Although bran is extensively used for feeding young stock of all kinds, its deficiency in calcium must be corrected by the inclusion of calcium rich foods or by the addition of finely ground steamed bone-meal, otherwise rickets are likely to result.

Pollards

Pollards are very similar in composition and nature to medium bran.

The finer by-products in the milling of wheat are known as shorts, middlings, third parings and topping, and contain less fibre and more nutrient ingredients than the bran, the percentage of protein increasing with the degree of fineness.

Wheat germ, which is largely removed in the bran, is rich in oil, and one of the best sources of the anti-sterility vitamin E.

In the milling of wheat by modern machinery much of the most nutritive parts of the wheat grain is removed and fed to animals, while the less nutritive part, deprived of valuable proteins, minerals and vitamins is converted into flour for consumption by human beings !

Wheat mamni

Wheat mamni consists mostly of wheat screenings, *i.e.,* small wheat grains mixed with weed seeds removed in the cleaning of wheat, and is frequently infested with earcockle. It is of variable character and composition, but being cheap is used a good deal by *gowalas* in towns. It should not be fed to any considerable extent, nor for long periods.

Barley

Barley contains more digestible carbohydrates than oats, and, according to some foreign analyses, less digestible protein.

The Lyallpur trials indicate that hard Indian barley, although slightly richer in total protein than oats, has a considerably greater proportion of digestible protein per 100 pounds, *viz.,* 6.7 against 4.4 for oats, and a nutritive ratio of 1 : 9.7 compared with 1 : 14.1 for oats.

The composition of barley therefore depends on the type used, especially with the best malting barleys which contain more starch, and the hard barleys which contain a higher percentage of protein and are better balanced for stock feeding. If the better types of barley are properly balanced with protein rich feeding stuffs they may be preferred to oats for pig meals, or may be given as an alternative to oats for horses, a given weight of barley replacing somewhat more than an equal weight of oats. If given in excess to horses, barley is liable to cause colic. However, provided account is taken of the type of barley used, it may be used in moderation for almost all classes of stock. Ground barley is of approximately equal value to ground maize for dairy cattle. Barley may be used for pigs, fattening cattle and sheep, but it should be crushed or ground, except for sheep, in order to help in its mastication.

Barley meal

Barley meal or *ardawa* is made by grinding the whole grain, but it is very variable in quality, as it is frequently made from inferior or damaged grain and does not keep well. Good quality barley meals are unrivalled for fattening pigs, promoting growth and producing firm well conditioned carcasses.

Paddy and rich kura or rice chura

Rice in the husk is generally known as paddy and is not as a rule fed to animals. The only areas in which small quantities of paddy are given to stock are parts of Madras, Sind and Assam. In the first and last named tracts paddy is given to elephants, while in Sind it is sometimes fed to riding horses. It has been estimated that not more than 2,000 tons in all are utilised for stock feeding.

Rice *kura* or *chura* (beaten rice) is one of the manufactured products of paddy obtained in the conversion of paddy into rice, and consists largely of the husk and a certain amount of the original grain. It is used as a cattle food in south and eastern India, and contains a higher percentage of protein than the original paddy, and also a very high percentage of fibre.

Maize grain

There are two chief types of maize grain, the dent or flat variety and the round or hard one. There is little or no difference in the composition of the two, but in the dent variety the starchy endosperm extends continuously from the base of the grain to the lip with layers of horny endosperm on both sides. Hence, when the grain dries, the starchy layer contracts and draws in the base, which therefore becomes indented.

The hard round variety contains far less starchy endosperm which does not connect with the horny base layer, and therefore does not become indented on drying.

The horny starch endosperm contains various yellowish or orange pigments which give yellow maize its characteristic colour. The chief pigment is carotene, the precursor of vitamin A. Yellow maize and meal are therefore of far higher nutritive value on this account than the white varieties which contain little or no carotene. Some stockmen prefer white maize meal on account of its more attractive appearance but it is lacking in the valuable growth promoting vitamin.

White maize is equal to the yellow in nutritive value if stock have access to green pasture or forage, or are fed a reasonable amount of well cured hay. If fed to pigs or poultry that are not so provided, the feeding of white maize instead of yellow may be disastrous. Maize with hulls of colours other than yellow may, or may not be rich in carotene, as this depends on whether the endosperm is yellow or not.

Maize is particularly rich in starch, and contains more than other cereals. That used in the Lyallpur trials contained over 77 per cent starch and 3 per cent fat. It is low in fibre and contains about 9.7 per cent protein, about the same as oats. An average sample of maize contains 70 per cent total digestible nutrients and 5.5 per cent digestible protein.

The chief protein of maize is zein, which is deficient in certain of the essential amino acids, *viz.*:—tyrosin and tryptophane, without which animal tissues cannot be built up. This deficiency can, however, be overcome by giving other feeding stuffs, such as leguminous fodders or hay, whose proteins contain these amino acids.

Maize contains very little calcium, from 0.04 to 0.09 per cent which means that one ton of maize contains less than 1.12 pounds of calcium. The maize used in the Lyallpur trials contained 0.06 per cent calcium, though there are many foreign samples which contain as little as 0.01 per cent, the figure given by Morrison. The phosphorus content, however, is very similar to that of oats. Maize is very extensively used for dairy cattle and other stock, and if due account is taken of its deficiencies, may partly or wholly replace oats for horses, provided the substitution takes place gradually, starting with not more than 2 pounds per day, and gradually leading up to the maximum. Maize and oats are not necessarily interchangeable pound for pound, as their respective total digestible nutrient contents are 70 and 66, and the digestible protein content 5.37 for maize and 4.4 for oats. These figures were determined from the Lyallpur experimental trials on heifers, but corresponding digestibility data for horses are not available. Maize should always be cracked before feeding to horses, as it is then easier to masticate. The dent varieties are to be preferred to the hard. Some horse owners prefer not to feed any maize to horses on the plea that it is heating and causes skin troubles, and a generally flabby and weak condition of the muscles. If given in judicious combination with other foods, however, no harmful results should accrue. It may also be used for pigs and sheep which relish it highly, probably on account of its palatability. Here again, however, the mineral and protein deficiencies of the maize must be properly balanced by other feeding stuffs for young or pregnant animals.

It is best not to grind maize too finely as this detracts from its palatability. Ground maize does not keep as well as ear or shell maize, and more than necessary should not be ground at one time. Some farmers prefer to grind the whole maize ear, *i.e.*, both the grain and cob, if they need a high fibre content which will prevent masses forming in the intestines when dry roughage is not available. This product is sometimes called ground maize and should not be confused with maize meal which is the ground grain without the cobs.

Maize meal

In the process of grinding the grain, the germ, which contains most of the oil, is crushed and the resulting meal tends to absorb moisture and is likely to become rancid on storing. In more refined methods of grinding the germ is removed with part of the starch, and is then crushed for extracting the oil known as maize germ oil. The dry meal thus produced, although poorer in fat or oil, will keep much better than the whole meal. As in the case of other meals, maize meal should be stored under dry conditions, and if kept in bags, these should be placed so as not to form too large a mass, and air spaces should be left at intervals to prevent undue heating, which may occur as a result of fermentation. Maize meal shares with maize the deficiencies previously pointed out.

BREWERY AND DISTILLERY BY-PRODUCTS

There are various by-products from breweries and distilleries which may be used in dairies and stock farms nearby. The chief of these are malt sprouts, brewers' dry grains and wet grains.

Malt sprouts

When barley is malted, about three pounds of malt sprouts are produced per 100 pounds of barley. These are generally mixed in the proportion of 1 : 10 with brewers' dried grains.

They are similar in composition to brewers' dried grains, and contain about the same amount of crude protein, much of which is not true protein, but less fat and somewhat more total digestible nutrients. They are bulky and absorb water readily and may therefore be mixed with dry molassed feeds, but ordinarily should not be transported as they readily become mouldy. On this account they must always be kept dry.

They are somewhat unpalatable and should not be fed in greater amounts than about 10 per cent in the concentrate part of the ration for dairy cows and other stock, as they are likely to taint the milk and cause colic due to their propensity to swell. Not more than 4 pounds should be fed to horses, 2 pounds to cattle and 1 pound to sheep. This material should always be combined in proper proportions with some other feeding stuff so as to obviate any harmful effects.

Wet brewers' grains

Wet brewers' grains constitute a bulky mashy material containing over 70 per cent water. They consist of all the husks of the barley, much of which is digestible, most of the proteins, and but little starch, most of which has been removed during the fermentation process. The composition of the wet grains varies very considerably, but 80 pounds may be taken as approximately equal to 20 pounds of dried grains. They are very palatable but rapidly decompose unless carefully stored. On this account they are usually fed near the brewery. They should be taken to the farm in as fresh a condition as possible, and if they must be stored, should be kept in air tight barrels or tanks, and inspected before being used to see that no decomposition has taken place. Wet brewers' grains are chiefly useful for dairy cows and may be given at the rate of 20-30 pounds per head daily. They may replace an equal weight of silage, or a part of the concentrates, and should be fed after milking to avoid any possibility of tainting the milk.

They may also be used for certain other classes of stock, such as fattening cattle and swine, but are as a rule too watery for horses and pigs. If fed to the latter they should not be given in excess of 15-20 pounds per head per day for horses, or over 2 pounds per 100 pounds live weight for pigs, and should be properly supplemented by other suitable feeding stuffs, otherwise scouring and digestive disturbances may result.

Dried brewers' grains

The wet brewers' grains when produced are watery in character and difficult to transport, and are usually dried and sold as dried grains. These may contain up to 20 per cent protein or more, 5 per cent fat, 14 per cent fibre and 40 per cent nitrogen free extract consisting largely of pentosans. They contain only about 60 per cent total digestible nutrients in comparison with 70 for wheat bran, and about 12 per cent digestible protein. Brewers' dried grains are slightly superior to wheat bran, but do not possess the laxative properties of the latter. They may safely be used for cattle and dairy cows up to one third of the concentrate mixture.

They are excellent for sheep, especially for suckling ewes and fattening lambs. Pigs, however, are not able to digest them properly. Dried brewers' grains may replace up to 50 per cent of the oats in horses' rations, pound for pound, although they tend to have a constipating effect and hence are not popular with horse owners on this account..

THE MILLETS

There are two chief types of millet namely, *bajra* (*Pennisetum typhoideum*), and *juar* (*Andropogon sorghum*), which are found throughout the drier regions of India and are fed to all kinds of stock except horses. They have somewhat hard seed coats, and should be ground or softened by water before being used for stock. They are useful for poultry, fattening cattle and lambs, but should not be used for young chickens on account of their hard coats.

In the Lyallpur trials *bajra* and *juar* were found to contain 9.75 per cent and 13.75 per cent protein and 4.86 pounds and 6.40 pounds digestible protein respectively per 100 pounds. The corresponding figures for total digestible nutrients were 54 pounds and 73 pounds per 100 pounds.

FEEDING STUFFS OF ANIMAL ORIGIN

The chief feeding stuffs of animal origin common in India are:—

Milk and its by-products.
Fish meals.
Bone meals.

Milk

Milk is the natural food of young animals from the time of birth until they are old enough to look after themselves. Nature has so ordained things that the young of any particular species thrive best on the milk of that species, but are able nevertheless to make satisfactory progress on the milk of others, which, though possessing the same type of nutrient ingredients, contains them in different proportions. Thus, cows' milk may with certain adjustments be used for rearing the young of the human species, young foals and lambs.

Milk is a bulky food and contains nearly 90 per cent water, but calculated on a dry basis, it is very rich in proteins of high biological value and high digestibility. All the constituents of milk are in fact almost completely digestible and exist in a form in which the digestive system of young animals can most easily deal with them.

Milk contains from 4.5 to 5.5 per cent fat, present in the form of a very fine emulsion in which it can be most easily digested, and about 4.5 per cent soluble carbohydrates in the form of lactose. Lactose differs from cane sugar, as it is not fermented by yeasts but is acted on by bacteria in the intestinal tract with the formation of lactic acid. The mineral matter comprises about 0.7 per cent of the whole milk and is rich in calcium and phosphorus and serves all the needs of the newly born animal for some considerable time. Milk, however, is deficient in iron and copper and if the young are suckled for too long a time they may suffer from anaemia and must then be treated with the iron and copper mixture described in Chapter XIV.

The chief permanent factor which influences the composition of milk is the breed of the animal, and the fat is the constituent which varies most. The next is undoubtedly feeding. Whilst it is not possible to get higher yields of good quality milk than the constitutional maximum for any particular breed by increased feeding, correspondingly it is not possible to get the constitutional maximum if the animals are not fed properly.

Milk if produced by animals which are given satisfactory rations is rich in vitamin A, but it contains little vitamin D, and still less vitamin C. Young animals are born with an ample supply of these vitamins in their bodies, sufficient to enable them to grow satisfactorily during the time that nature has ordained that they should feed on their mothers' milk. Prolonged feeding on their mothers' milk beyond this period should be watched with care as symptoms of rickets may develop if the young animals are deprived of the essential vitamins.

Ordinarily, cows' milk is too expensive to be fed to farm stock other than young calves until such time as the latter can take solid food. It should not usually be necessary to continue whole milk feeding for more than four or five weeks if plenty of skim milk is available.

The table on page 200 adapted from Heinemann's 'Milk' as given by Linton [1927] shows the chemical composition of the milk of different animals.

The milk from Sahiwal cows when being fed from 2 to 6 pounds of cottonseed per head per day, with bran, *bhusa* and green fodder has the following composition:—

Water.	Protein.	Fat.	Sugar.	Ash.
86·98	3·80	4·50	4·00	0·72

The table on page 201, given by Linton [1927] and adapted in part from 'The Chemistry of Cattle Feeding and Dairying' by J. A. Murray, shows the percentage composition of the various by-products obtained from milk.

It is not proposed to discuss here the problems connected with the production of milk, which will be dealt with in another manual, beyond saying that rationing schemes should be in accordance with scientific standards and that milk should be produced under the best hygienic conditions and stored until ready for use under similar conditions.

Separated and skim milk

Separated and skim milk are the products left after most of the fat has been removed by mechanical methods in the case of separated milk, and by hand methods in preparing skim milk.

TABLE IV

COMPOSITION OF THE MILK OF DIFFERENT ANIMALS

Animal.	Specific gravity.	Water.	Casein.	Albumin.	Total proteins.	Fat.	Sugar.	Ash.	Total solids.
Woman.	1029	87·58	0·80	1·21	2·01	3·74	6·37	0·30	12·42
Cow.	1032	87·27	2·88	0·51	3·39	3·68	4·94	0·72	12·73
Cow colostrum.	1042	75·07	4·19	12·99	17·18	3·97	2·28	1·53	24·93
Ass.	1032	90·12	0·79	1·06	1·85	1·37	6·19	0·47	9·88
Sheep.	1035	83·57	4·17	0·98	5·15	6·18	4·17	0·93	16·43
Goat.	1030	86·88	2·87	0·89	3·76	4·07	4·64	0·85	13·12
Mare.	1034	90·58	1·30	0·75	2·05	1·14	5·87	0·36	9·42
Rabbit.	1047	69·50	—	—	12·00	13·50	2·00	2·50	30·50
Sow.	1038	83·94	—	—	7·23	4·55	3·23	1·05	16·06
Bitch.	1035	75·44	—	—	11·17	9·57	3·09	0·73	24·56
Elephant.	1031	79·30	—	—	2·51	9·10	8·59	0·50	20·70
Camel.	1042	86·57	—	—	4·00	3·07	5·59	0·77	13·43

TABLE V

COMPOSITION OF COWS' MILK AND ITS BY-PRODUCTS

	Whole milk.	Cream.	Skim milk. (Deep set.)	Separated milk.	Butter milk.	Whey.
Water.	87·5	70·00	90·35	90·61	90·97	93·26
Fat.	3·5	22·86	0·35	0·07	0·80	0·24
Nitrogenous matter.	3·4	2·71	3·51	3·52	3·60	0·73
Milk sugar.	4·85	3·86	5·01	5·02	4·06	4·[illegible]
Ash.	0·75	0·57	0·78	0·78	0·75	0·44
Specific gravity.	1032	1010	1036	1037	1030	—

Skim milk contains a little more fat than separated milk owing to the more efficient separation of the fat in the latter.

Calculated on a dry basis, skim milk is very rich in protein, as only the fat has been removed, hence it has a narrow nutritive ratio of only 1 : 1.5 aganst 1 : 4 for whole milk. Hence in feeding to young stock, there is little need to give other, protein concentrates in addition. It is especially suitable for calves, young pigs and poultry, but as nearly all the vitamin A has been removed in the cream, the efficient use of skim milk will depend on a satisfactory substitute for the cream being found. Ground or crushed oats or linseed meal are excellent for young calves and they may be used as a part substitute for skim milk; one pound of crushed oats replacing approximately a gallon of skim milk.

In the case of pigs a suitable mixed meal should be given with skim milk in order to obtain a properly balanced ration suitable to the animals' age, skim milk being most suitable for young pigs. Generally speaking skim has a higher feeding value per unit of protein contained in it than an equivalent of feeding stuff from other animal sources, such as fish meal.

Butter milk

Butter milk is the residue left from whole milk after churning, and the percentage of fat left in it will depend on the efficiency with which churning is done. It generally contains more fat than either skim or separated milk, but all the protein and carbohydrates of the whole milk, and somewhat less lactose than skim milk, due to a mild degree of fermentation of lactose to lactic acid. On this account butter milk has a greater laxative effect than skim milk and should be given to calves at a somewhat later age than the latter. It is suitable for pigs and chickens, but it should not be kept in dirty containers as it is liable to ferment and become rancid, in which condition it is dangerous.

Whey

Whey is the residue left from the manufacture of cheese and chhana during which most of the casein and fat of the milk are removed; hence the residue left in the whey consists only of the lactalbumin, lactglobulin, the lactose and the ash.

Its composition is variable but it is far less nutritious than other milk products. It contains only about 0.8 per cent protein, 5 per cent lactose and 0.25 per cent fat.

Whey is chiefly suitable for maturing pigs owing to the high quality of the remaining protein (albumin), and pigs over 100 pounds in weight will put on weight on a ration of whey and barley, or wheat, or gram, without any other concentrate, but the whey is only half as valuable for pigs as skim milk.

Whey is not satisfactory for young calves as it needs to be supplemented by fish liver oil, and protein from other sources to make it equal in nutritive value to an equal amount of whole milk. It is not always easy to feed rich supplements in the amount necessary to balance the ration.

Ellis [1937] recommends the following possible combinations with one gallon of whey:—

1. Five parts bean meal, 4 parts linseed cake meal.
2. Two parts linseed cake meal, 2 parts flaked maize, 1 part white fish meal, 1 part oats.
3. Two parts linseed cake, 1 part bean meal, 1 part superfine weatings.

He advises one pound of mixed meal to every gallon of whey, but the combination is not even approximately as good as whole milk. Whey is a good poultry feed owing to its protein, and vitamin B content.

Colostrum

Colostrum is the first milk secreted by the cow after calving and forms the link between the inter-uterine nutrition of the young from its mother's blood system, and extra-uterine nutrition when the calf's digestive system is capable of taking whole milk. It is much richer in protein and minerals than ordinary milk, but contains less milk sugar, and has a sickly smell. It has a slightly laxative effect and seems to be a provision of nature to clear out the digestive tract of the newly born calf. It is also said to exert a protective effect against the ingress of harmful bacteria to the intestines, thus preventing scour. Cows' milk is not suitable for human consumption until about five days after calving when it becomes normal. Colostrum is 35 times richer in vitamin A, and 65 times richer in carotene than normal milk.

FISH MEALS

There are two types of fish meal obtainable, only one of which, the best quality white meal, is fit for feeding to stock; the other, which is really fish guano, is only fit for manure. White fish meal is produced from the heads, bones and some of the fish of white types of fish which are cooked by dry heat in containers surrounded by steam at a temperature of 280°F for about 10 hours. It is then cooled and ground, when a fine meal, which should not contain more than 10 per cent water, is obtained.

The second type, or guano, is made from other fish including offal and entrails, and is similar in chemical composition to the white meal but contains less oil. It possesses a penetrating odour, and if fed to animals such as cows or pigs, will taint the milk and flesh for long after its use has been discontinued. Great discrimination should therefore be used in purchasing fish meal for stock, and the meal should be guaranteed high quality white meal.

This contains about 55 per cent protein of high nutritive value and 25 per cent mineral matter, chiefly in the form of calcium phosphate from the bones, and should not contain more than 5 per cent fat or oil. Four per cent common salt should be added as a preservative. Fish meal is rich in vitamins A, B and D, although if dried at too high a temperature these may be deficient.

The best grade fish meal is a very highly concentrated protein and mineral feeding stuff, 90 per cent of the protein of which is, according to Orr, Crichton and Green [1922] digestible.

Fish meal can also be used as an economical protein supplement for cattle and sheep, although it does not appear to have for these animals any superiority over protein feeds of vegetable origin, such as oilseed cake meals, and is sometimes unpalatable.

As it is highly concentrated, only a small supplement of fish meal is needed as a rule to balance ordinary rations, but it produces a marked effect on an animal's capacity to digest its food as a whole, and is of special value when only cereal concentrates or their by-products, which are deficient in calcium, are available.

Orr and colleagues [1922] have recorded the results of trials conducted at the Rowatt Institute to ascertain the extent to which high grade fish meals do or do not taint the milk and flesh of animals which eat them. They experimented with cows, pigs and poultry. Cows were given up to 1 pound of fish meal per day without any harmful effect on the milk. Hens and cockerels were fed fish meal with other meals and no taint could be found in the eggs or the flesh. In the case of pigs, the trials were started when the animals were 8 weeks old, and continued for 3 months. The ration consisted of fish meal, potatoes, oats and maize, in the proportion of 1 part of meal to 6 of the rest, and in no case could any taint be observed in the flesh of the animals after slaughter.

Other trials were also conducted with unguaranteed and not specially selected herring meal and it was found that, if the fish meal was discontinued a fortnight before slaughter, no taint could be detected in the flesh, but if the fish meal was continued up to the date of slaughter, the bacon had a fishy taste and the fat was of a yellowish colour, although the pork appeared to be unaffected.

Fish meal does not keep well in storage and should only be bought in small quantities at a time and kept perfectly dry and as cool as possible. If any signs of rancidity or decomposition appear it should on no account be used.

BONE MEAL

There are two grades of bone meal. The first is made by grinding bones to varying degrees of fineness, and is only fit to be used as a manure. In India, even this meal is frequently far from being a meal in the proper sense of the term and may be of a very coarse type.

The second type, and the only one which should be fed to stock, is steamed bone meal. This is prepared by very thoroughly cooking the raw bones under steam pressure which removes most of the fat and protein. The residue is then dried, finely ground and sold as steamed bone meal. The cooking process sterilises the meal and removes any possibility of infection with Anthrax, which might be conveyed to stock by the use of raw or uncooked meal.

Meal thus prepared naturally varies in composition, but contains on an average 30 per cent calcium, 15 per cent phosphorus, but only about 7 per cent protein and 3.0 per cent fat. It is thus essentially a well balanced calcium-phosphorus mineral supplement, and is particularly useful when sufficient high calcium phosphorus feeding stuffs are scarce, or even as a supplement when they are not. It may be used for young growing stock and high yielding dairy cows. The meal may also be fed to pigs and poultry, and a small amount, varying from 4 to 6 ounces per head per day, may conveniently be given to colts or horses when on poor pasturage, or whose concentrate is mainly oats.

High yielding dairy cows may be given up to one pound per day, calves when fed on milk substitutes up to one ounce per day, and sows and young pigs up to half an ounce a day.

The necessity for bone meal will depend on the nature and quality of the rations available, and if given in the above mentioned or smaller quantities, should prove beneficial.

What is sometimes called raw bone meal must be distinguished from the uncooked variety used as a fertiliser. Strictly speaking it is not raw, but is prepared by boiling the bones in open vats instead of under steam pressure. This method does not remove as much protein and fat as the steam process, so the meal does not have as high a calcium-phosphorus content as the steamed variety. It is otherwise much the same, except that, as a rule, it is more coarsely ground and more suitable for poultry than for other stock.

Stock owners may be inclined to feed rock phosphate instead of bone meal but as pointed out in Chapter II, this is not to be recommended, as rock phosphate may contain up to 3 or 4 per cent fluorine, which has a most deleterious effect on the teeth and bones. A safe plan therefore is to avoid rock phosphates as a mineral supplement and never use them for stock.

BLOOD MEAL

Blood meal is not much used in India, but the blood from slaughter houses may be dried by various processes and ground to form a meal or dried blood, as it is called.

This material is very rich in protein, containing over 70 per cent, but it is low in calcium and phosphorus and thus differs from meat scraps. Blood meal may be fed up to one or two ounces a day to calves and young pigs, though tankage is to be preferred for the latter. It may also be given to chickens or horses which have become run down or debilitated. Its special value is due to its high protein and lipoid content, and to the iron contained in the haemoglobin. It may also be used for dairy cows but only in small quantities not exceeding one pound per head per day. Owing to the temperature employed in drying, the protein is not highly digestible, a fact which to some extent detracts from its value. It is also considered by some people to be useful as a supplement for increasing the quantity of wool on sheep.

MOLASSES AND MOLASSED FEED

Molasses

The value of molasses as a feeding stuff lies almost entirely in its carbohydrate content, and to some degree also in its high phosphorus content. It is a well known fact that when a meal is taken consisting of protein and carbohydrates, a high percentage of carbohydrates has a tendency to depress the digestibility and proper utilisation of the protein.

Trials carried out at Lyallpur with heifers and working bullocks on normal rations consisting of wheat *bhusa* and gram indicated that the addition of from four to six pounds per head per day of molasses to the ration considerably depressed the digestibility of the protein. Therefore, whatever the value of molasses may be in terms of total digestible nutrients, its inclusion in the ration beyond a certain amount is not likely to increase the value of the ration proportionately; on the other hand it may diminish it. The evidence so far available indicates that up to two pounds molasses per head per day may safely be fed to dairy cattle as a part of the total ration with beneficial effects, but beyond this amount its value tends to diminish and may ultimately become negative.

Molassed feed

During recent years various types of molassed feeds prepared by adding a certain percentage of molasses to other feeding stuffs have appeared on the market. Some of these contain up to fifty per cent molasses or even more. The author has conducted feeding trials at Lyallpur on heifers for a period of six months on a molassed feed obtained from the Imperial Institute of Sugar Technology, India, (Cawnpore), which contained two parts bagasse screenings, three parts mustard cake and four parts molasses. Some difficulty was at first experienced in inducing animals to eat this material, but after a time they became accustomed to it. The chemical composition of the molassed feed was similar to that of wheat bran, although its digestibility was considerably lower. For comparison the chemical composition and the feeding values of the molassed feed and bran respectively are set forth in tables VI and VII below:—

TABLE VI

	Moisture. %	Dry matter. %	Ash. %	Fat. %	Crude fibre. %	Protein. %	Nitrogen free extract. %
Molassed feed.	12·20	87·80	10·29	2·71	13.01	13·44	48·35
Wheat bran.	9·24	90·76	4·83	3·79	11·53	14·1	56·48

TABLE VII

	Total digestible nutrients per 100 lb. of the feed. lb.	Digestible protein per 100 lb. of the feed. lb.	Albuminoid ratio 1:
Molassed feed.	50·47	7·00	5·6
Wheat bran.	50·59	10·60	4·1

TABLE VIII

CALCIUM AND PHOSPHORUS CONTENT

	Calcium as CaO per cent.	Phosphorus as P_2O_5 per cent.
Molassed feed	1·240	0·429
Wheat bran	0·356	2·000
Oat hay	0·330	0·294
Wheat *bhnsa*	0·358	0·060

The molassed feed was rich in both calcium and phosphorus. The corresponding calcium and phosphorus figures for wheat bran, oat hay and wheat *bhusa* are shown in table VIII above for sake of comparison. It may be noted, however, that the ratio between calcium and phosphorus in the molassed feed is not well balanced, as an ideal ratio is roughly 1 part calcium to 0.9 phosphorus. Whilst therefore such feeds have an undoubted value in certain conditions, they have their limitations and drawbacks. As they are highly saccharine in nature they are inclined to ferment when stored, particularly under damp conditions. If stored in bulk there is a danger of fermentation progressing to such a degree as to cause spontaneous combustion. Such feeds are also very bulky, which is a disadvantage if they have to be transported any considerable distance.

BIBLIOGRAHPY

1. Ellis, J. C. B. 1937. The Feeding of Farm live stock. p. 145. (London: Crosby Lockwood).

2. Jones et al., Tex. Bul. 379. Tex. Report, 1935.

3. Kammlade and Mackey, Ill. Bul. 260 and Carmichael, Ohio. Bul. 245.

4. Kellner, O. 1908. The Scientific Feeding of Animals, p. 208.

5. Lander, P. E. 1942. Indian grazing conditions and the mineral content of some Indian fodders. I.C.A.R. Misc. Bul. No. 16.

6. Lander, P. E. & Dharmani, L. C. 1930. Mem. Dept. of Agri. India, Chem. Ser. 10, No. 6, pp. 181—192

7. Lander, P. E. & Dharmani, L. C. 1931. Ind. Jour. of Vety. Sci. and Animal Husbandry. Vol. I, Part III, 177—191.

8. Lander, P. E. 1935. The Indian Jour. of Vety. Sci., and Animal Husbandry, Vol. V, Part VI, pp. 343—349.

9. Linton, R. G. 1927. Animal Nutrition and Veterinary Dietetics. p. 183—(Edinburgh; W. Green and Sn. Ltd.)

10. Linton, R. G. 1927. Animal Nutrition and Veterinary Dietetics. p. 183—(Edinburgh; W. Green and Sn. Ltd.) page 183.

11. Orr, Crichton & Green. 1922. Scottish Jour. of Agri. April, Vol. V, No. 2.

12. Progress Report of the Physiological Chemist, Dacca, for 1941—42. Submitted to the I.C.A.R.

13. Symonds, 2. J. 1886. Indian Grasses. (Bangalore.)

CHAPTER VII

THE PREPARATION AND STORAGE OF FEEDING STUFFS

In the previous chapter some of the common feeding stuffs used in India have been reviewed, and the preparation and methods of storage of some of them described. We can now review in somewhat more detail the methods employed in preparing and storing some of the more common feeding stuffs in India.

HAY

Hay is the product made from grass and other fodder plants after having been carefully dried for future use. It should not contain more than about 12—14 per cent moisture, so the original grass must be dried sufficiently rapidly in order that fermentation and decomposition do not take place before it is ready for stacking. The best plants to make into hay are grasses, oats, clovers and other leguminous crops such as lucerne and berseem. The thicker stemmed crops, such as maize, are not suitable for hay making owing to the length of time their stems take to dry.

Good pasture hay should form an essential part of the rations for all farm stock, except poultry and swine.

Hays vary very widely in composition and nutritive value according to the pastures from which they are produced.

In India grass is still collected largely by recognised grass cutters and coolies who cut it wherever and whenever they can find it, their chief object being to get the maximum bulk, and little or no attention is paid to quality which is governed largely by the stage of growth at which the grass is cut.

Factors influencing the composition of hay

There are wide variations in the composition of grasses and crops from which hays are made, at different stages of growth, and the quality of hay will depend on the stage at which the grass from which it is made is cut. Different grasses naturally vary in the extent and nature of these variations, but in general, all grasses in their very early stages of growth are less nutritious than later on, because they contain a large proportion of their nitrogen in the non-protein form. The protein content then rises to a maximum at what may be termed the milk stage, and then falls again. If, therefore, grass is cut for hay when fully ripe much of the protein will have passed into the seeds and be lost in hay making, as well as a considerable part of the dried leaves. In the case of the millets, sorghums, maize and soybeans, this rule does not strictly hold, if both the seed and the straw are taken into consideration, and such crops are usually richer in total digestible nutrients at a more mature stage than earlier, owing to the storage of nutritive material in the seed. Fig. 8 shows the variations in the protein content of certain crops from month to month as growth progresses.

Early cut hay is reputed to contain a greater percentage of ash, chiefly in the form of calcium and phosphorus, but the author's investigations on many different grasses indicate that this does not necessarily always hold good. Very great variations do occur, but these may not be constant for the same grass in the same locality from year to year. Fig. 9 which represents respectively *anjan, palwan* and *dhub* grass from Sirsa in the Punjab, also illustrates these variations over the growing period.

The ultimate value of hay will be chiefly determined by the stage in growth of the grass when it is cut. Prior to the flowering stage the nutrient content of the leaves and culm is not fully developed, but on the other hand if cutting is deferred until the seed is ripe the latter contains most of the nutrients of the plant and falls off during the operation of cutting and hay making.

The best time to cut grass for hay is when the bulk of the principal grasses have flowered. Naturally, except under controlled conditions, a variety of grasses may be found in natural pastures, and these may not always flower at the same time; such hays are called composite hays and may present wide differences in composition and nutritive value.

Making hay

When grass is ready for cutting it should be cut in the cooler hours of the day, but not before the dew is off. The time of

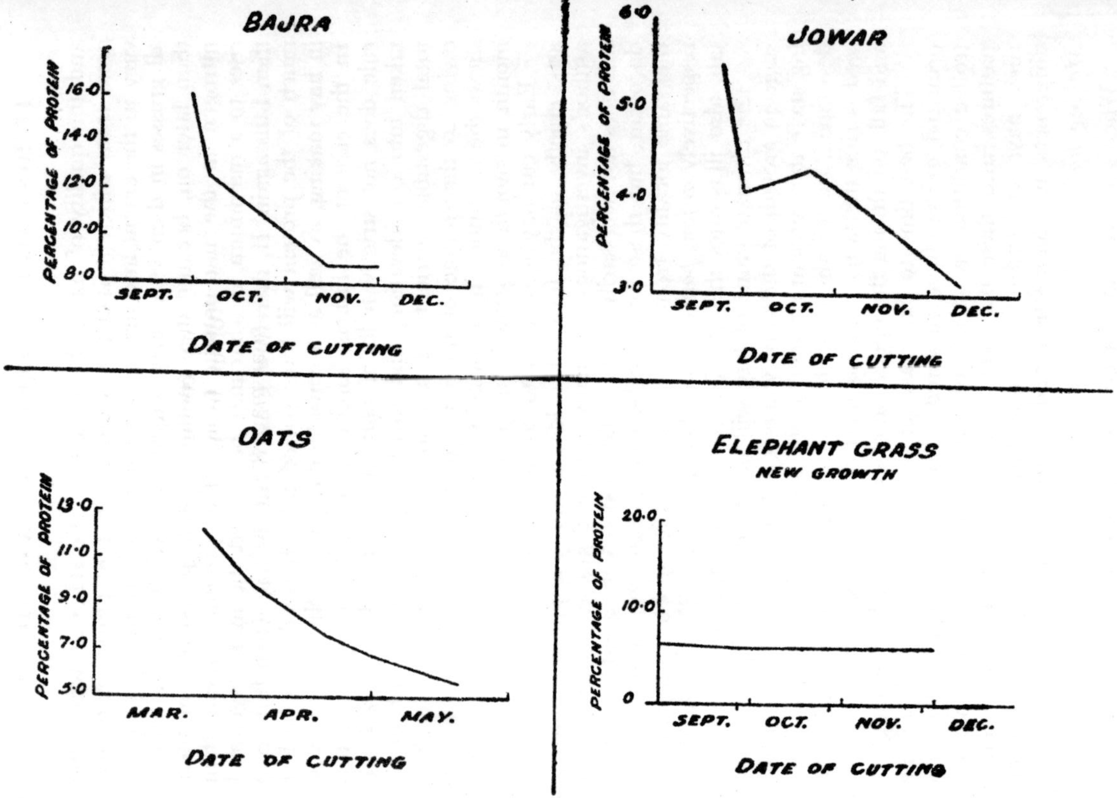

FIG. 8. SHOWING VARIATIONS IN PROTEIN CONTENT OF CULTIVATED CROPS DURING GROWTH.

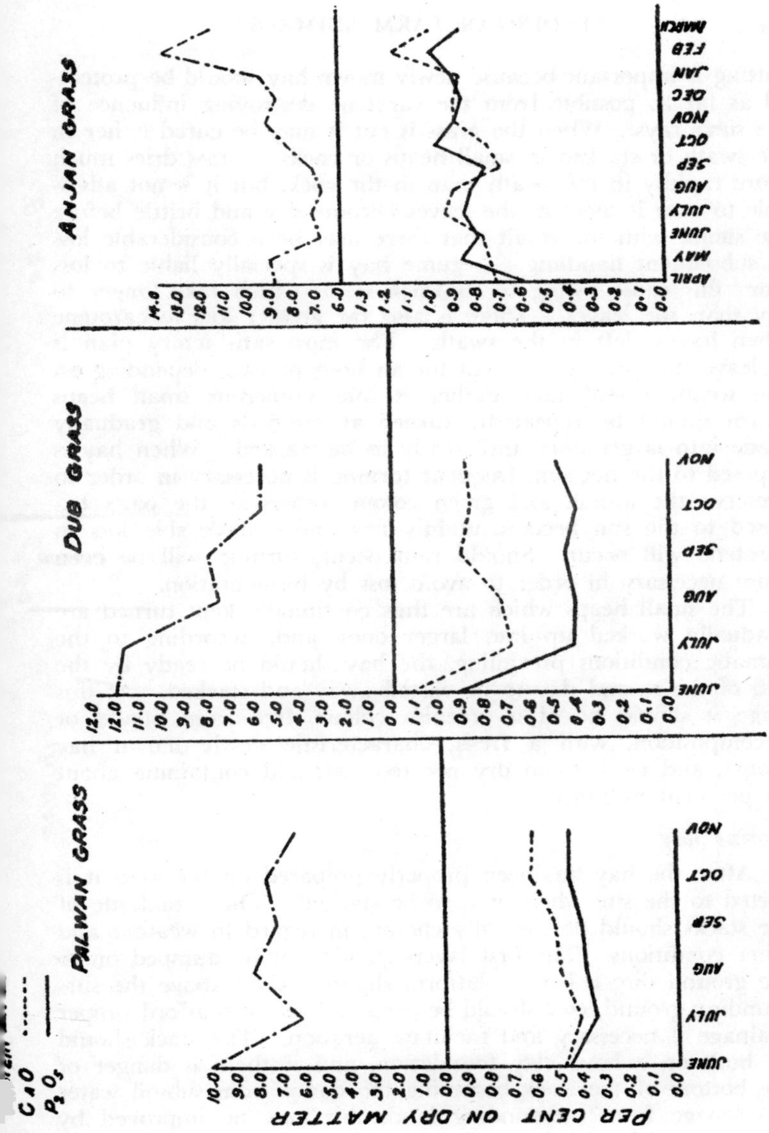

FIG. 9. SHOWING VARIATIONS IN THE PROTEIN, CALCIUM AND PHOSPHORUS CONTENTS OF PALWAN, DHUB AND ANJAN GRASSES DURING GROWTH.

cutting is important because newly mown hay should be protected as far as possible from the carotene destroying influence of the sun's rays. When the grass is cut it may be cured either in the swath or stacked in small heaps or cocks. Grass dries much more rapidly in the swath than in the cock, but it is not advisable to cure it there as the leaves become dry and brittle before the stems, with the result that there may be a considerable loss in subsequent handling. Legume hay is specially liable to loss from this cause owing to its thick stems which take longer to dry than the leaves. There is also far greater loss in carotene when hay is left in the swath. The most satisfactory plan is to leave the grass as it is cut for an hour or two, depending on the weather, and then gather it into numerous small heaps which should be repeatedly turned at intervals and gradually made into larger ones until ready to be stacked. When hay is exposed to the hot sun, frequent turning is necessary in order to preserve the aroma and green colour, otherwise the parts exposed to the sun become unduly dry and considerable loss in carotene will occur. Should rain occur, turning will be even more necessary in order to avoid loss by fermentation.

The small heaps which are thus continually kept turned are gradually worked up into larger ones, and, according to the climatic conditions prevailing, the hay should be ready by the end of the second day to be carted away and stacked. At this stage it should be of a greenish colour, free from mould or decomposition, with a fresh, characteristic newly mown hay aroma, and neither too dry nor too wet, and containing about 12 per cent moisture.

Storing hay

After the hay has been properly prepared on the field it is carted to the site where it is to be stacked. The actual site of the stacks should be carefully chosen, in regard to weather and other conditions. The first layers should not be dumped on to the ground direct, but a platform slightly raised above the surrounding ground level should be prepared, so as to afford proper drainage if necessary and facilitate aeration. The stack should be built on a hard clay foundation, and if there is danger of the bottom of the stack becoming too damp from subsoil water or seepage from surrounding land, this may be improved by

Barn for hay, Brucepur, W. Punjab.

cementing the surface with clay with which a little sodium carbonate has been mixed. The main task of constructing the stack now remains. The dimensions on which it is desired to build the stack having been fixed, the hay on arrival should be well shaken out and evenly packed over the entire surface, which is gradually raised until the desired height is attained. (Legume hay needs special care in this respect and should not be shaken out in the same way as grass hays owing to the liability to loss on account of the brittleness of its leaves, see also Chapter XII.) Special care is needed in well distributing the hay and not packing it in compact bundles in the stack when it arrives from the field.

It is important both for practical and economic reasons to locate the stacks as near as possible to where the hay will eventually be required, and not to mix hays of different quality in the same stack. Every stack should be roughly earmarked as regards quality, otherwise if, for example, dairy cows are suddenly changed from a high grade stack to a low grade one their milk yield may be adversely affected.

The question is sometimes asked whether hay should be stored in barns (Plate XXII) or some permanently covered structure. In places with high rainfall the latter might be desirable, but, as a general rule, hay should be stacked in the open with a sloping thatch preferably made from paddy straw, or reeds, tied into small bundles and carefully placed so as to form a water-tight roof. A reasonable amount of air will not harm hay, and if the stack is a large one and more aeration is needed, this may be attained by placing a line of small six-inch drain pipes at intervals from the outside to the centre of the stack, or a thick bamboo at various levels. An occasional layer of dry paddy straw or wheat or millet straw will also prove beneficial and prevent undue heating or fermentation.

If hay is carefully stacked in this way it should be in good condition and ready for use after a year or eighteen months. After this time it gradually deteriorates in nutritive value. It is an advantage if a little salt or pepper is sprinkled at intervals as the stack is being built up, to act as a preservative and a protection against insects.

Brown hay

If hay is stacked in too moist a condition, it may become

overheated and considerable fermentation ensue, with the result that what is called brown hay is produced. As a result of the oxidation of starch to sugar and alcohol, and losses from these due to further fermentation to carbon dioxide and alcohol, brown hay will be of considerably less nutritive value than green hay.

Sometimes a mild degree of fermentation is encouraged as horses, for example, develop a taste for such hay, and may even prefer it to green hay, but if fed in large quantities it is likely to cause scouring. Excessive fermentation due to bad curing and stacking in a damp condition may cause fermentation to proceed so far that the temperature rises to a point when spontaneous combustion occurs. This should not happen in a well prepared stack.

Computing the weight of hay in a stack

The amount of hay in a stack depends on a number of factors, such as the type of hay, degree of moisture when stacked, and methods of stacking. The following methods of computation are those recommended by the U.S. Department of Agriculture, Morrison, [1936]:—

For lucerne hay after being from 30 to 90 days in the stack, 458 cubic feet per ton (28 Punjab maunds).

For mixed hay, 640 cubic feet per ton.

For wild hay (Indian equivalent is grass collected at random), 600 cubic feet per ton.

For hay over 90 days in the stack.

For lucerne hay, 470 cubic feet per ton of hay.
For mixed hay, 625 cubic feet per ton of hay.
For wild hay, 450 cubic feet per ton of hay.

Computing the volume of hay in a stack

It is somewhat difficult to determine accurately the volume of hay in a stack, as the shape of the stack may vary considerably. The following method, however, is that recommended by the U. S. Department of Agriculture [Morrison, 1936] after detailed investigation:—

The volume of a rectangular or oblong stack is:—

For low, round topped stacks: $(0.52 \times O) - (0.44 \times W) \times W \times L$.
For high, round topped stacks: $(0.52 \times O) - (0.46 \times W) \times W \times L$.
For square, flat topped stacks: $(0.56 \times O) - (0.55 \times W) \times W \times L$.
For volume of a round stack: $(0.04 \times O) - (0.012 \times C) \times C^2$.

In these formulae, O is the 'over,' *i.e.*, the distance from the base on one side of the stack over the stack, and to the base on the other side. W is the width, L is the length, and C is the circumference or distance around a round stack. In measuring a round stack the 'over' should be measured in two portions at right angles to each other.

When to feed hay

The best time to feed hay is when it is about a year old. Up to this time if curing and stacking have been well done, a mild degree of fermentation occurs which helps to make the proteins more digestible, and also breaks down some of the more indigestible parts of the fibre. After a year, however, hay tends to diminish in nutritive value and becomes dry and dusty, and if exposed to bad weather may become mouldy and develop a bad odour which makes it unsatisfactory for stock, particularly horses. When hay has become damaged in this way it should not be fed to dairy cattle or horses, but may be given in small quantities to working bullocks. New hay should not as a rule be given to horses.

Effect of soil on hay produced

The nature of the soil on which a crop is produced for hay affects not only the yield, but also the composition of the crop.

Soils which are deficient in phosphorus, such as some of the Bihar soils, may produce crops in which the phosphorus content is low. Calcium has also been found to be notably below normal in the crops growing in Kangra in the Punjab where there is a very high rainfall which tends to wash the calcium out of the soil. Osteomalacia, rickets and goitre are all very prevalent in such areas among human beings, and the animals also show signs of calcium-phosphorus deficiency.

Losses incurred in hay making

A certain amount of loss in nutrients is inevitable in hay making, but these can be reduced to a minimum with proper care.

Loss is incurred as a result off:—

1. Breaking off the leaves and finer stems.
2. Loss in carotene (vitamin A) by exposure to the sun and air, and by fermentation.

3. Loss of carbohydrates by fermentation.
4. Loss by leaching in heavy rain.

(1) Loss is liable to occur by the breaking off of leaves and flowering heads, or parts of the stems at various stages of hay making. As these are the parts which contain the greatest percentages of nutrients and vitamins, particular care should be taken to avoid such loss.

Legume hays are particularly liable to loss on this account as the leaves dry much sooner than the thick stems, and it is not uncommon to find legume hays thereby reduced 50 per cent in value.

(2) Loss of carotene (vitamin A) is likely to be great if the hay is left long exposed to the air and the direct rays of the sun. The carotene content of a hay may be judged largely by its colour, which in a well cured hay should be a greenish brown. A hay that is a dead brown colour may be taken to be poor in vitamin A. Therefore hay should be kept continually turned as far as possible while in heaps in the field prior to being carted for stacking.

(3) A certain amount of fermentation occurs under the most favourable conditions during which some of the starch is changed into sugar, which may be further oxidised to carbon dioxide and water, thus causing loss of valuable nutrients. Loss from this cause is usually relatively small with well cured hay. Fermentation may also cause destruction of carotene.

If hay is allowed to stand too long in heaps in the field without being turned, very considerable loss extending to as much as 40 to 50 per cent may take place from fermentation. Excessive heating due to fermentation also contributes to the loss.

(4) When hay is subjected to heavy rain during the curing process while in the swath, considerable loss may result from the leaching out of nutrient material. Such loss is less when the hay is stacked in the field in small heaps, but whenever thorough soaking occurs from rain and weather permits, steps should be taken to minimise loss from leaching which may in some circumstances be very considerable.

Drying Grass Artificially

The sun in India usually suffices for drying hay before being stacked, but the practice of artificially drying grass is common in some countries where drying by the sun is difficult. The green grass is first chopped into small lengths and then passed through a drying chamber where it is exposed to hot air and the gases from the furnace. A temperature of about 600°F is usually employed, and if the operation is properly conducted there should be no loss and the green colour of the grass is preserved. When grass is dehydrated in this way almost the whole of the original nutritive value is preserved, although there may be considerable loss in vitamin D. If the grass is carefully dried the carotene content should be practically undiminished, whereas in the best sun dried hay losses in carotene may amount to as much as 50 per cent or more in the process of hay making.

The digestibility of dehydrated grass is not materially impaired if a temperature up to 600°F is employed, but higher temperatures may considerably diminish its value due to lowered digestibility.

The method of artificially drying grass is not used in India to anything like the extent that it is in some other countries, where damp or rainy conditions prevail during the grazing season. The necessary appliances are too costly for the average farmer, but on large farms and Government estates where pasture lands can be scientifically and systematically controlled there is much to be said for the drying of grass by this method.

Some Typical Indian Composite Hays

Emphasis has been previously stressed on the need for cutting grass at the proper time for preparing good quality hay. Some years ago the writer carried out an investigation by means of digestibility trials on Sahiwal heifers on the feeding value of certain composite hays obtained from a number of military grass and dairy farms in the Punjab [Lander, 1942], and representative of the districts from which they were obtained. The results obtained show how hays prepared in different districts vary widely in chemical composition and nutritive value. This may be due, partly to the type of grass collected for making the hay, and partly to the fact that many of the grasses were not cut at the optimum nutritive stage. Table I shows the districts from which the hays were obtained and gives a summary of the nutritive data relating to them.

Table II, in which some data relating to non-Indian fodders taken from Orr's 'Minerals in Pastures' are also given for comparison, shows the complete analytical data of these hays. Those marked 'M' in the remarks column proved to be maintenance rations for three-year-old heifers, as judged from the conservation of body weight and the daily nitrogen balances found in the digestibility trials which have been fully described by the author in a recent publication [Lander, 1942].

A study of Tables I and II reveals many interesting features a few of which may be selected for discussion.

If the Sialkot (plains) and Muree (hills) hays with respective protein contents of 4.83 per cent and 3.68 per cent (Table II) are compared, it will be noted that, although these figures are not widely divergent, yet the digestibility coefficients for the respective proteins contained in them are, being 40.18 per cent and 15.84 per cent respectively (Table I).

This phenomenon may be explained by saying that it relates entirely to the nature of the proteins present. It is interesting, however, to note the respective phosphorus contents, *viz.*:—0.43 per cent for the Sialkot hay and 0.11 per cent for the Murree hay (Table II).

Again it will be noted from Table III that in the Sialkot hay, with the higher phosphorus (P_2O_5) content, the digestible protein per 100 pounds of hay is 2.17 pounds compared with 0.56 pound in the case of the Murree hay with a low phosphorus content.

In the case of the Ambala hay (phosphorus as P_2O_5, 0.67 per cent) and the Dalhousie hay (phosphorus as P_2O_5, 0.15 per cent) an even more pronounced difference in the availablity of the digestible protein will be noted, the figures being respectively 2.26 per cent for the Ambala hay (plains) and 0.35 per cent for the Dalhousie hay (hills) (Table III). The same relationship between the phosphorus (P_2O_5) content and the digestible protein per 100 pounds of hay is seen if the other hays rich and poor in phosphorus content, such as, for example, the Ferozepur and Kasauli hays, the Ambala and Multan hays, and the Jhelum and Multan hays, are compared. It may be possible that the

TABLE I

FEEDING VALUES AND DIGESTIBILITY COEFFICIENTS

Feed.	Dry matter	Total Digestible Nutrients (lb.)	Digestible protein (lb.)	Albuminoid ratio 1:	Digestibility coefficients.						
		per 100 lb of feed			Dry matter	Ash	Fat	Fibre	Protein	Nitrogen free extract	
Jutogh Hay ..	93.56	23.28	0.60	74.2	48.23	8.62	27.28	58 06	17.24	48.69	
Sialkot Hay ..	90.72	33.67	2.17	22.4	50.57	18.36	29.25	54.83	40.18	50.04	M
Jullundur Hay ..	92.60	23.35	0.96	45.7	50.73	5.77	33.61	64.65	25.00	43.63	
Murree Hay ..	93.40	20.09	0.56	78.9	44.29	..	36.67	60.98	15.84	44.88	
Rawalpindi Hay ..	91.50	20.11	0.66	66.9	48.23	9.00	47.73	64.81	21.02	40.23	
Ambala Hay ..	91.15	23.92	2.26	18.1	48.45	17.74	38.19	56.58	44.10	50.00	M
Lahore Hay ..	89.90	26.29	1.51	30.0	52.65	24·52	26.14	64.02	36.43	51.52	
Jhelum Hay ..	89.85	33.60	1.58	30.3	56.16	18.24	34.72	64.29	34.24	60.84	M
Kasauli Hay ..	85.50	22.34	0.72	69.0	45.94	5.83	40.00	59.65	19.29	45.54	
Multan (Geneva) Hay	94.30	21.85	Neg.	∝	46.26	5.87	21·43	59.66	Neg.	47.67	
Multan (Musel) Hay	93.40	24.57	0.35	71.4	48.41	17.47	16·67	62.20	12.01	49.51	
Ferozepur Hay ..	94.89	27.18	4.99	8.1	51.84	38.57	36·94	58.01	58.41	50.20	M
Dalhousie Hay ..	93.60	23.89	0.35	14.26	49.91	24.71	30·44	63.65	11.22	48.82	

TABLE II

GRASS HAYS

Description.	PERCENT ON DRY CROP.																Remarks.
	Soluble ash.	Ash	Fat	Crude fibre	Protein	Soluble carbo-hydrates.	Insoluble residue.	Iron (Fe O)	Aluminium (Al O)	Calcium (CaO)	Magnesium. (MgO)	Sodium (Na O)	Potassium (K O)	Phosphoric acid (P O)	Sulphates (SO)	Water soluble. Chlorine	
Jutogh hay. .	3.11	7.33	1.40	39.37	3.74	48.16	4.22	0.08	0.16	0.80	0.35	0.22	1.30	0.42	0.11	..	
Sialkot hay. ..	3.57	10.09	0.97	38.65	4.83	45.46	6.52	0.12	0.19	0.54	0.19	0.33	1.84	0.43	0.39	..	M
Jullundur hay. ..	3.64	8.15	0.99	40.11	3.51	47.24	4.51	0.06	0.12	0.57	0.22	0.30	1.55	0.41	0.10	..	
Murree hay. ..	2.74	7.35	1.91	43.89	3.68	43.17	4.61	0.09	0.02	0.91	0.26	0.49	0.81	0.11	0.12	0.10	
Rawalpindi hay. .	3.00	8.65	1.89	45.91	3.42	40.13	5.64	0.12	0.04	0.92	0.28	0.43	0.86	0.13	0.14	0.10	
Ambala hay. ..	6.26	10.72	1.42	35.84	5.55	46.47	4.46	0.06	0.18	0.57	0.24	0.87	2.53	0.67	0.54	0.49	M
Lahore hay. ..	4.61	9.76	1.40	38.76	4.38	45.70	5.15	0.04	0.14	0.67	0.20	0.48	1.81	0.35	0.44	0.45	
Jhelum hay. ..	5.25	9.63	1.21	29.93	5.15	54.08	4.38	0.19	0.33	0.67	0.38	0.46	1.63	0.32	0.79	0.40	M
Kasauli hay. ..	3.84	8.76	1.74	36.90	4.19	48.41	4.92	0.14	0.29	1.01	0.29	0.26	0.69	0.21	0.12	0.07	
Multan hay (*Janewa*).	4.44	10.68	1.23	39.20	2.98	45.91	6.24	0.08	0.16	0.60	0.17	0.27	1.07	0.20	0.45	0.27	
Multan hay (*Musel*).	4.46	11.22	0.97	36.37	3.15	48.29	6.76	0.13	0.26	0.83	0.18	0.18	1.09	0.22	0.43	0.16	
Ferozepur hay. ..	7.27	11.56	1.04	32.93	8.96	45.51	4.29	0.07	0.10	0.71	0.32	0.82	2.28	0.56	1.21	0.77	M
Dalhousie hay. ..	2.51	9.33	0.88	38.63	3.41	47.75	6.82	0.25	0.38	0.76	0.19	0.15	0.44	0.15	0.09	0.05	
From Minerals in Pastures' by J. B. Orr.																	
Falkland Island. ..	4.56	..	..	29.3	12.19	..	..	..	..	0.29	..	0.31	2.20	0.54	..	..	
Scotch Pasture. ..	5.49	..	..	25.2	15.88	..	..	..	..	0.56	..	0.41	2.60	0.60	..	..	
Natural Pasture. ..	5.85	..	..	24.5	15.63	..	..	..	..	0.65	..	0.37	2.66	0.67	..	..	
Timothy. ..	..	..	..	..	..	..	..	..	..	0.11	0.060	..	1.56	0.41	..	..	
Orchard grass. ..	..	..	..	..	..	..	..	..	..	0.12	0.086	..	2.24	0.35	..	..	
Clover red. ..	..	..	..	..	..	..	..	..	..	1.50	0.390	..	2.10	0.40	..	..	
Clover alsike. ..	..	..	..	..	..	..	..	..	..	1.26	0.350	..	2.72	0.51	..	..	
Alf alfa. ..	..	..	..	..	..	..	..	..	..	1.76	0.340	..	2.25	0.56	..	..	
Vetch. ..	..	..	..	..	..	..	..	..	..	1.21	0.300	..	2.06	0.58	..	..	

calcium-phosphorus ratio affects the digestibility of the protein of the grass in some way at present obscure.

TABLE III

District of origin of hay.	Protein percentage of dry hay	Digestibility coefficient of protein.	Digestibility protein per 100 lb of the hay.	P_2O_5	CaO/P_2O_5	
Sialkot.	4·83	40·18	2·17	0·43	1·3	M
Murree.	3·68	15·84	0·56	0·11	9·0	
Multan (*Musel*).	3·15	12·01	0·35	0·22	3·8	
Dalhousie.	3·41	11·22	0·35	0·15	5·0	
Jhelum.	5·15	34·24	1·58	0·32	1·9	M
Ferozepur.	8·96	58·41	4·99	0·56	1·3	M
Kasauli.	4·19	19·29	0·72	0·21	4·8	
Ambala.	5·55	44·10	2·26	0·67	0·9	M

These data show that great variations occur in what is commonly called hay.

If, however, good grass is cut at the correct stage, the grass itself and hay made from it constitute a protein rich feeding stuff much richer in total digestible nutrients than most roughages.

Where grass is carefully grown for hay as in Remount Department farms, the number of varieties found in a paddock is considerably less as a rule than on ordinary grazing grounds, and is more comparable to seed hay in western countries ; consequently a more uniform type of hay is produced.

The question may arise in regard to land primarily devoted to producing grass for hay, whether to allow the grass to grow to the stage when it is ready for cutting for hay or whether to cut it at frequent intervals throughout the year. In the latter case the total yield of dry matter may be much less than in the former, but on the other hand the total yield of protein may be greater when the crop is cut frequently, or grazed under the system known as rotational grazing, than if the grass is allowed to grow to the stage required for hay. According to Morrison [1936], grass in general, if cut at intervals of 2-3 weeks will yield only 50 to 65 per cent as much dry matter, and 60 to 75 per cent as much total digestible nutrients during the season, as when it is harvested for hay at the usual stage. In investigations carried out at Sirsa, the author has shown [Lander, 1942] that successive cuttings of various selected grasses may give greater yields than when the grass is left till maturity. A safe rule to follow is always to cut the grass for hay at the proper nutritive stage just after flowering, when the maximum yield of nutrients is likely to be obtained.

It is not only desirable to cut grass for hay before the seed is ripe, on account of the higher protein content, but also because the protein is more digestible when the grass is cut at this stage. The digestible protein and the total digestible nutrients are therefore both invariably higher when grass is cut at the right stage. Even a delay of a few days when the seed is ripening may make a difference in this respect, and consequently alter both the cash and nutritive value of the hay.

Bhusa or Straws

Bhusa and the straws of other cereals are the residue left after the ripe crop is threshed for removing the grain. In India, rice is generally threshed by banging the crop with sticks after it is cut, but the most common method of threshing other grains is to allow bullocks to trample on the gathered crop which is continuously turned over and broken up. The *bhusa* then consists of the dried leaves and stalks which are stacked in the field at suitable places by one or more of the following methods:—

(1) *In kups*

A *kup* (Plate XXIII) is a structure resembling in shape the domes on mosques, and is made by packing straw or *bhusa* tightly together and then thatching with *bhusa* in such a manner that there is, so to speak, a succession of overlapping eaves from the top of the dome to ground level as shown in the photograph. The thatch protects the *bhusa* in the *kup* from rain and is particularly suitable for wet districts. If *bhusa* is stacked in this way it may be kept for a year or two.

(2) *The palla*

The *palla* is a somewhat simpler structure than the *kup* and is common in villages. It is made by packing *bhusa* in the space made by placing a number of ordinary charpoys set up

Bhusa stored in *kups*.

on end so as to form squares or rectangles. The top is then covered by a thatch of unbroken wheat straw and plastered over with mud.

(3) *The dhar*

The *dhar* is a primitive method and merely consists of a long heap of *bhusa* covered over with mud. It is perhaps the most wasteful method of storing *bhusa.*

ENSILAGE

If certain green fodder crops are cut and dried in the sun they may be preserved as hay as described above.

If, however, they are cut and left exposed in heaps on the ground they will soon become mouldy and decompose and become totally unfit for feeding to stock. Decomposition is due to the combined action of bacteria (some of which act best in the presence of air and are called aerobic, while others act best without air and are anaerobic), fungi, moulds and the various plant enzymes present in the cells of plant tissues.

If these processes could be checked and controlled, it is clear that decomposition could also be checked, and the plant material kept for a considerable time on much the same principle as canned foodstuffs, which can be kept for a long time if they are properly prepared.

The process by which green fodders may be preserved is called siloing or ensiling, and the resultant product is known as silage or ensilage. In the process of siloing, fermentation is never actually stopped, but continues in varying degree according to the method employed. Various substances are thus produced from the partial decomposition of the fibre, cellulose, starch and proteins of the plant material. Some of these products are carbon dioxide, water, acetic acid, lactic acid and butyric acid, and their amount and type will depend on the nature of the material siloed and the method of siloing it. It is clear that the type of silage produced will also depend on the method employed. The process of siloing has developed greatly in those countries where wet weather may hinder the making of hay, or where large quantities of green succulent feeds are required to carry stock through the winter, as in the U.S.A. where practically no farm in the cattle belts is without its silo tower.

In India siloing is the exception rather than the rule, partly owing to the cost of construction of tower silos in areas where these can be utilised, and partly because in irrigated tracts green crops are usually available throughout the year. The practice, however, might be made more use of to preserve fodder which goes to waste in the monsoon as a reserve against famine conditions, or on large estates where the rotations followed necessitate keeping a reserve of green fodder in silos. The average zamindar, however, will need considerable persuasion to induce him to construct simple pit silos where they could advantageously be used.

Fermentation changes occurring in silage

When green fodder is cut there is always a certain amount of air occluded in the tissue cells of the plants, and when placed in the silo, the cells continue their respiratory activities for a time. Some of the fibre, cellulose, starch and protein become partially digested and broken down into carbon dioxide and water and simple nitrogenous products, with the evolution of a considerable amount of heat. These activities involve the loss of a certain amount of the dry weight of the original material and in some cases may be as high as 40 per cent. Air occluded in the spaces between the plant material, and bacteria, also assist these changes, and if a silo is very loosely packed in a damp condition, a very high temperature may be attained with the result that a considerable amount of the silage may be destroyed.

Preparation of silage

The great art in silage making is the control of quality so as to prepare silage which is palatable, with the minimum amount of loss through fermentation or decomposition. The changes which take place during the preparation of silage from the raw material are brought about by the agency of micro-organisms, some of which are beneficial, such as the lactic acid and acetic acid producing organisms which operate best with occlusion of some air. Acetic acid is produced by a minute organism known as *Bacterium acetii* from starch or sugar contained in the fodder, while lactic acid is produced by the lacto bacilli or *Streptococcus lactis*.

The presence of the acids produced by these organisms prevents the development of certain undesirable bacteria which are responsible for decomposing the proteins of the material ensiled. There is a considerable number of these organisms, and one group is known as the butyric acid bacteria (*Clostridium butyricum*) as it produces butyric acid which is very undesirable in silage. These bacteria are most active in anaerobic conditions, *i.e.*, with a minimum or total exclusion of air. The amount of acetic and lactic acids produced in a well made silage should constitute approximately 2 per cent by weight of the total material. In making silage, therefore, the lactic and acetic acid producing organisms have to be encouraged by means of proper packing and a judicious occlusion of just the right amount of air to encourage their activity.

In western countries these activities are sometimes increased by inoculating the material to be ensiled with the proper type of organism.

In order to produce silage of good quality the material to be ensiled should be allowed to wilt slightly before being placed in the silo and should be reasonably tightly packed, but not too tightly so as to exclude all air. When packing the silage special precautions are necessary to see that air spaces are not left at the sides of the pit. The best silage is made by employing bullocks to tread down the material and even elephants, when available, may be used in a pit silo. In the process of maturing, the relatively high temperature of 100°F may be attained and this helps to inhibit the action of undesirable proteolytic bacteria, such as those which form butyric acid. Lactic and acetic acid producing bacteria are active in such conditions.

If the material has been too loosely packed with much air included, oxidation processes are more intense and a temperature of 120°F may be attained, which will still further tend to inhibit the development of the proteolytic organisms. In such conditions there will be a tendency for the development of moulds. When properly prepared both lactic and acetic acids are present, the former in greater amount, and the silage has a sweet or fruity taste. Such silage is very palatable to animals.

An intensive study of the quality of silage has been made at Cambridge[7] by (a) inspecting silage from a large number of silos and ascertaining the conditions under which each was made by the farmer concerned, and (b) by making accurate observations of the conditions of the crop as ensiled, and subsequently observing the types of silage produced. In this way it has been possible to distinguish five distinct types of silage:—

1. Sweet, dark-brown silage.
2. Acid, light-brown or yellow-brown silage.
3. Green fruity silage.
4. Sour silage of several forms.
5. Musty silage.

The chief characteristics of these are as follows:—

(1) *Sweet, dark-brown silage*

Sweet, dark-brown silage is a good feeding stuff but is not the best type of silage. It has a sweet pleasant smell similar to that of over-heated hay, but it is generally rather dry, although it is very palatable and readily eaten by cattle. This type of silage is generally produced when the temperature rises above 113°F, but is not produced below that temperature. It frequently happens that a shallow layer of sweet silage is found between 6 inches and 2 feet below the surface in tower silos filled with almost any crop other than maize. This is because air has more ready access to the surface of the silage than to the lower layers and so encourages fermentation and heating. Sweet silage is therefore produced when conditions are such that fermentation is facilitated by the presence of sufficient air, so that the temperature rises above 113°F. This frequently occurs when dry crops or those which have been allowed to dry after being cut are ensiled. Such crops facilitate fermentation because they cannot be very tightly packed, and also because the heat generated by fermentation has less moisture to heat, hence the temperature rises. The production of sweet silage is also facilitated by making the silage in intermittent stages, thus allowing fermentation to occur in each layer before the next layer is put on. Considerable wastage occurs with this type of silage due to excessive fermentation although the product is both attractive and nutritious.

(2) *Acid, light-brown or yellow-brown silage*

This type of silage is generally produced when oats or seed crops are used, especially if they are wilted shortly before being ensiled, so that the crop contains from 25 to 30 per cent dry matter when ensiled. The maximum temperature attained varies between 86°F and 104°F. Not much juice is usually lost in making silage of this type. Brown silage is brown or yellow-brown in colour, and the more yellow the colour the better is the silage. It has a pleasant, acid smell due to the presence of acetic acid. This silage is readily eaten by stock which thrive upon it, and is one of the best types of silage.

(3) *Green fruity silage*

This type is not very common, but may be produced if crops such as oats are cut before maturity at the milk stage before the seeds are properly formed. To produce this type the crop should be cut and ensiled without delay. The maximum temperature attained is about 86°F and the resulting silage, which has a green or olive green colour, has an attractive smell, is neither sweet nor sour, and is greedily eaten by stock. It is also highly digestible.

This type of silage suffers from the practical disadvantage that considerable quantities of juice containing nutritive material are likely to drain to the bottom of the pit or tower. This juice can be collected from tower silos and fed to cattle or pigs but involves loss and deterioration of the bottom layers of the silage in under-ground silo pits.

(4) *Sour silage*

Sour silage generally results when very immature and succulent crops are ensiled, as the watery fodder packs down very closely in the silos and excludes air so that little rise of temperature is possible. It is generally dark brown and has a pungent and unpleasant smell, due chiefly to the presence of butyric acid. If immature crops such as maize are siloed, sour silage is likely to result. This will also happen if crops are siloed in wet weather. This type may also be produced if crops such as oats, which have been cut and allowed to remain in a damp condition, are used, or if rain falls and some rotting at the base of the stems occurs. In such cases the crop should be dried somewhat before being ensiled.

(5) *Musty silage*

Musty silage is frequently found at the top of silos and at the sides if the silage has not been properly packed, due to the fact that owing to access of air normal fermentation cannot take place and moulds consequently develop. This gives the silage a musty or ammoniacal smell which makes it distasteful to stock. If crops such as oats or maize have become over-matured before being cut for silage they should be ensiled at once and not allowed to dry further in the field or there will be a danger of musty silage being produced.

Types of Silo

Various types of silo may be used, depending on local conditions. The best type is the tall cylindrical tower which may be made of wood, concrete, or bricks, and should be constructed so that the walls are air tight, smooth and perpendicular, in order that the fodder as it is placed in the tower can settle down into a compact mass. The silo should be provided with a roof and a window, and a series of doors from top to bottom which can be opened so as to permit workers to pack or remove the silage at different levels. When the silo is to be filled, the fodder should be cut up in a chaffing machine on the ground, and then blown up a chute to the top where it falls through a distributor so as to ensure even packing. A worker must be in the tower to place the fodder evenly and compactly all round the circumference in such a way that as it settles no air spaces are left. If the fodder used is too dry for it to be closely and evenly packed, water should be added in sufficient quantity to enable this to be done. The top layers should be carefully tramped down before the roof is finally put on. Good silage may be obtained by this method, and although there will be a certain amount of spoilage at the top, the loss in a 30-foot deep silo should not usually be greater than about 10 per cent for well prepared maize or oat silage, if proper methods have been adopted.

The size of the silo will depend on the amount of fodder usually available, and the number of animals to be fed from it, but its height and diameter should be so designed that from 2 to 4 inches of silage are removed daily according to the season,

Silo pits at Coleyana Estate, W. Punjab.

more being necessary in hot weather than in cool owing to spoilage, which may occur by daily removal of small amounts in hot weather. Morrison [1936] gives the following data as a guide. Two inches in depth of ordinary maize silage weigh about 5 pounds per surface square foot near the top of the silo, and 7 pounds near the bottom, averaging about 6.5 pounds in a silo filled to a depth of 30 feet. To use 2 inches daily from the surface, approximately the following amounts must be fed from silos of various diameter: Diameter 10 feet, 510 pounds silage; 11 feet, 615 pounds; 12 feet, 735 pounds; 14 feet, 1,000 pounds; 16 feet, 1,300 pounds; 18 feet, 1,655 pounds; 20 feet, 2,045 pounds; 22 feet, 2,470 pounds; 24 feet, 2,940 pounds; 26 feet, 3,450 pounds. The following table shows the approximate capacity of silos of different depth and width when the silage has settled down for a month:—

CAPACITY AFTER SILAGE HAS SETTLED DOWN FOR ONE MONTH OR MORE

Inside diam. in feet	Depth of silage in feet											
	8	10	12	14	16	18	20	22	24	26	28	30
10 ..	11	14	17	20	23	26	29	33	36	39	43	46
12 ..	16	20	24	29	33	38	42	47	52	56	61	66
14 ..	21	27	33	39	45	51	58	64	70	77	83	90
16 ..	28	35	43	51	59	67	75	84	92	100	109	118
18 ..	35	44	54	64	75	85	95	106	116	127	138	148
20 ..	43	55	67	80	92	105	118	130	144	157	170	184

Ground silos

A cheaper form of silo (Plate XXIV), more suitable for India, may be made by digging a cylindrical pit in the ground comparable in some ways with the above-ground silo tower. The dimensions of the pit will be governed partly by the nature of the soil, but the above calculations will serve as a guide. Where the sub soil water is very low, the silo may be 15 feet deep if the soil around the surface is banked up and kept dry. The loss from such a silo is naturally greater than from a tower silo.

The writer has used *kachcha* silos 6 feet and 8 feet in diameter and 12 feet in depth, and a *pukka* cement silo of the same type with a sloping floor at the bottom through which a pipe passes to allow seepage to escape into a chamber underneath, access to which is gained by means of a sloping stairway. These silos when filled, held from 125 to 150 maunds of green fodder. Experience proved the necessity of extremely careful packing and the necessity of some form of lining to a *kachcha* pit silo to prevent silage at the circumference from becoming spoiled. This disadvantage was absent in the cement pit, but a cheap lining can be made by cementing the sides of the silo with a good clay made into a cement by the addition of sodium carbonate.

Imperial Chemical Industries, London, have recently published an excellent little booklet dealing exhaustively with silage, which is available from the Company's branches in India. Whilst the general principles of making silage are much the same in all countries the practical details necessary in India will depend on climatic and soil conditions. In India the most suitable and practicable form of silo is undoubtedly the ground silo. It is also cheaper to construct. It merely consists of a properly prepared cylindrical pit in the ground of approximately the abovementioned dimensions, and the top may be constructed so as to project somewhat above the surface of the ground. This will afford protection from surface water draining into the pit. The photographs show a series of such simple silo pits constructed at the Coleyana Estate, Okara (Punjab).

Crops suitable for silage

Forage crops vary in their suitability for silage, and the best for Indian conditions is perhaps maize, although the author has made very satisfactory silage from berseem, *juar* and oats, and also from *shisham* leaves as an experimental famine fodder. Good pasture grass may also be made into silage when hay making conditions are not satisfactory, preferably mixed with other fodder crops, although it is not usually grown for that purpose. The sweet sorghums and grain sorghums, or a mixture of oats and beans in the proportion of two of oats to one of beans, makes an excellent and nutritious silage. Conditions vary so much in different parts of the country that no fixed rule can be laid down and the purpose for which the silage is needed will govern the type of crop used.

When siloed alone, the legumes are not as satisfactory as

maize and the small grain crops, because they contain but little carbohydrates. This diminishes the intensity of acid fermentation, although this defect may be overcome to some degree by allowing them to wilt before siloing, or by mixing them with forage of higher sugar content.

The best time to cut maize and oats and the sorghums for silage is at the milk or flowering stage before much fibre has developed.

A.I.V. silage

A special type of silage has been developed by A. I. Virtanen of Finland (hence A. I. V. silage) by which the degree of fermentation and breaking down of nitrogenous compounds is said to be considerably reduced. The essential feature of this method is the spraying of successive layers of silage as they are placed in position in the silo with dilute hydrochloric or sulphuric acid, which helps to prevent the growth of undesirable bacteria or moulds. It is necessary to drain off the seepage some days after filling the silo and a certain amount of calcium carbonate should be added to the fodder when used for feeding to neutralise any unpalatable effects of the acid. Another method sometimes employed to increase acidity is to spray the fodder as it is filled in with molasses, which assists in the production of acetic acid.

Silage for Stock

The feeding value of silage will depend on the nature of the materials from which it is made, and the manner in which it is made, but good silage has certain desirable qualities quite apart from its nutrient content, and retains its vitamin A content better than hay. It is very palatable and stock will usually eat more roughage when fed silage and roughage than when fed dry roughage alone; this may make a considerable saving in the amount of concentrates needed. A certain amount of the hay or roots of rations may be replaced by silage.

Woodman [1925-26] has carried out some trials with green fodders and the corresponding hay and silage made from them, and he has recorded that the total digestible nutrients in 100 pounds of green oats, oat hay and oat silage, were 9.14 pounds,

50.71 pounds and 14.87 pounds respectively. Corresponding figures from some preliminary trials at Lyallpur have shown that 100 pounds of green oats, oat hay and oat silage contained 10.04 pounds, 50.98 pounds and 14.35 pounds respectively of total digestible nutrients. On the basis of these data one pound of hay is approximately equivalent in digestible nutrients to from two to three pounds of silage, or in round figures a hundred pound of silage is equal to thirty-four pounds of hay. In the case of silage from mixed crops the proportion is approximately one hundred pounds of silage to forty pounds of hay.

It is difficult to say how far silage can be substituted for roots. Silage on a dry basis is fairly rich in protein, and comparable with the fodder from which it was produced, whereas roots are non-protein feeding stuffs and consist of little more than a solution of carbohydrates and salts in water. On the other hand, roots contain little or no fibre, and hence, calculated on a dry basis they contain more total digestible nutrients than silage, and they are also more digestible. Hence if roots are entirely replaced by silage the ration will, when calculated on a dry basis, contain too much dry matter and insufficient total digestible nutrients; therefore the hay part of the ration should be reduced somewhat to accommodate the extra crude fibre contained in the silage.

Silage is not a complete substitute for either hay or roots, but if properly made there should not be great loss in nutrient content compared with the green fodder, and its digestibility should be not less than that of the green fodder from which it was made.

Silage is chiefly used for feeding dairy cattle, especially milch cows, which will generally produce more milk on a ration consisting of concentrates with hay and silage than on concentrates and hay alone, unless the hay is high quality legume hay. Silage is especially valuable if no green fodder is available and the roughage is of inferior quality.

Morrison [1936] recommends the following as the amounts of silage which can safely be fed to various classes of stock as a part substitute for other parts of the rations:—

Dairy cows 30 to 50 pounds for those in milk, with somewhat less for dry cows; dairy heifers 12 to 20 pounds; beef breeding cows 30 to 50 pounds; fattening 2-year old steers 25 to 30 pounds, at the beginning of the fattening period the allowance gradually decreasing until only 10 to 15 pounds are fed; fattening calves from 10 to 20 pounds at the beginning of the period 8 to 10 pounds or less at the latter part; brood mares and idle horses 15 to 30 pounds; breeding ewes 2 pounds per 100 pounds live weight; and fattening lambs 1.5 to 3.0 pounds per 100 pounds live weight.

SPECIAL PREPARATION OF FEEDING STUFFS

It is a vexed question with stockmen as to whether it will pay to incur the expense involved in special treatment of feeding stuffs, such as crushing, grinding, soaking or cooking. There may be circumstances in which feeding stuffs should be specially treated, but it is not correct to assume that such operations will necessarily save an animal unnecessary expenditure in energy, which could be more profitably utilized for other purposes. The operation of mastication gives a certain amount of exercise, which is necessary, although grinding may be advisable for certain classes of stock such as sick animals, or, in the case of certain grains which may otherwise escape proper mastication. The value of special preparation will depend on the type of material and the type of animals to which it is fed.

Grinding and crushing grains and seeds

Grinding and crushing may be an advantage for young animals before their teeth are fully developed, or for old ones with imperfect teeth, or when the seeds are so hard that they cannot be thoroughly chewed. Hard grains such as flint maize, gram, peas, beans, wheat and cotton seeds should be ground or crushed before being fed. Calves will chew maize and oats fairly well up to about 8 months of age, after which it is advisable to have them crushed. Pigs will masticate ordinary grains fairly well up to a certain age, but they usually do better on crushed grain, especially after the stage when they are being finally prepared for the market.

Kellner [1908] has referred to the results of 18 series of feeding trials carried out on 280 pigs in the U.S.A. over a period of ten years, in which ground maize was compared with whole maize, and the increases in weight recorded.

The total food consumed and the total increase of live weight during this time were:—

	lb.		lb.		lb.
Whole maize	46,736	Bran	22,590	Increase	13,828
Ground maize	50,647	Bran	24,189	Increase	15,891

Thus in the whole maize and bran series, it took 501 pounds of food to produce 100 pounds increase in body weight, while only 471 pounds were necessary for the same increase with the ground maize, which shows that 6 per cent more maize would be needed if fed whole to produce the same result; the cost of grinding also has to be taken into account in making the balance sheet. Pigs, which have become accustomed to ground grain may be upset if changed on to whole grain. Grinding appears to increase the palatability of grains, and animals will frequently eat more when crushed than when uncrushed.

Grains and seeds should invariably be crushed or ground for dairy cows, but unless they are exceptionally hard, grinding is not necessary for sheep, which masticate their food more thoroughly than most animals. Horses and mules are usually given crushed grain although it is often a matter for debate whether this is really necessary.

Kellner [1908] records trials in which 64.6 per cent of the dry matter of oats was digested when chopped hay was fed with whole oats, when the oats were crushed 68.6 per cent was digested, and 72.7 per cent when the oats were fed coarsely ground. In other trials it was found that 5 to 16 pounds of oats could be saved per 1,000 pounds of grain by crushing.

Linton [1927] has recorded trials conducted by him to ascertain the relative time horses take to consume whole and bruised oats respectively, and he found with 28 horses that 3 pounds of whole grain were consumed on an average in 17.3 minutes, and the bruised grain in 16.1 minutes by the same horses.

Another series of tests with horses and mules showed that 12 horses ate 3 pounds of whole grain in 17.4 minutes, while 11 mules took an average of 19.1 minutes. The same horses eating 1.5 pounds of whole oats mixed with 1.5 pounds of bruised oats took 14.9 minutes to finish, while the mules took 17.4 minutes, thus showing that bruised oats were consumed in less time than whole grain. These trials indicate that the bruised oats were eaten more quickly than the whole, and therefore were not so thoroughly masticated. Oats should always be ground for dairy cattle and pigs and the husks should be removed for young animals.

Suddenly changing an animal over from crushed or ground grain to whole grain is likely to involve digestive disturbances and should therefore be avoided as far as possible. Crushed grains and meals absorb moisture readily and will not keep as well as the whole grain, and they should be prepared in small quantities at a time when actually needed.

Soaking feeding stuffs

Many small seeds and grains which cannot easily be ground, and such grains as hard maize, hard wheats and barley are improved if they are soaked for some time before feeding. Sufficient warm water should be added to soften the hard seed coat, but not an excess to cause them to swell.

Cottonseed, which has a hard outer shell, should always be crushed and then soaked for some hours before feeding. It should then be mixed with other concentrates in the form of a damp mash but not a slop. The benefits of soaking feeding stuffs are not as great as is sometimes thought. In feeding pigs there is a danger of considerable waste if the food is fed dry, and so it is generally fed wet, although dry feeding has been found to produce just as good results. Soaking grain for horses should be the exception rather than the rule, and in any case only very small quantities of soaked grain should be used, as the ingestion of any considerable quantity is likely to lead to considerable fermentation in the digestive tract. Merely moistening food may be useful to prevent waste of finely ground material and allow of better mixing, but it has no effect on digestibility.

Cooking and Steaming Food

Until comparatively recently, the idea was prevalent that cooked and steamed foods were good for animals, and even today many stockowners believe that it is an advantage to cook food. This idea, however, is becoming out of date. Cooked food is the practice for human beings in order to make it more attractive and palatable, but there is no valid reason why the practice should be followed for animals, except in the case of sick animals, who could not otherwise be induced to eat, and for pigs, which should not be fed on such materials as raw potatoes. Cooking undoubtedly destroys bacteria, and causes starch grains to burst, and induces partial dextrinisation; fats are also hydrolysed to some extent, but the general effect is to render the food, particularly the proteins, less digestible.

In the case of animals there should be no necessity to make the food appear attractive or palatable by cooking except in special circumstances. On the whole cooking is more likely to prove deleterious than otherwise.

Ladd and Hornberger [1885] have shown that cooking and steaming grain and roughage does not increase the digestibility of their protein but reduces it, while trials at Ohio [1936] have proved that cooking lucerne hay and maize for fattening cattle reduced the gains made by the animals, and the profits.

Experiments have also been conducted in Ireland [1916] to ascertain the effect of cooking meals for pigs when fed over a considerable period. Half the pigs employed in the trials were fed meal which had merely been damped, and half fed meal which had been boiled or steeped in boiling water. The trials revealed that the dry meal fed pigs invariably made greater gains in daily weights than the cooked meal fed pigs.

Cooked feeds should not be given to horses. A horse coming in after work or exercise is likely to be attracted by the flavour and aroma of the cooked grain, and in consequence is liable to eat too fast and bolt the food without proper mastication and salivation, with the result that proper digestion is impeded and fermentation may be set up with considerable production of gas, and gastric tympany may result.

Malting and sprouting grain

It has been claimed from time to time that if grain is malted and allowed to sprout, this has a beneficial effect on stock to which such grain is fed. The general consensus of foreign opinion is that such claims cannot be substantiated although there are many firm believers in it in India. Morrison [1936] has recorded the results of numerous trials which indicate that feeding grain, which has been allowed to sprout, does not result either in increased production or saving of food. Beneficial results are said to have been obtained by feeding sprouted oats to horses, but this may have been caused by the fact that when grain is allowed to germinate a considerable amount of vitamin C is produced, and this in itself is an advantage.

Cutting straw and green fodders

Green succulent fodders, such as maize, are best fed to dairy cattle and other stock after having been chopped into small pieces of from one to two inches long, by a chaff cutter. This saves considerable waste and ensures a more even feed. The various straws should also be fed after being chaffed. Straw should be cut into small lengths of from one to one and a half inches in length for cattle, and somewhat longer for sheep. If fed to horses, straw should be cut into the same lengths as for cattle. Cutting straw prevents waste by scattering in the mangers, and also makes mastication easier, and if the straw is mixed with the grain, helps to prevent the latter from being bolted too quickly by greedy animals and ensures the straw being eaten along with the rest of the ration.

Cutting fodder does not increase its digestibility, but, on the other hand, if cut too finely it may be swallowed without proper mastication and cause colic.

It is not necessary to cut hay for horses, which should always be provided with a certain amount of long hay to keep them occupied during idle hours in the stable. This should be given as an aftermath to the ration proper. Some people prefer not to cut straw, but to leave animals to select the finer leaves and stems at leisure, rather than compel them to eat the whole straw including the less digestible fibrous stems.

The storage of roots

If root crops such as turnips have to be kept for any length of time they are best stored in pits or clamps. Clamps are made by preparing a long rectangular foundation somewhat higher than ground level, and then packing the roots on top as closely as possible until the required height of from 3 to 4 feet is reached. As each length is completed the roots are carefully covered with straw and then the whole clamp is closed with a roofing of earth having sufficient slope and compactness to keep out rain. When roots are stored in this manner they continue to undergo respiratory activity, which may continue for sometime; hence some ventilation is necessary. This may be provided by leaving small holes at suitable distances apart in the earth roof with a clamp of straw projecting chimney-wise. These may then be closed in like the rest of the roof after appropriate intervals depending on local climatic conditions.

If roots are stored in pits the above general principles hold good, but pits are not as satisfactory as clamps, except in a very dry soil with a deep water table; otherwise they may become damaged from a damp sub soil.

Owing to the initial respiration which may continue for sometime, some of the starch in roots is converted by enzyme and respiratory activity into sugar during storage. This action is diminished after a time as tissue oxygen becomes used up, and almost ceases when the roots are finally closed in. Roots should not usually be kept in store for a longer period than is needed for animals on the farm before the next crop is ready.

Storage of bran

Bran is very liable to be attacked by weevils and should be stored in loose heaps on the floors of properly constructed rooms and not packed in bags. As bran contains much of the germ of the wheat grain it is liable to absorb moisture and become rancid if kept in damp conditions. It should not be stored in larger quantities at a time than are needed for actual feeding.

The storage of grains

There are certain fundamental principles which should be observed in storing all types of grain if damage and deterioration are to be averted. The individual grains are still alive, and respiratory activities, although slight, are still going on. Therefore, if grain is stored in large quantities in badly ventilated barns, a considerable amount of carbon dioxide may accumulate. Hence storage rooms should always be well ventilated. Grain should be dry when placed in store and the softer varieties should be partially dried and must not contain too much moisture. When in store the grain must be kept dry.

When cereals, and especially oats, are allowed to become damp they become musty and develop toxic properties. No

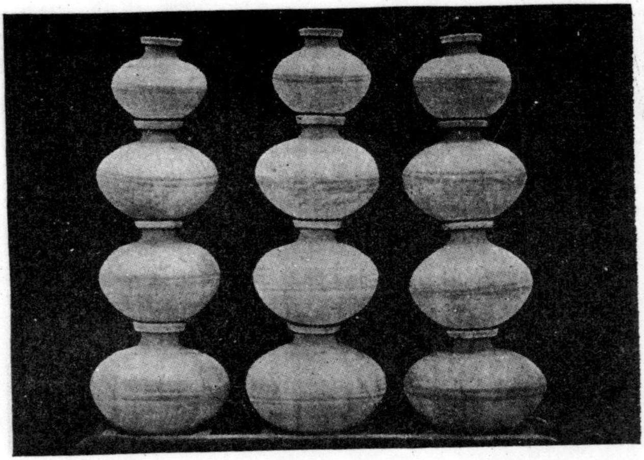

FIG. 1. Earthen pitchers used for storing grain.

FIG. 2. Earthen bins (*Bharola*) for storing grain.

grain which has thus become damaged should be fed to animals. Horses are specially liable to suffer from indigestion and wasting if given such grain.

All types of grain are liable to attack by insects, such as weevils. Strict cleanliness is therefore essential and barns should be thoroughly cleaned before the new crop is introduced, because unsuitable barns and containers may harbour these pests from one season to another. Holes in walls and woodwork should be carefully filled in, and where necessary all wooden bins, containers, and the woodwork of the barn should be scrubbed with a strong solution of soda in hot water.

Even under the best conditions stored grain may become damp or infested with insects; hence it should be taken out whenever possible and spread in the sun occasionally, preferably after the monsoon has started, as this is the period when insects become most active. This will enable insects, or defective grain which may have become damaged to be periodically removed.

The methods of storing grain will depend upon the tract in which it is stored, the class of people storing it, the type of grain which is stored, and the purpose for which it is stored.

Some of the more common methods employed are the following:—

(i) Earthen containers (Plate XXV, Fig. 1). These are very common among the poorer classes and are prepared as a rule by potters. They vary in shape and size and are used for storing grain, flour, pulses, jowar, rice and oilseeds in small quantities which are needed for seed. The amount which is stored in such containers may vary from a few seers to over a maund.

(ii) Earthern bins (Plate XXV, Fig. 2). Bins may also be of various designs and shapes. They may be (1) small cylindrical ones called *bharoli* in the Punjab, or large ones called *bharola*, (2) rectangular ones called *kothi* or (3) square ones called *chauras*. They are usually made of clay mixed with *bhusa*, and are finished off by being coated with a thin layer of cow dung. These storage utensils are generally built on a platform inside the cultivators' living quarters. They are provided with an opening at the top which can be closed with a lid, and a small side hole near the base through which grain can be removed. *Kothis* are generally supported on mud supports and do not have an opening at the top, but have one in the front wall sufficiently large to enable a man to enter. The larger type of *kothi* may be divided into compartments for storing different types of grain. All types of grain such as wheat, rice and maize are stored in these containers, the amount varying from very small quantities, up to from 100 to 200 maunds, depending on whether grain is being stored for household purposes or for seed.

Grain stored in such bins is liable to become badly attacked by insect pests particularly at the top and near the side opening.

(iii) Bamboo bins. Where bamboo is available it is often employed for constructing bins known as *pairoo* (Plate XXVI, Fig. 1). These are made by matting together thin plaited strips of bamboo and may be either oval or cylindrical in shape with an open top and a hole at the bottom, and, like the earthen containers, are usually placed on a support inside the house. The top is closed with a lid which is then plastered over with mud. In spite of closing the top by mud or cow dung, insect pests invariably gain access to these containers and cause considerable damage.

(iv) Bukhari. The *bukhari* is a home-made device formed by fencing off a corner in the house by erecting two walls of wheat-straw, cotton sticks, or bricks and mud, perpendicular to two corner walls. Any surplus grain is then stored for future use.

The methods so far described are chiefly employed by small cultivators and the poorer classes. Large scale farmers have to employ more commodious devices, the chief of which are as follows:—

(i) Open heaps (Plate XXVI, Fig. 2). Grain may be stored in bulk in a heap in any convenient and suitable enclosure or godown which affords protection from the weather, and is the method common in many parts of the country. In some markets, special godowns are constructed whose walls are tarred to a height of 2 to 3 feet from the ground, the walls and floors generally being *pukka*. Underground chambers may also be employed. Damage from insects may be considerable in grain stored in bulk.

(ii) Gunny bags. Gunny bags are extensively used for storing grain of all kinds (Plate XXVII). When full

PLATE XXVI

FIG. 1. *Pairoo* made of bamboo for storing grain.

FIG 2. A grain market. Heaps of grain temporarily protected by gunny bags.

PLATE XXVII

Grain stored in gunny bags placed in tiers.

they are piled up in godowns, open sheds and verandahs, or even in the open, covered with tarpaulin. This is an unsatisfactory method of storage for any length of time, as the grain is exposed to attack by insects from all sides. Insects gain access through the interstices between the strands of the fibre, but in spite of this, the method is very popular owing to the ease with which the grain thus stored can be moved and transported.

(iii) Kachcha khatti. The *kachcha khatti* is an underground pit with a narrow opening at the top, the walls and floor of which are plastered with mud and cow dung. It is very common in some districts of the Punjab, the United Provinces and Central India, and after being filled with grain the mouth is closed and carefully plastered with mud. These receptacles may hold up to 300 maunds and, although grain stored in them is considered to be comparatively immune from insect attack a certain amount of damage is generally caused at the bottom and along the walls on account of moisture. This constitutes one of the chief defects, as a certain amount of grain becomes spoiled and unfit for feeding purposes. Such bins can only be satisfactorily employed where the water table is very low.

(iv) Various types of metal and concrete bins (Plate XXVIII Figs. 1, 2) are also employed in different parts of the country. The metallic bins are made of iron sheets, and built in a series along the walls of storage chambers. Each bin is fitted with a separate lid. They are common on some larger estates and government farms, and their dimensions vary from 120 to 250 cubic feet. Such bins are not insect-proof, and grain stored in them may suffer considerable damage. They possess, however, the advantage of being easily and efficiently fumigated either when empty or full.

Reinforced concrete bins are not very common, but are occasionally used for storing wheat and gram. They are cylindrical, rectangular or hexagonal in shape, and each is fitted with an exit hole at the bottom and a circular or square opening at the top, the latter being closed with an iron lid. Their dimensions may vary from 400 to over 4,000 cubic feet, but unless the lids fit perfectly, insect pests easily gain access through the top opening and the exit holes. One advantage of these is that they can easily be cleaned and disinfected before being filled, or if the grain becomes infested while in storage, they can be cheaply fumigated.

Storage of maize

When harvested, maize contains moisture varying from 12 per cent to 20 per cent, which is a safe limit for storing. If it contains more than this amount, as it may if gathered before it is properly ripe, it is liable to be spoiled when stored in the shelled condition.

Maize is best stored in the ear as there is less likelihood of damage from fermentation, or by the ubiquitous weevil, the enemy of all stored grains and meals. Maize always shrinks somewhat on storing due to loss of moisture, the rate of shrinkage depending on the humidity of the atmosphere, but after attaining a level of about 10 per cent to 12 per cent of moisture, further loss does not occur except in very dry conditions.

Ground maize, especially if it contains the germ, is liable to become rancid on being stored, and should not be kept longer than absolutely necessary, especially if it contains more than 12 per cent water. Maize should not therefore be ground too long before being used.

Storage of oilseed cakes and meals

Oilseed cakes and meals need special care as they are more liable than most feeding stuffs to suffer from bad storage conditions, especially lack of ventilation and damp. For this reason larger quantities than are needed for comparatively short periods should not be kept in store, but consignments should be bought from the mills as and when required. It may be necessary sometimes to store moderately large quantities and certain safeguards should be undertaken to keep the cake in good condition. In the first place every bag should be inspected before it is stored, and when bags are torn, or the cake is broken up to any extent, or is in a powdery condition, such bags should be kept separate from the rest. The whole cakes in intact bags should then be stacked in such a way that access of air is possible. The bags should not be packed too close to walls, and if the store room has a concrete floor the bags should be raised from ground level by means of cross bars on which the bags are placed, as concrete is liable to sweat and the lower bags may

FIG. 1. Iron container used at the Agricultural College, Lyallpur.

FIG. 2. Reinforced concrete bin used at the Agricultural College, Lyallpur.

become musty on account of dampness. The bags should be placed in tiers of two, three or more bags in thickness, according to circumstances, with sufficient space between the tiers for periodic inspection.

Meals need special care and frequent inspection if they are left for any length of time in large heaps. It is essential that cakes and meals should be absolutely dry when stored, and kept so. If any bags become wet they should be removed from the dry ones and the contents dried or used at once. Wet cake may prove toxic or even fatal to farm stock. Needless to say, all necessary precautions should be taken to keep down rats and mice, and this will be easier if proper spaces are left between the tiers.

BIBLIOGRAPHY

1. Amos, A., & Woodman, H. E. Ensilage. Ministry of Agriculture and Fisheries. Bul. No. 37.
2. Bohstedt, Bell and Gerlaugh, Ohio Special Cir. 10.
3. Hosterman, U.S. Dept. of Agri. Leaflet 72 & Tech. Bul. 239.
4. Hosterman, U.S. Dept. of Agri. Leaflet 72 & Tech. Bul. 239.
5. Jour. of Dept. of Agri. & Tech. Ireland. Vol. XVI, No. 3, April, 1916, pp. 419-21.
6. Kellner, O. 1908. Scientific Feeding of Animals, p. 139.
7. Kellner, O. 1908. Scientific Feeding of Animals, p. 138.
8. Ladd, N. Y. (Geneva) Rpt. 1885; Hornberger, Landw. Jahrb., 8, p. 933.
9. Lander, P. E. 1942. Indian Grazing Conditions and the Mineral Content of some Indian Fodders. I.C.A.R. Misc. Bul. 16.
10. Linton, R. G. 1927. Animal Nutrition and Veterinary Dietetics, pp. 56-57.
11. Morrison, F. B. 1936, Feeds and Feeding. p. (i) 223. (ii) 223. (iii) 206. (iv) 240-241. (v) 231. (vi) 63. (vii) 62.
12. Woodman, H. E. 1925-26. The Chemistry of Silage. Jour. of the Min. of Agriculture, Vol. XXXII, No. 2, pp. 124-131.

CHAPTER VIII

SOME HARMFUL FOODS AND FAMINE FODDERS

There are many plants in India which are poisonous in varying degree when eaten by animals, although it is likely that some which are reported to be poisonous are not actually so.

A poisonous plant is one which when eaten by animals causes varying degrees of indisposition, with narcotic or irritant symptoms which may have serious or fatal results, either very rapidly, or even after considerable time has elapsed, due to cumulative effects.

These effects are brought about by poisonous substances actually present as such in the plant, or which may be liberated in the animal's digestive tract as a result of chemical change.

Many of these substances are highly toxic and constitute the essential principles in what are known as medicinal plants, and are used extensively in medicine, but from the animal's or the stockman's point of view they are merely poisonous.

There is a very large number of such poisonous plants in India, and Chopra and Badhwar [1940] in a recent publication have given an extensive list of many of the better known ones reputed to have toxic properties, although, as they say, there are over 2,000 such plants known to exist in the country, but the majority have not yet been fully investigated nor their toxic principles determined.

It is very difficult sometimes to differentiate clearly between a poisonous plant and one which is not, as some which may be harmless in certain circumstances of soil or climatic conditions may be toxic in others. There is no doubt, however, that poisonous plants are very widely distributed, and that the losses incurred by animals eating them are considerable. It is desirable therefore that the stockowner should know something of the extent to which some plants are actually poisonous, *i.e.,* capable of producing indisposition, illness or death when eaten in sufficiently large quantities.

Conditions in which poisoning occurs

Many poisonous plants are shunned by animals on account of their pungent odour or disagreeable taste, which form a natural protection, as animals, as a rule, have an instinctive aversion to eating them. It is in times of fodder scarcity and actual famine, however, that the danger is great, for then animals will frequently eat anything green that they can find.

Animals vary considerably in their predilection for different types of plants, and some are more liable to danger from poisoning than others. Sheep, for example, have a special liking for the soft and succulent green parts of plants which make luxuriant growth, while cattle and horses will eat the coarser and larger plants more readily. Again, certain individual animals will tend to seek out certain types of plants, while others of the same species will avoid them; similarly, the same animal may at certain times show partiality for a special plant and at others reject it. This phenomenon is particularly noticeable with sheep. Different species of animals may be differently affected by poisonous plants; some, such as pigs, may remain unharmed and be able to reject any dangerous or unpleasant plant eaten, while another type, such as the horse, may be harmed through its inability to do this

The stockman should be on guard against poisonous material being introduced into the rations of his stock. Wheat screenings, for example, often contain a considerable admixture of weed seeds of different kinds which may be poisonous. Cereals may be contaminated by poisonous seeds such as the seeds of some of the Euphorbias, for example, *Euphorbia dracunculoides,* which are liable to occur with grain of low quality, or poisonous material such as the Java bean may find its way into them.

Another source of danger is the presence in cereals of broken or inferior grains damaged by harmful fungi, such as the ergot of rye, and the various fungi which attack cereals. Oil cakes and meals which have become damp are particularly liable to develop moulds and fungi, and these, although not necessarily toxic in themselves, produce decomposition products in the foodstuffs.

Some fresh green plants may be toxic in the young condition but lose their toxicity when older and dry, as, for example, certain members of the N. O. Ranunculaceae, or plants whose toxic principles consist of certain volatile oils.

On the other hand, some plants such as the sorghums are harmless when actively growing under normal conditions, but develop poisonous properties under drought conditions when they become stunted and dry. It may happen also that such plants may not lose their toxic properties when dried, and may find their way into hay and be fed to stock in a form in which they are not readily detectable.

It may not always be possible to pronounce an opinion on a fodder from chemical analysis. The author has frequently received samples of fodder for analysis alleged to have caused hydrocyanic acid poisoning, but no trace of the acid could be found.

It is an almost general rule that plants which are toxic in the wild state lose their toxicity under cultivated conditions. The gourds are an example, and, as is well known, cane, which is the cultivated offspring of the sorghums, is not toxic in any stages of its growth. The nature of the soil and the amount of moisture it contains, exert considerable influence on the toxicity of plants, and a plant may be toxic in one part of the country but not in another. This susceptibility to environment is specially characteristic of the N. O. Solonaceae, and Dunstan [1906] has shown that *Hyoscyamus muticus,* grown in Egypt, may contain from two to three times as much hyoscyamine as the same species grown in India. Many plants, for example, *Nerium odorum,* are poisonous throughout, including roots, stems, leaves and flowers, while in others, such as *Datura stramonium,* only the seed may be toxic, or the leaves, as in the sorghums, while in others the root may be the most toxic part.

Chemical nature of the toxic substances in plants

Chopra and Badhwar (*loc-cit*) have classified the chemical constituents in plants responsible for poisoning into five main groups:—(1) The first consists of the class of bodies known as the alkaloids which have an intense physiological action and are very poisonous. They are responsible for the poisonous character of many fungi of the mushroom type, and the pungent odour of many weeds. These properties protect animals to a large extent as they will avoid them, and fortunately they do not as a rule occur in the grasses, but they do occur in certain other natural orders of plants which are widely distributed. Typical examples are strychnine in *Nux vomica* (N. O. Loganiaceae), aconitine from Aconitum of which there are several species, *e.g., Aconitum napellus* (N. O. Ranunculaceae), atropine and allied alkaloids from *Atropa belladona* (N. O. Solonaceae), nicotine from tobacco (N. O. Solonaceae), and morphine from the poppy (N. O. Papaveraceae).

(2) The second class consists of the cyanogenetic glucosides, thus named from the fact that on hydrolysis they split up into a sugar (glucose), and hydrocyanic acid. A typical example is amygdalin which occurs in the bitter almond and cherry kernels. When fermentation sets in this is decomposed by the ferment emulsin contained in the almonds, as follows:—

$$\underset{\text{amygdalin}}{C_{20}H_{27}NO_{11}} + 2H_2O = \underset{\text{benzaldehyde}}{C_6H_5CHO} + \underset{\text{hydrocyanic acid}}{HCN} + \underset{\text{glucose}}{2C_6H_{12}O_6}$$

A similar reaction takes place in the digestive tract of animals.

Other toxic glucosides occur in the Oleander family (Apocynaceae) and Digitalis (Scrophulariaceae).

Another group of glucosides is the saponins which occur in at least 400 plants belonging to 50 different families, and are so named from the fact that when mixed with water they produce a soapy foam. When plants containing them are eaten by farm animals they cause vomiting and diarrhoea and haemolysis of the blood. Examples of plants containing saponins are the corn cockle (N. O. Caryophyllaceae) and the soaproot of the same natural order.

(3) The third class consists of certain essential oils such as those occurring in the Eucalyptus or in Absinth, but cattle do not as a rule feed on plants containing toxic essential oils.

(4) The fourth class is known as the toxalbumins which occur in the castor and croton seeds. These poisons cause considerable losses among livestock, although immunity may be developed if the seeds are continuously eaten in small quantities. This class chiefly affects the blood, causing disintegration of the haemoglobin of the red blood corpuscles.

(5) The last group consists of certain resins such as are found in wild members of the cucumber family, and in the rhododendron family, toxic oils such as croton oil, and picrotoxin, a convulsant poison found in *Anamirta cocculus,* a climbing shrub with poisonous berries found in many forests in India. All these resins are toxic to animals.

Some of the more common types of poisonous plants and feeding stuffs which the stockman may meet will now be described. These may be dealt with under the following groups:—

1. Plants and plant preparations.
2. Seeds and cakes.
3. Roots and tubers.
4. Miscellaneous.

Plants and Plant Preparations

The Sorghums and Maize

The sorghums and maize belong to the N. O. Gramineae, which includes all the ordinary grazing grasses, and a considerable number of the members of this family develop the glucoside responsible for hydrocyanic acid poisoning. Very little is known about the incidence or extent of the glucoside in ordinary grasses, but the sorghums and maize are well known to contain it under certain unfavourable climatic conditions such as drought. It may occur also in very young plants, as well as in the more mature stunted plants. It is not known precisely what function the glucoside fulfils in the general plant economy, but as it generally occurs in conditions when nitrogen metabolism is disturbed or incomplete, it appears to play some intermediate part in the formation of proteins. The soil in itself does not appear to play an important role, but where soil nitrogen is deficient, or where nitrogen metabolism in the plant is interfered with, as may happen on highly saline soils, such conditions conduce to a higher glucoside content.

The chief varieties of sorghum which are known to cause hydrocyanic acid poisoning are: *Andropogon sorghum* or *Sorghum vulgare* (commonly known as *juar*) which is cultivated both as a cattle fodder and for human consumption, and *Sorghum halepense,* or Johnson grass, a tall perennial grass with long rhizomes which is found throughout India. This is also known as *baru* in the Central Provinces, *kala macha* in Bengal, and *dadam* in the N.W.F.P.

These varieties, particularly the latter, have been responsible for considerable mortality among cattle in various parts of India. The hydrocyanic acid producing glucosides of very young succulent plants are particularly liable to cause fatal results. As the plants mature the amount of hydrocyanic acid, or rather the glucoside, decreases continuously until the flowering stage when little or none is found under normal conditions. [Ghose, 1919]. On the other hand, if normal development is arrested as described above either the diminution of the glucoside does not take place, or fresh accumulations occur due to disturbed nitrogen metabolism, and a toxic condition is maintained or established. When grown under ordinary normal conditions, the sorghums are a good source of fodder, but when normal growth has been interrupted by drought, frost, trampling or other causes, hydrocyanic acid may develop to a point where the plant becomes toxic.

If plants which contain the glucoside are thoroughly dried they are less liable to be harmful, and if siloed for a few weeks, even stunted and harmful fodders may safely be fed to cattle [Taylor, 1916].

The leaves of the sorghums contain the greater part of the glucoside present, and calculated on a dry basis a greater percentage of the total than the roots and stems.

Acharya [1933] has recorded observation on the diurnal variation of the glucoside content in *sorghum vulgare* in which he finds that it is lowest in the morning, increasing up to midday, followed by a slow decline to evening, and a rapid decline during the night. Feeding in the early morning and late in the evening would therefore diminish the likelihood of poisoning in conditions where it might otherwise occur.

Ordinary drying does not necessarily reduce the glucoside below lethal limits, and thorough heating in the sun is necessary for the hydrocyanic acid to become totally dispersed.

Cases of poisoning from maize are not as common as from the sorghums, but a dry or stunted maize grown under drought conditions should not be given to farm stock.

The symptoms of poisoning by cyanogenetic plants consist of

uneasiness, rapid respiration, depression, stupor, convulsions, cyanosis of the mucous membranes, paralysis and ultimately death. The entire process may require only a few minutes or may last considerably longer. Hydrocyanic acid is a rapidly acting poison and offers little opportunity for remedial treatment. Recent work has shown however, that if the victim can be reached in time a combination of sodium thiosulphate and sodium nitrite given intraperitonially or intravenously by a veterinarian is an effective remedy.

Deadly Nightshade. (ATROPA BELLADONA)

The N. O. Solonaceae contains many examples of plants which are poisonous, *e.g.,* the deadly nightshade is highly toxic to human beings, but much less so to horses, while pigs, sheep and goats are little affected. Large ruminants are more likely to suffer than small ones which are but little affected.

In slight cases of poisoning the symptoms are dilation of the pupils, nausea, general muscular weakness and inability to coordinate movements or to rise after having fallen, followed by coma. In more severe and fatal cases these symptoms are intensified and appear more quickly. Violent vomiting occurs and complete blindness may ensue, the pulse gradually weakens with a fall in temperature, and death is preceded by a short period of violent convulsions.

The Lupins. (LUPINUS)

These plants which are common in gardens belong to the legume family and are very nutritious; some species are harmless, at least at some stage of growth, and are excellent for grazing animals; others are dangerous at certain times, and some are toxic at all stages of growth. The most harmful one found in India is the yellow lupin (*L. cuteous*), which has caused many deaths among sheep, though horses, cattle and pigs may also be affected. Sheep are particularly susceptible and may succumb rapidly, or a chronic form of poisoning known as Lupinosis may result, depending on the amount of poison ingested. In the acute form of the disease the sheep suddenly become ill, with acute cerebral congestion causing them to rush about, apparently with no self-control and great irregularity of movement. Muscular spasms occur followed by exhaustion and death which may result within an hour or less. In the chronic form of the disease considerable swelling of the head may appear as in the acute form, and severe digestive troubles which last for a considerable time, during which the sheep displays no appetite and appears to be in a semi-comatose condition.

The amount of the alkaloid found in different plants varies considerably and not all are necessarily poisonous, and considerable amounts of the fodder are usually required to produce toxic symptoms, although these may appear even after a single meal. All parts of the plant contain the alkaloid, and the toxicity of the whole plant increases considerably by the time the seed is ripe due to the excessive toxicity of the latter. Drying does not remove the alkaloid, but if the plants are cut before the pods are formed toxicity is very much less.

If lupins are fed at all, the safest plan is to give only small quantities at a time as the effects are not cumulative. The toxicity may also vary considerably according to the soil on which the plants are grown, and various other indefinite climatic factors, and sometimes a very small quantity of lupin grown on one soil may suffice to kill a sheep, whereas a similar species grown on another may produce no ill effects. The toxic character of the lupins is due to alkaloids of a special group having the peculiarity that slight alterations in the chemical structure of the molecule may convert a very toxic alkaloid into a comparatively non-poisonous substance, and *vice versa.* These alterations are such that they may readily take place within the plant itself by the natural processes of plant chemistry, which may account for the great variations that occur in the toxicity of different plants.

Sastry and co-workers [1942] have recently recorded the results of some work in Madras on the use of the seeds of *Lupinus angustifolius* as cattle food and have summarised their findings as follows:—

1. The alkaloidal content of the lupin seeds received from the Agricultural Research Station, Nanjanad, was found to be 0.1302 per cent.
2. The alkaloid lupanine has been isolated from the seeds in yields varying from 1.5 to 2 gm. from 8 kilogrammes of the material.

3. Debittered seeds fed to the goat did not produce any untoward symptoms.
4. Seeds fed with the husk intact and without being debittered also did not produce any bad symptoms when given in quantities up to 12 ounces a day, but showed satiation on the 4th day after the feed was raised to 1 lb. a day.
5. 25 to 50 milligrammes of the alkaloid, administered intravenously to dogs weighing about 10 to 15 lb., caused a definite fall in the blood pressure, and a dose of 100 to 150 milligrammes ($1\frac{1}{2}$ to $2\frac{1}{4}$ grains) proved lethal, and it is inferred that the alkaloid contained in 60 to 70 lb. or even more of the seeds would be required to prove lethal to a bullock weighing about 300 to 350 lb.
6. The alkaloids of the seeds do not appear to produce any cumulative effects.
7. Debittering of the seeds does not appear to be necessary and they may, with safety, be fed raw (even with the husk intact)—powdered, crushed or in any edible form—to the extent of 3 lb. a day for a bullock with an average weight of about 300 lb.—and in proportionately smaller quantities for smaller animals.

The Thorn Apple. (DATURA STRAMONIUM)

The thorn apple has an unpleasant taste and aroma, and livestock usually avoid it, though it may sometimes find its way into hay and cause poisoning. Drying does not destroy the toxicity which is due to the narcotic alkaloids, hyoscyamine and atropine.

Rhododendron nilagiricum

This is a large shrub found in the higher Himalayas and the Nilgiris where it is known as *billi* or *pumaram*. The leaves are poisonous to goats and ruminants, the symptoms produced being tympany and severe diarrhoea. If neglected, fatal results ensue.

Ficus elastica, or the India rubber tree

This is known as *bor* in Bengali. The young leaves of this tree are poisonous to most animals, producing profuse sweating, paralysis and ultimately death.

Bambusa arundinaceae or Bamboo

This is found all over India and is known as *bans* in Hindi and Bengali, *mungil* in Tamil, and *veduru* in Telugu. The young shoots form a common article of human diet all over India and the leaves and twigs are often used as animal fodder and are particularly relished by elephants. The stiff hairs or bristles on the sheaths of the shoots, however, are supposed to be responsible for cases of poisoning, the symptoms of which develop slowly and may ultimately cause death. The active principle is not known and no curative treatment has yet been discovered.

Croton tiglium

This is known as *jaypal* in Bengali, *jamal-gota* in Hindi, *nervalam* in Tamil and *nepalavittu* in Telugu. It is a small tree cultivated in many parts of India, the seeds of which act as a powerful purgative and in large quantities produce acronarcotic poisoning which is usually fatal.

Anamirata cocculus

This is known as *kakamari* in Telugu and *kakkai-killivirai* in Tamil. It is a large climbing shrub with a rough bark, the berries of which when eaten by animals cause colic, nausea with tetanic convulsions, and ultimately death. The active principle is a substance known as picrotoxin.

Crinum defixum

This is a bulbous plant with succulent leaves found in swamps and on river banks in Madras and Bengal, and is known as *sukdarshan* in Bengali, *kasar chettu* in Telugu and *visha mungil* in Tamil. The flowers are large sessille, white and fragrant at night time. The bulbs, however, are poisonous to cattle and if eaten cause death in a short time through severe gastro-intestinal irritation, asphyxiation and heart failure. Affected animals show profuse salivation and suffer from diarrhoea and congestion. In less acute cases breathing is laboured and is followed by drousiness and stupor. In more severe poisoning respiration is difficult and death is very rapid. No remedial or antidotal measures have yet been discovered.

Abrus precatorius

Known as *gaungchi* in Hindi, *gunch* in Bengali, *guriginja* in Telugu and *kundumani* in Tamil. This is a beautiful climbing plant found all over India, of which there are three chief varieties:—

1. Those with rose coloured flowers and red seeds having black eyes.
2. Those with dark coloured flowers and black seeds having white eyes.
3. Those with white flowers and white seeds.

The seeds contain a poisonous albuminous substance known as abrin which loses its activity on being boiled. If eaten in any quantity they are highly poisonous, producing symptoms of narcotic poisoning which may end fatally. The seeds are used in the criminal practise of *sui* poisoning, by being made into a paste and injected subcutaneously.

Andrachne cordifolia. Vern. *kurkni, gurgul* or *kukuli*

A small shrub found in the north-western Himalayas up to 8,000 feet from the Indus to Nepal. The twigs and leaves are said to have fatal results if eaten by cattle on an empty stomach early in the day.

Avena fatua. The wild oat. Vern. *kulgud, gandal, jei, ganergei, kasamm*

This is a plant which occurs as a weed amongst cereal crops in the plains and is said to produce symptoms of poisoning among cattle.

Buxus sempervirens. The box-wood tree. Vern. *chikri, papri papra. paprang, shamshad*

This is an evergreen shrub found in the Suliman or Salt Range of the north-western Himalayas. It thrives on calcareous soil and in sheltered places, and on alluvial deposits on the banks of streams. Goats browse on the leaves, but other animals will not do so except in times of scarcity, and it is known to have produced fatal results in camels and cattle which eat it.

Daphne oleoides. Vern. *kutilal, shing, kagsari*

A small well branched shrub found in the western Himalayas from Garhwal to Murree. It is poisonous to cattle, producing violent diarrhoea, and camels will not eat it except when driven to do so by hunger.

Euphorbia tirucalli. Milk hedge or milk bush. Vern. *sehud, sehur*

A small tree indigenous to Africa which has become naturalised in the drier parts of Bengal, the Deccan, south India, and Ceylon. It is often cultivated for hedges. Goats and camels eat both the leaves and the bark, and while it is not strictly speaking poisonous, the juice causes great pain if it gets into the eyes.

Kalanchoe spathulata. Vern. *tatara, haiza-ka-patta, rugnru*

This is a succulent perennial plant generally found between 1,000 and 4,000 feet in the lower Himalayan foothills, but near Simla may be found at higher levels. Goats occasionally eat it although it is poisonous to them, but cattle do not, as a rule, touch it.

Ranunculus arvensis. Corn butter cup. Vern. *chambal*

This is a small, pale green annual plant, found in the foothills of the Himalayas from Kashmir to Kumaon, and in the north-western Punjab plains. Goats and sheep frequently eat it, although it produces in them symptoms of irritant poisoning.

Sophora mollis. Himalayan laburnum. Vern. *kun, malan, kathi*

A small plant with pale green leaves five to ten inches long which often do not appear till the flowers are in bloom. It is found in the plains and lower hills in north-western India and is poisonous to cattle. Goats, however, appear to be immune.

Taxus baccata. The Yew. Vern. *birmin, tung, thunu*

The leaves of the yew, which is found in the temperate Himalayas, are often eaten by goats and sheep with impunity, although when dry they are highly poisonous to horses and cattle.

SEEDS AND CAKES

Paspalum scrobiculatum

In addition to the foregoing, there is an extremely large number of members of the N. O. Gramineae which are toxic in one form or other. An important example is the *kodri* plant, or *Paspalum kora,* a wild variety of *Paspalum scrobiculatum.* The latter is widely cultivated in India, especially in the United Provinces, the Central Provinces and the Bombay Presidency. In different provinces the plant is known as either *kodra, kodri, kodon, kodi,* or *kodam.* It is a simple millet which does well on poor soil for which reason it is widely cultivated. Several varieties are recognised by cattle owners and in the Bombay Presidency in particular, two types called respectively, wholesome and unwholesome, are well known. The former has a small pale grain, while the latter has a deep brown seed-coat and is known as the black kodra.

Although used for food by a very large number of people in India, the grain can by no means be considered a wholesome article of diet, and unless special precautions are taken it is liable to act as a narcotic poison both for human beings and animals.

The grain, especially the husk and testa, is the only part of the plant which is poisonous, and the smaller grains are proportionately more toxic than the larger ones. For this reason they are generally carefully separated before the larger grain is used for food.

Watt, in his Dictionary of Economic Products in India, records the statement of Surgeon-Major Pyrie, "the outer coat or husk has a dark outline of a fungus-like character, on the internal surface of which are a number of minute round cells containing dark sporules", and it is these which are the source of the poisonous principle.

Boiling does not entirely destroy it, but if the grain is kept for a number of years it becomes much less. The seed-coat should always be removed whether the seed is used for human beings or for animals.

The symptoms of kodan poisoning, which occur within two hours of the seed being fed to horses or cattle, are stiffness of various parts of the body with violent tremors of the voluntary muscles, difficulty in moving the limbs, extension of the neck, protruding eyes, and laboured respiration. The animal is unsteady on its feet and when it falls down cannot get up again. Most cases end fatally unless treatment is quickly given. The nature of the poisonous principle is at present unknown.

Cases of kodan poisoning are sometimes treated by homely remedies, such as an administration of ghee and pickles, or some other sour or acid food.

Cattle and buffaloes readily eat the young leaves, and the straw is sometimes used as fodder, but animals should on no account be allowed to eat the crop when it is ripening. An interesting case of kodan poisoning is reported in the Indian Forester for 1934 on page 570, in which thirteen elephants who raided a cultivated field of *Paspalum scrobiculatum* in Madras died from its poisonous effects.

Java beans and Burma beans

These beans should always be regarded with suspicion, particularly the many coloured Java varieties which may be dark red, black, purple or mottled in colour, and have been found to yield as much as 9 grains of hydrocyanic acid per pound of beans, two pounds of which may be fatal to a bullock, assuming that 20 grains is the lethal dose. These beans seldom appear on the market, but the new Burma beans (*Phaseolus lunatus*) are a common feeding stuff for animals and may frequently contain appreciable quantities of the hydrocyanic glucoside, although much less than the Java variety. The white Burma bean, however, may be regarded as safe as it rarely contains more than a trace of the glucoside.

It is never possible for the stockowner to know the extent of poisoning which Burma beans may cause as this depends primarily on the climatic conditions under which the crop was grown, and while a crop grown under favourable conditions of rainfall, etc., may be harmless, a similar crop grown under adverse conditions may be very much the reverse.

Charlton [1928] has shown that the hydrocyanic acid content is much more affected by adverse conditions than by the particular type of bean, and it may be very high. He has also found it impossible to reduce the hydrocyanic acid content by careful plant selection over a number of years, and as unfavourable seasons always occur at frequent intervals it is not likely that a hydrocyanic acid free strain can be evolved.

Effects of storing the beans

The quantity of hydrocyanic acid present in Burma beans depends on their age and on the method employed in storing them. Enzyme activity continues after the normal ripening process is complete, and when stored in hot damp climates this may cause a considerable increase in the hydrocyanic acid content, although diminution will occur on removal of the beans to colder localities. During storage the amount of hydrocyanic acid which can be extracted by hot water, *i.e.*, the glucosidal hydrocyanic acid (as distinct from the enzymic hydrocyanic acid which may be liberated continuously by the activity of enzymes in the seed), increases to a maximum during the first year of storage, and then decreases to almost *nil* in a good sample. On the other hand, the total hydrocyanic acid content increases after being stored for a year, and remains high for some years afterwards.

As the stockowner is primarily concerned with the total hydrocyanic acid potentialities of the beans, and as the enzymic and glucosidal moieties may vary considerably in relative amounts, he should always take care in studying an analysis to be sure how much hydrocyanic acid is glucosidal or removable by hot water, and how much is enzymic, or liable to be continuously produced. It is advisable, however, to avoid the beans altogether unless he has reasonable grounds for believing them to be safe.

The Linseed plant and Linseed cake

The flax plant (*Linum usitatissimum*) from which linseed is produced contains a cyanogenetic glucoside, linamarin, which is present in the leaves and the seed, the latter generally containing about 1.75 grains per pound.

This glucoside develops to a maximum very early in the plant's growth and disappears entirely later on, persisting only in the seed. It is thus dangerous for cattle to eat the immature or wilted plants.

Before the cake is fed to animals, it should be treated with boiling water in order to destroy the enzyme which liberates the hydrocyanic acid from the glucoside. The cake should *not* merely be soaked in cold water and left till feeding time as this will merely increase the production of hydrocyanic acid by increased enzyme activity. The hydrochloric acid of the gastric juice is liberated gradually in the stomach and the increasing acidity tends to slow down the production of the poison. Hydrocyanic acid is not a cumulative poison and it is both quickly absorbed into, and quickly eliminated from, the system; therefore, assuming an average glucoside content of 1.75 grains per pound, a bullock would need to eat 11 lb. to acquire a lethal dose. Hence, before the animal could eat this quantity (an amount far beyond that which would be fed) any effects of the gradual production of hydrocyanic acid would be eliminated. Hence, with a ration containing from 4 to 6 lb. of linseed cake for horses, bullocks, or fully grown cows there should be little danger.

In feeding calves, however, it is sometimes the practice to mix warm water with the cake before feeding. This facilitates the production of hydrocyanic acid, and, as the calves eat, weight for weight, considerably more cake than a fully grown animal, there may be considerable danger if the cake is prepared in this manner. A safe plan for calves and young stock is always to scald the cake well with boiling water before feeding, and arrange for it to be eaten slowly and not bolted.

Indian Mutter peas or Lathyrus Peas. (LATHYRUS SATIVUS)

There are several species of Lathyrus, but *khaseri dhal* or the Indian mutter pea, is the one which causes most trouble in this country, and is responsible for much sickness both among human beings and animals. It is essentially a cold weather crop and is generally grown on land which is too poor for any other type of crop and may be sown broadcast. Two types are common in India:—a large variety grown on the dry wheat lands, and a smaller variety on the damp wet lands of Bengal and south India. It is cultivated principally for fodder, but as it is easily grown and very cheap it is also extensively used by the poorer classes especially in time of famine or scarcity. It is also used as an adulterant of the *dhals* from which it is difficult to distinguish it.

When human beings take it in their diet for any length of

time, they are liable to suffer after an intervening period, which may amount to some months, from varying degrees of paralysis which are now considered to be caused by the presence in the peas of a volatile alkaloid or neurotoxin.

Stockman [1917—1919] has given an interesting account of the symptoms of this disease which is known as Lathyrism; according to this observer there is considerable pain, pricking and numbness, a diminution of tactile, heat and pain sensations, and cramp. The motor symptoms are the most conspicuous features however, the legs being particularly affected; the gait is jerky and irregular and only small steps can be taken. The degree of these symptoms affords some measure of the extent of the disease. In slight cases the patient has to walk with the aid of a stick, then with increasing severity, with two sticks, and so on until finally he is only able to move about on his hands and feet in a sitting posture. These symptoms are caused by degenerative changes in the motor fibres and their nerve cells in the nervous system, caused by the neurotoxin. It has now been established by Mellanby [1934] that the responsible neurotoxin can only be effective in the absence of vitamin A, a matter of cardinal importance to the stockman, as if his stock are fed with lathyrus peas plus plenty of green fodder or good legume hay, there is little danger of the disease developing. The fact that the alkaloid is volatile is also important, as it is unlikely to be present in well cooked grain or in thin pressed cake made at a high temperature, though in food cooked at lower temperatures such as *dhal* or curries, it may remain and produce the disease if such food is eaten for long periods. Furthermore, there is a great variation in the amount of the alkaloid present in different peas, and some varieties are more poisonous than others. The effects of the grain when fed to animals are similar to those produced in man. The horse is the most susceptible of all farm animals, and in cases of poisoning there is weakness of the lumbar muscles, and paralysis of the laryngeal nerves which causes the well-known roaring noise which is always a feature of horses suffering from lathyrism. Ultimately the animal finds great difficulty in breathing and death may supervene from dyspnoea.

Walker [1925] has stated, however, that he was unable to produce toxic effects with experimental feeding of *L. sativus* to horses or with the vetch seeds separated from *L. sativus*. Swine fattened on the meal of the peas are reported to lose the use of their limbs and grow fat lying on the ground, whereas sheep are said not to be affected, and poultry will not as a rule eat the meal or the peas. Cattle are also said not to suffer any apparent harm from eating Lathyrus [Watt, 1889]. These apparently conflicting results may be due to the fact that the animal body is able to store varying amounts of vitamin A, which may exert a modifying or inhibiting effect on the neurotoxin and limit its capacity to produce symptoms of the disease.

Mustard and rape seed and cakes

Mustard and rape seeds contain glucosides which may cause poisoning if eaten by stock due to the liberation of volatile oil by the action of an enzyme. The cakes which are made from the seeds may also contain the glucoside and produce chronic irritant poisoning and prostration if fed for any length of time. Mustard appears to be the more dangerous but if the cake is boiled or has boiling water mixed with the ground meal the enzymes are destroyed and the cakes rendered harmless.

Castor seeds. (RICINUS COMMUNIS)

Castor seeds are obtained from the plant known as *Ricinus communis* (N. O. Euphorbiaceae) and are liable to give rise to intense forms of poisoning. Castor oil is extracted from the seeds either by one of the pressure methods, or the extraction method. In the former case the temperature attained is insufficient to destroy the poison in the seed, while in the latter case, owing to the high temperature employed, the poison is rendered inactive. The cake obtained by pressure should never be used for animals, and owing to the chance of some poison still remaining in the extracted cake it is advisable not to use it either. Both cakes should be employed as manures only.

There are many cases on record in which husks of the castor seed have been found in other cakes owing to the presses not having been properly cleaned after pressing castor seeds, and occasionally cereal or other grains may also be accidentally contaminated.

The adulteration with castor seed of ordinary oil seed cakes used for feeding may be detected by heating a small sample of the cake with dilute acids and then alkalis and examining the residue, when any castor seed present will be conspicuous as small black flakes, which will show up against the rest of the cake which is light or brownish in colour.

The poison is present in the endosperm and embryo of the seed, and not in the seed-coat or husk, but the fact that the seed-coat is present should be sufficient warning that the endosperm of the crushed seed is likely to be present also.

Further confirmation of the presence of the poison may be obtained by various agglutination and precipitation tests.

The poison present in the castor seed is Ricin which belongs to a class of very virulent vegetable toxins, of which Crotin found in the croton seed is another example. It is also very similar to certain bacterial toxins and has some features in common with snake venom. It differs from the latter however in being capable of being absorbed directly through the healthy alimentary epithelial cells, while snake venom is destroyed by the digestive juices.

Ricin is extremely powerful in its action and very small quantities are sufficient to cause fatal results, three grains being reported to be the fatal dose for man. The symptoms of poisoning after eating the castor seed or cake are not immediate. They are characterised by loss of appetite, abdominal pain and constipation, shivering and coldness of the extremities but with high temperature, and death may follow in a few days.

Cotton seed

Cotton seed cakes are coming into increasing prominence in India as feeding stuffs for stock but they cannot be fed with impunity to all kinds on account of a toxic substance which is present, known as gossypol. This is often present in appreciable quantities both in the cakes and the raw seed but the actual amount varies considerably according to climatic and soil conditions.

When subjected to considerable heat this substance is converted into an oxidation product D. gossypol whose toxicity is considerably less, so that in the ordinary cooking processes during the production of oil from cotton seed, and in the expeller process, much of the original toxicity is lost. It is not entirely removed however, and ordinary cotton seed cakes and meals cannot be fed to swine which are particularly susceptible, but if the meal or cake is subjected to steam heat under pressure the toxin is completely destroyed and the cake or meal may be fed to all classes of stock. Animals which suffer from this form of poisoning become lame, are irregular in their movements, and suffer from ocular discharges which may often result in blindness and death. Different species of animals vary widely in their susceptibility, pigs being most affected, while cattle do not appear to suffer at all even after taking large amounts of cake or seed for long periods. In the feeding trials with experimental animals at Lyallpur, cotton seed cake and whole cotton seeds have been fed for many months to heifers, milch cows, and buffaloes with no ill effects.

Milch cows of 800 lb. weight and giving an average of 20 lb. of milk a day have been fed with raw cotton seed at rates varying from 2 to 6 lb. per head per day with no deleterious effects whatsoever either on the animals or the milk yields. Buffaloes of 1,200 lb. and giving 24 lb. of milk a day have also been fed as much as 8-10 lb. of cotton seed per day with no ill effects, and working bullocks up to 4 lb. per head per day with no ill effects.

It is inadvisable, however, to feed calves younger than four months more than small amounts, as they are more susceptible than adult cattle. Horses and sheep are more susceptible than cattle and may be affected by much smaller amounts than are the latter. Poultry appear not to be affected but even the best prepared cakes from the mills produce toxic effects with swine, and no cake should be fed to them unless it has been thoroughly steam heated.

Various explanations have been put forward as to the nature of the harmful effects which are produced by gossypol.

According to Withers and Canreeth [1915-1918] the harmful effects of gossypol are due to its withdrawing the iron from the haemoglobin of the blood. He attempted to remedy this by adding ferric sulphate to the cake fed to pigs and also a copper-iron mixture (described in Chapter XIV) with some beneficial effect, but the pigs still died under this treatment.

On the other hand M'Gowan and Crichton [1924] dispute this explanation and state that:—"Cotton seed meal injury or poisoning in pigs in all probability is not due to a toxic substance in the cotton seed meal. The same condition can be brought about by feeding pigs on other substances concerning which there is no suspicion of toxicity. In all probability cotton seed injury is brought about by feeding young pigs, which have already suffered as sucklings to a greater or lesser extent from iron deficiency, on an incomplete diet. The incompleteness of the diet would appear to be a general one and not having specific reference to the absence of vitamins. Iron in the shape of ferric oxide would appear to have a specially beneficial effect in preventing symptoms from arising."

Hence scientific opinion differs, and it may be that after all, the toxic effects are produced in part by a vitamin deficiency acting in conjunction with the toxic gossypol.

The point of practical importance to the stockman however, is that he should avoid cotton seed cakes or meals which have not been steam heated for pigs, and feed calves only small amounts of the ordinary market cakes and be sparing in the amount fed to horses.

An alleged harmful effect of a different type is that which is stated to be caused to cattle which are fed any considerable quantity of the fuzzy or linted American cotton seed cakes now abundant in India. Considerable prejudice has arisen because the lint is supposed to cause mechanical obstruction in the animal's intestinal tract and cause harmful results. The author and Dharmani [1937] have shown, however, as a result of extensive trials at Lyallpur that there is no foundation for this prejudice, and that the feeding of considerable amounts of the fuzzy seeds and cake over extended periods produces no ill effects whatever on this account.

The Corn Cockle. (AGROSTEMMA LYCHNIS GITHAGO)

The corn cockle is a very common weed in wheat fields in some countries but is not extensively found in India, and although the green plant is generally harmless to animals, the seeds are highly toxic. They may find their way into threshed wheat if the latter has been imperfectly cleaned, and they are also likely to be present in wheat screenings, and mixed with inferior grains.

The seeds of the corn cockle are rather large, and often difficult to separate on account of the numerous small spikelets projecting from the surface. They are dark in colour, and when ground with wheat for flour may impart a dull brown colour to the latter and an unpleasant odour to bread made from it. Such bread should not be regarded as fit for human consumption, as the toxic substance present is not destroyed by heat.

When the corn cockle is mixed with feeding stuffs it is likely to cause sickness or death to animals eating it, although different animals respond differently.

Long [1917], quoting the findings of Cornevin on experiments with pigs, calves, dogs and poultry gives the following amounts of corn cockle which will produce fatal results in these animals:

Calf	0.25 lb.	per 100 lb. live weight.
Pig	0.10 lb.	
Dog	0.90 lb.	
Fowl	0.25 lb.	

and quoting further from the studies made by Pesch states:—

"Under certain conditions corn cockle is injurious to domestic animals. The amount of the poisonous substance in the seed is variable, depending probably on the season and the soil. Animals become accustomed to it so that amounts of the seed, which at first cause sickness, later have no injurious effects. The susceptibility to the poison varies both with the species and the individual. Young animals are more readily affected than older ones. It is believed that rodents and sheep are not susceptible, and, as far as is known, grown cattle are only slightly affected by the poison. Calves, swine, horses and especially dogs are more or less susceptible. Concerning birds and fowls there is some doubt. Although animals are reported to become tolerant of the poison if the cockle is taken in only small regular doses, yet there appears to be a chronic form of poisoning due to this cause and termed Githagism, while there is an acute form of poisoning due to the ingestion of large quantities of the seed which may cause death in 24 hours or less"

The balance of evidence is therefore to the effect that the corn cockle is definitely dangerous if taken in more than very small quantities, and the best guard against it is to prevent animals grazing in fields where it is present and to see that their grain is properly cleaned. The poison responsible is the glucoside saponin. The vegetative portion of the cockle plant may contain small quantities of saponin but it occurs in large quantities—up to nearly 7 per cent in the seed.

The chronic form of poisoning is rarely found in animals except, perhaps, in the pig, and is characterised by diarrhoea, a general wasting and decline in strength, and nervous irritation.

Long (*loc. cit.*) describes the symptoms, as recorded by Cornevin, for horses, cattle and pigs as follows:—

"In the horse if a small quantity only is taken there is yawning, heavy colic, stamping and evacuation of rather soft faeces. If larger quantities are taken the symptoms which commence in about an hour are salivation, frequent yawnings and burning of the head, colic, pale mucus, hurried and weak pulse, a rise in temperature and accelerated respiration. Some time later there are muscular tremors, succeeded by pronounced rigidity, and the faeces are diarrhoeic and foetid. The animal lies down and getting up is painful, it falls into a kind of coma, stretches itself to the utmost and death takes place without convulsions."

"In cattle the symptoms observed one hour after eating are restlessness, salivation and grinding of the teeth. Excitement and colic are followed sometimes by coughing, this state lasting from five to eight hours. There is then a period of coma, characterised by permanent decubitis, repeated foetid diarrhoea, hurried and plaintive respiration, accelerated and gradually weakening pulse, a gradual loss of sensory or motor powers and a progressive decline in temperature. Death occurs in 24 hours."

"In the case of pigs the animal grunts, lies down and remains thus with its snout in the straw. There is vomiting, more or less violent colic and diarrhoea, the evacuation consisting of bad-smelling spumous faecal matter. At times there are colonic contractions. Young pigs are most susceptible".

Roots and Tubers

Potatoes. (Solanum tuberosum)

Potatoes which have become green and started to sprout develop a very toxic alkaloid known as solanine, not only in the haulm, the leaves and the flowers, but also in the peel. Such potatoes should always be thrown away and on no account be fed to stock of any kind.

The Oleanders. (NERIUM ODORUM)

Known as *karber* or *kaner* in Hindi, *alari* in Tamil and *ganneru* in Telugu, are cultivated plants with red or white flowers found in gardens throughout India. The roots from which two bitter toxic substances have been isolated are highly poisonous to animals, and a poisonous substance of a resinous nature is also known to be contained in the leaves, bark and flowers. These substances are powerful nerve-depressants and heart poisons, and are found in greater quantity in the wild than in the cultivated varieties.

Thevetia neriifolia

This is the Yellow or Bastard Oleander, and is commonly used in cases of criminal poisoning, particularly the seeds. It is known as *pila-kaner* in Hindi, *pachai-alari* in Tamil and *pacha-ganneru* in Telugu, and is a common bush or hedge plant in gardens throughout the plains of India. The milky juice is highly poisonous, and the seeds, in the kernel of which is a powerful acro-narcotic poison, are very bitter. If cattle eat them they die from violent purging and emesis. An active principle called vetine is considered to be the responsible agent.

Tapioca. (MANIHOT UTILISSIMA)

This is known as *maravalli* in Tamil and *marachini* in Malayalam. It is cultivated chiefly in Travancore in south India, and its roots when properly prepared are a common article of diet for human beings and are also sometimes given to cattle. The green tops are also used as cattle fodder. Fatal poisoning, however, may occur if the raw roots are eaten.

The roots of sweet varieties contain less than 0.005 per cent hydrocyanic acid and are harmless, but the roots of the bitter varieties may contain ten times this amount, and if eaten will produce hydrocyanic acid poisoning.

Exposure to high temperature will generally render the roots harmless.

MISCELLANEOUS

Fish Meal

Some account has been given in Chapter VI of the distinction which must be drawn between good white fish meal and the inferior variety of guano, which is only fit to be used as manure. Stockmen are continually advised to use fish meal for various classes of stock but should only buy guaranteed whitemeal which is made from the bones, tails and heads, with some white flesh, of white fish only, while the inferior type may be made from whole fish of varying types, including fish offal.

If the latter is fed to stock it invariably taints the produce, making it unpalatable and inferior in quality. Another feature is that the taint cannot be got rid of when once established, even after discontinuing the use of the food. Thus the flesh of the pig will remain tainted for months after being fed inferior fish meal, and the taint increases on cooking.

Similarly, poultry thus fed will produce eggs which are tainted; cows will produce milk with a fishy smell, and the butter produced from it will carry on the taint.

While therefore, inferior fish meal cannot be said to be actually poisonous, it impairs the quality of the produce of animals fed on it in every way, and should on no account figure in the rations for any class of stock.

Ergot. (*Claviceps purpurea*)

Ergot is a fungus which attacks rye, barley and other cereals and is visible as a mass of dark brown or blackish dust covering the grain on the flowering heads. This dark mass consists of the mature sclerotia of the ergot, each minute sclerotium consisting of a horn shaped body which may vary considerably in size from the others. Some are small and some are large and much contamination may occur without being easily noticed.

When grain infected by ergot is eaten by animals or man it causes a disease known as ergotism, the causative agent being an alkaloid known as ergotinine.

This toxin causes certain degenerative changes in the central nervous system which manifest themselves particularly by contraction of the blood vessels.

There are two types of the disease known as convulsive ergotism and gangrenous ergotism. The first form is the rarer of the two and is characterised by dullness or delirium due to contraction of the cerebral blood vessels.

In the second—the gangrenous form—the chief symptoms

are extreme pain in the various extremities of the body, such as the hoofs, eyes, ears and limbs, followed by a dry gangrenous condition due to failure of the blood supply. Owing to the universal contraction of the blood vessels in the intestinal tract, nausea, vomiting and diarrhoea may ensue with general wasting of the whole body.

The disease is generally found in the chronic form, and the degree of severity and the effects which follow, will naturally depend on the amount of ergot eaten. If only small quantities are taken nothing worse than mild digestive disturbances may occur. Abortion is said to be the result of ergot poisoning, although recovery generally occurs unless the disease develops into the gangrenous condition which is usually fatal.

Until comparatively recently little was known either of the cause of, or cure for the disease, and it was generally treated by giving the affected animal emetics and sudorifics. The researches of Mellanby (*loc.cit.*) however, have thrown a new light on the outstanding conditions associated with this disease. In the first place, all the symptoms of convulsive ergotism can be experimentally produced in dogs and rabbits by feeding them on a cereal diet deficient in vitamin A or carotene, without the presence of any ergot at all. The effect of such a diet causes degeneration in certain parts of the nervous system, which are considerably increased if ergot is introduced into the diet. Ergotism may therefore be said to be essentially a disease caused by a vitamin A deficient diet lacking in all the protective foods, and augmented by the presence of the toxin in rye infected by ergot.

If, therefore, animals are fed on a ration which is deficient in carotene or vitamin A, and at the same time have access to infected rye, they are likely to develop the disease in an acute form. The stockowner can protect his stock by carefully examining the pastures his stock feed on, or the fodder they obtain for the characteristic curved ergots, and, taking care that they do not get dirty or contaminated wheat screenings or grain, and to see that grain which may be contaminated is properly cleaned. It is possible that ergotism may be present in cattle without its being suspected, but any unusual blanching of the ears or extremities, or unusual nervous symptoms should give the stockowner a clue as to what is amiss. Once the disease is fully established and nervous degeneration set in, the disease is almost incurable, but if taken in time it can be cured in some months by proper feeding.

Decomposed and mouldy foods

Feeding stuffs are liable to undergo damage and deterioration in certain circumstances, the conditions most favourable being heat and humidity, although damage from various other causes may occur in dry cold weather.

If a feeding stuff is kept in a condition of refrigeration no decomposition can result because the low temperature inhibits the agencies responsible for it, but at ordinary or high temperatures with the assistance of moisture, these agencies come into operation. Firstly, all feeding stuffs are, to some degree, alive and contain what are known as intracellular enzymes or ferments which, according to the temperature, effect some degree of breakdown of the cellular tissue; for example, starches may be partially hydrolysed to sugar, proteins may be slightly digested, and so forth. Secondly, all feeding stuffs are, by the very nature of circumstances, infected to some degree by the spores of fungi, and when circumstances are favourable these spores develop into moulds. Two of the most common moulds found in feeding stuffs are the well known *Aspergillus niger* and *Penicillium*; others are *Botrytis* and *Oidium*, and mouldy cheese or bread are typical examples of the effect they produce. The third agency or group of agencies is constituted by the numerous and ubiquitous bacteria. Both moulds and bacteria belong to the lowest forms of life, and they derive the food they need from the hosts on which they happen to be. In the process of obtaining this food they cause a breakdown of the fats, carbohydrates and proteins of the food, and decomposition products, which are responsible for the smell and nature of the changed food, result. These lowly organisms are not necessarily poisonous in themselves or injurious to animals, but the products they produce are, and any feeding stuff which has become badly infected should not be used.

Certain types of feeding stuffs are specially liable to damage from this cause when kept in damp conditions; oilseeds and cakes and other protein rich feeds are particularly dangerous, and Linton records cases of horses which had died as a result of eating beans which had become mouldy. Under such unfavourable conditions, poisonous products are derived from the proteins, and from the oil which becomes rancid.

The cereals are specially liable to damage from these agencies under damp conditions, and when fed to stock, particularly horses and sheep, may produce fatal results. Oats are perhaps more susceptible than other cereals, and horses have a greater record of sickness on this account than other stock.

Only the best oilcakes should be purchased, as cake made from damaged seed, even if it shows no outward signs of decomposition, may be extremely dangerous.

Some suggestions have been given in Chapter VII on precautions which should be taken to guard against damage to feeding stuffs, but the best one is never to feed anything which shows signs of mould or decay.

Insects are a contributory cause, as grain damaged by them is more liable to suffer than whole grain, due to the more ready ingress of fungal spores and bacteria.

Newly harvested grains

If cattle or horses are fed newly harvested grains after being accustomed to old grain, they may suffer from digestive disturbances. New grain should therefore always be stored for a time before being used, although oven drying may prove to be beneficial if necessity demands its immediate use.

Newly harvested oats are very likely to produce colic in horses if fed in excess or as a sudden change over from old grain.

FAMINE FODDERS

Certain parts of India experience periodic droughts and famine when very little fodder is available for cattle and other animals in the areas affected. The parts of the country most likely to be affected by drought and famine are well-known arid tracts such as the Hissar district of the Punjab, parts of Sind and the Rajputana States, which are not served to any extensive degree by canal irrigation. Similar circumstances arise in other parts of India, for example, in the United Provinces, Bombay and the Central Provinces, when rainfall is deficient or fails to a major degree. In such conditions of drought and fodder scarcity the problem which confronts the zamindar is frequently not one of feeding his animals with a maintenance, still less a production ration, but of keeping them alive at all. He is thus confronted with the necessity of obtaining any fodders not normally eaten there may be in the locality; not always an easy problem. The method of dealing with such periodic shortages of fodder appears to resolve itself into two lines of attack. Firstly, creating reserves of fodder in those areas which experience has shown are most likely to suffer, and secondly, earmarking the leaves and twigs of those species of trees, shrubs and plants which can be commandeered as a last resort.

Fodder Reserves

The grazing conditions over a large part of India, even in the best of times, are not such as will enable any very large stocks of fodder to be accumulated as reserves. There are tracts, however, in which this could conveniently be done, and it is bound up with the wider problem of curtailing the large numbers of non-descript animals which roam at large over the country side, and also with regional improvement of grazing conditions. These factors have been discussed in some detail by the author in a recent publication 'Indian Grazing Conditions and the Mineral Contents of Some Indian Fodders' [Lander, 1942]. It is suggested, therefore, as a first step towards providing supplies of fodder during periods of drought that these two problems must be tackled, and, along with them, reserve stocks of hay should be accumulated in good seasons. Hay can easily be kept for two years, and if a proper system were adopted there might at least be a moderate reserve of fodder always at hand. A second method which might be adopted where circumstances permit is the construction of silage reservoirs in famine areas. These could be built in the form of long silage pits, say, 3-4 feet deep, 6 feet wide and as long as necessary, and silage prepared of whatever surplus fodders may be available according to the description given in Chapter VII.

The writer has described [Lander and Dharmani, 1924] some digestibility trials on Hissar bullocks with fresh *shisham* leaves and pointed out that these could only be taken by the animals under trial in a limited quantity up to about 2 lb. per head per day, owing to the disturbing effect which they had on the digestive system when fed in large quantities. He [Lander and Dharmani, 1927] has also described the value of siloed *shisham* leaves as a feeding stuff to Sahiwal cows and bullocks, and shown that siloed *shisham* leaves alone do not constitute a maintenance ration, but that when *shisham* leaves are siloed with *bhusa,* or siloed alone, and then fed with *bhusa,* a considerably improved ration is obtained. It is well known that animals which have been habitually under-fed can subsist for a considerable period on a ration which would not be even a maintenance ration for better-fed animals. It appears therefore, that if *shisham* leaves are siloed alone, or still better, with a certain amount of *bhusa,* they might provide a valuable source of fodder in times of emergency for cattle, which are not ordinarily fed well balanced rations, and if sufficient of this material were available a reasonable period of famine might be tided over. However, even if reserves of the common fodders, either as hay or as silage, have been established, and considerable amounts of *shisham* leaves are available, either fresh or siloed, there would still be a demand for supplementary feeding stuffs.

There is a large number of trees and shrubs growing throughout the length and breadth of India, both in the plains and in the hills, which could be fed to animals for this purpose in time of famine.

Watt, in his 'Dictionary of the Economic Products of India', Vol. III 1890, gives a long list of hill and plains fodder trees and plants which are eaten by cattle and could be used as famine fodders. Laurie, Silviculturist of the Forest Research Institute, has also given a considerable list, in a short pamphlet entitled 'Fodder Trees in India', 1939, the leaves and shoots of which can be eaten by cattle in times of scarcity or other times. A list is also given in an article entitled 'Some common fodder yielding trees in the Madras Presidency', by S. N. Chandrashekhar Iyyer and T. Venkataramana Reddy, published in the Indian Forester, Vol. LXVIII, No. 8, p. 435, and No. 10, p. 536. Another list, containing more than 400 names of trees and plants which can be used as fodder, is given in a publication entitled 'Fodder plants in Orissa' by M. L. Sen Gupta, available from the Superintendent, Government Press, Cuttack, Orissa (1942).

The Monthly Bulletin of the Bangalore Dairy Cattle Society,

Vol. I, No. 10, October, 1942, also gives two lists comprising good and medium types of fodders, taken from the report of the Mysore Fodder and Grazing Committee. Space does not permit the reproduction of the details of the very large number of trees which are included in these lists, but some of the more common and useful ones are given below, with a few notes on those of particular importance, or more uniform distribution in India.

Good Fodder Species

Acacia arabica. Gum arabic tree. Vern. *babul, kikar*

This is a common plant found in the Punjab, Bihar, the western part of the peninsula and in Ceylon. It is cultivated throughout the greater part of India, except in the moist humid regions on the coast, and the extreme north-west beyond Jhelum. It is one of the commonest plants of the Deccan, and covers most parts of Surat and Gujrat, the upper Godavari regions, and is very abundant in the Shewan district of Sind. Although plentiful it is not indigenous to the Punjab or to Madras.

The gum pods with tender shoots and leaves are fed to cattle, sheep, goats and camels and are specially useful as a famine fodder. In ordinary seasons goats are largely fed upon the pods which may explain the rapidity with which the plant becomes diffused over the country, springing up self-sown on the banks of tanks, rubbish heaps and walls. It is remarkable that sheep and goats eat greedily a substance so rich in tannin.

Artocarpus integrifolia. Jack-fruit tree. Vern. *kanthal, kathal, panas*

This is a large tree cultivated throughout India and Burma, except in the north, and grows wild in the mountain forests of the western and eastern Ghats. It has dark coloured leaves 4 to 8 inches long, smooth and shining above and rough beneath, which are considered to be fattening for cattle who greedily eat the rind of the ripe fruit.

Bauhinia variegata. Vern. *kachnar*

A moderate sized, deciduous tree, found in the sub-Himalayan tracts from the Indus eastwards and throughout the forests of

India and Burma. It is common everywhere and grows up to 4,000 feet in altitude but prefers the low hills. It is also largely cultivated throughout the plains as an ornamental tree which is often completely covered with large purple and white flowers which appear in the beginning of the hot season. The leaves are broad, being usually about 4 to 6 inches in length. The branches are often lopped for fodder.

Bauhinia purpurea. Vern. *koliar, koilari, sona*

This is a moderate sized deciduous tree found in the sub-Himalayan tract from the Indus eastward, in central and south India and in Burma. The leaves are from 3 to 6 inches long, a little longer than broad, split to the middle into two halves, and are used as fodder for cattle.

Cordia myxa. Vern. *lasora, lasura, gordi*

A moderate sized deciduous tree met with in the Salt Range and the sub-Himalayan tract from the Chenab to Assam, ascending to 5,000 feet. It is also found in the Khasia Hills, in Bengal, Burma, Sind and in western, central and south India. The leaves which are 3 to 6 inches long by 2 to 4 inches broad are leathery and smooth above and below, except when young, and are used as a cattle fodder.

Dalbergia latifolia. Blackwood or rosewood of southern India. Vern. *shisham, sissu, shisao*

This is a deciduous tree which attains a large size in south India. It is also found in Oudh, eastern Bengal, Bihar, central India, the Deccan and the western Peninsula, and in the hills up to nearly 4,000 feet. It grows equally well in dry deciduous forests with teak, and in the moist evergreen *sholas,* and it is often associated with the bamboo. The leaves which are $\frac{1}{2}$ to 3 inches long are used as fodder.

Bassia latifolia. Butter or mahua tree. Vern. *mahua, mahwa*

This is a large deciduous tree indigenous to the forests of the Central Provinces, and extends from Kangra, Kumaon and Oudh through the Central Provinces and Chota Nagpur to the Western Ghats. In many parts of Bombay and Gujrat it forms scattered and isolated forests. It thrives on dry, stony ground, and sheds its leaves from February to April; the cream coloured flowers, clustering near the ends of the branches, appear in March and April and are soon followed by new leaf buds. The fruit is sometimes eaten by human beings, but the principal edible parts are the succulent ripe flowers which are eaten either raw or cooked. They afford both food and drink to a large number of people for the greater part of the year. One tree may yield from 6 to 8 maunds of flowers during the season. Cattle eat the leaves, flowers, and fruit, and the flowers are said to be fattening.

Eugenia jambolana. (SYZYGIUM JAMBOLANUM). Vern. *jaman, jammnu, phalinda*

A moderate sized tree, found wild or cultivated over the greater part of India from the Indus eastward, and in the extreme south of the Madrās Presidency. It is found up to 3,000 feet in the Punjab Himalayas, and up to 5,000 feet in the Kumaon. Its smooth, shining leaves, which are 3 to 6 inches long, are often eaten by buffaloes.

Ficus glomerata. Vern. *gular, lelka*

A large tree found in the Salt Range and in Rajputana, and along the sub-Himalayan tract as far as Bengal, also in central and south India and in Assam and Burma. The fruit is greedily eaten by cattle, and the leaves are collected as cattle and elephant fodder.

F. hispida. Vern. *gobla, daduri, kat gularia*

A moderate sized tree or shrub, common throughout the lower Himalayas from the Chenab eastward, ascending to 3,500 feet in Bengal and in central and south India and Burma. The leaves which are 4 to 12 inches long are lopped as fodder for cattle and elephants.

F. infectoria. Vern. *wan, jangli, pipal, pilkhan, palikh, khabar, pakhar*

A large, widely spreading fast growing tree with leaves 3 to 6 inches long, found in the Suliman and Salt Range, the outer Himalayas, the plains and hills of India, Bengal, Assam, Burma, central India and especially the western coast forests. It is a cultivated tree and is rarely met with wild. The leaves make good elephant and cattle fodder.

F. religiosa. Vern. *pipal, asvattha*

A large glabrous tree with leathery shining broad based long pointed leaves, about 3 to 4 inches long and a little less in breadth, found wild in the lower Himalayan forests, in Bengal and in central India. It is extensively cultivated in most provinces of India. The leaves and branches make a good elephant and buffalo fodder and are extensively lopped for this purpose.

Gmelina arborea. Vern. *kumbbar, gumbhar, kumar, gambri, siwan*

A large deciduous tree, met with in the lower Himalayan tracts from the Chenab eastward, and also throughout India. The leaves which have a leathery yellow skin and are 4-8 inches long and 3-6 inches broad, are used as fodder, and are much browsed by deer and wild animals. Cattle also eat the fruit.

Melia azedarach. The Margosa tree. Vern. *neem, vepilai*

The neem tree is common to the moister parts of India, and is widely cultivated in those parts as a shade tree. The leaves are bitter and are generally collected for goats, but cattle are not fond of them except in times of scarcity.

Morus alba. White mulbery. Vern. *tut, tul, chimi, chun*

A deciduous, monoecious tree, cultivated in the Punjab, the north-west Himalayas, and in western Tibet. The leaves which are from 2 to 4 inches long, are largely used for feeding silkworms, but are also considered to be an excellent cattle fodder.

M. indica. Vern. *tut, tutri*

This is a moderate sized, deciduous tree or shrub, found in the temperate and sub-tropical Himalayas from Kashmir to Sikkim, and ascends to 7,000 feet. It is met with wild, and is cultivated in Bengal, Assam and Burma. In north India the tree is leafless during the winter, the new leaves appearing about the middle of February, or even later in March or April. It flowers in March and April and the fruit ripens in May and June, or later at high elevations. The fruit is inferior to that of *M. alba* and is used only in times of scarcity. The leaves are used for feeding silkworms and are also useful as fodder.

M. serrata. Vern. *himu, nimu, shahtut, karun, kart-tul*

A large deciduous tree found in the north-west Himalayas between 2,500 and 9,000 feet. The twigs and leaves are extensively lopped for cattle fodder.

Quercus dilatata. Vern. *moru, tilanga, brain, banni*

This is a gregarious evergreen tree, with large smooth leaves, and sometimes attains a height of 80 feet. It is met with in the north-west Himalayas between 4,500 and 9,000 feet. The leaves and shoots are extensively lopped as fodder for sheep and goats. Unlopped forests are rare.

Q. incana. The grey oak. Vern. *banj, ban* or *bán*

A large gregarious, evergreen tree, found in the temperate Himalayas from the Indus to Nepal, generally between altitudes of 3,000 to 8,000 feet. It can also be grown on the Punjab plains, and is found in the Shan States of Upper Burma. In spring it becomes of a purplish colour, owing to the appearance of a fresh burst of soft tomentose leaves, which when young are pink and woolly, and when mature, dark green and smooth above and white or grey below. The acorns are greedily eaten by bears, which may account to some extent for the sparse natural reproduction of the tree in spite of its profuse seeding. The leaves are extensively lopped for cattle fodder in the hills.

Sesbania aegyptiaca. Vern. *jayanti, jait, jainter, sewri, shevari*

A soft short-lived wooded shrub found throughout India from the Himalayas, where it ascends to an altitude of 4,000 feet, to Ceylon and Siam. The leaves, which are about 3-6 inches long, and the young branches are lopped for fodder, which are readily eaten by cattle.

Terminalia tomentosa. Vern. *asumi, sein*

A large deciduous tree, which may attain a height of 80-100 feet. It is common throughout the moister regions of India. It does not generally lose its leaves until February or March, or even as late as April, and is one of the latest trees in dry forests to come out in fresh leaf. The leaves which are 4 to 8 inches long are smooth above and woolly beneath, and are generally lopped for cattle fodder.

Zizyphus jujuba. Indian jujube. Vern. *ber, baer, beri*

A small tree, both wild and extensively cultivated throughout India. The leaves which are 1 to 4 inches long are nearly round, and are dark green and smooth above and grey and velvety beneath, and are a good fodder for cattle and goats.

Z. nummularia. Vern. *jhar, beri*

A prickly shrub found in the Punjab up to 3,000 feet (though less common in the southern Punjab), in the N.W.F.P., Sind and Baluchistan, also in Gujrat and the western Peninsula. The fruit is small, round and sweet and is eaten by the poorer people especially in times of scarcity. The leaves are ½ to 1 inch long, dark green and velvety above, pale and densely woolly beneath, and form a most valuable fodder for camels, goats, buffaloes and cows. They are highly esteemed in the sandy districts of Sind, the Punjab and Baluchistan, and are usually stored for winter use by allowing the cut branches to dry, beating the leaves off and gathering them into heaps. They may be given to cattle alone, or with some form of chaff, straw or *bhusa*, and are supposed to be heating and to promote the secretion of milk.

Medium Fodder Species

Acacia catechu. Vern. *khair, khair-babul, katha*

This is a moderate sized, deciduous tree, common in most parts of India and Burma, extending to the sub-Himalayan tract westward to the Indus and eastward to Sikkim. It is found up to 3,000 feet and occasionally as high as 4,000 feet. Cattle eat the leaves and the lower and smaller branches.

Albizzia lebbek. The Siris tree. Vern. *siris, siras, lasrin*

This is a large, deciduous, spreading tree which grows wild or cultivated in most parts of India and in the mixed forests in the sub-Himalayan tract from the Indus eastward up to 5,000 feet, in Bengal, Burma, central and south India. The leaves and twigs are gathered as fodder for camels and other animals.

It is often cultivated for fodder in Mysore. The tree grows rapidly and flourishes on almost any soil, specially on canal embankments and roadsides, and provides both fodder and fuel.

Bombax malabaricum. Silk cotton tree. Vern. *semul, semal, sumbal*

This is a very large, deciduous tree, with horizontally spreading branches, and is met with throughout the forests of the hotter parts of India and Burma. It is abundant in eastern India and is found up to 4,000 feet in the hills. It is the largest and most characteristic tree of Rajputana, and is a striking object in the spring with its immense buttressed trunks and large red flowers, six inches in breadth, clustered on the leafless branches. The leaves which are 2 to 3 inches long are felted with star shaped hairs. These with the twigs are severely lopped for fodder in Bombay, the Punjab, the Central Provinces the United Provinces and in Assam.

Butea frondosa. Vern. *dhak, palas, chichra*

This is a moderate sized deciduous tree, found throughout India and Burma, and in the north-west Himalayas as far as the Jhelum. It is one of the most beautiful trees of the plains and lower hills of India and is popularly known as the 'Flame of the Forest.' Ordinarily it is a shrub, but when preserved grows to a considerable height with large handsome red flowers, which blossom in March and April. The trees are often mercilessly lopped and stripped of the leaves which are used as fodder for buffaloes and elephants.

Cassia fistula. Indian laburnum or purging cassia. Vern. *amaltas*

This is a moderate sized, deciduous tree of the sub-Himalayan tract common throughout India and Burma, and is found up to 3,000 feet in altitude. It is very common on the lower mountainous tracts skirting the Himalayas (from the Khasia Hills to Peshawar) and from Chota Nagpur and central India to Bombay. It occurs chiefly as a small spreading tree not more than 20 feet in height which is leafless in March, the long pendulous racemes of bright yellow flowers and fresh green leaves appearing together in April. The leaves are about one foot long, and with the twigs are lopped for cattle fodder in Oudh and Kumaon.

Dalbergia Sisso. The Sisso. Vern. *shisham, sisam, tali*

This is a large deciduous tree found in the sub-Himalayan tract from the Indus to Assam, up to 2,000 feet. It is found throughout the plains of India and also in Baluchistan and Afghanistan. It may be found in every district of India, largely as the result of efforts to extend its cultivation, though it is often self-sown. It thrives best on a light soil and requires a considerable amount of moisture. The old leaves turn reddish brown and begin to fall in December, but continue to be shed up to February, when the young foliage appears progressively until April. The leaves, which are rather small, are a good fodder. Cattle are very fond of the young shoots and leaves, and will browse freely on them if allowed to do so. Experiments carried out at Lyallpur have shown that the leaves can be made into a good palatable silage.

Erythrina indica. Indian coral tree, mochi wood. Vern. *pangra, pharad, mandara*

This is a medium sized, quickly growing tree, with a straight trunk, which is usually armed with prickles when young. It occurs throughout India from the lower Himalayas and also in Burma. It is often grown in gardens, and the leaves are used as a cattle fodder, specially in Madras, Bombay and Assam.

Garuga pinnata. Vern. *ghogar, kaikar, tum*

This tree attains a height of from 30 to 40 feet, and is found in the sub-Himalayan forests from the Jumna eastward, up to 3,000 feet. It is also found in central and southern India, Chittagong and Burma. It flowers from February to March and the fruit ripens in June and July. The shoots and leaves are collected for fodder, specially for elephants.

Prosopis spicigera. Vern. *jhand, jand, jandi*

A moderate sized, deciduous, thorny tree, found in the arid, dry zones of the central Punjab, Sind, Rajputana, Gujrat Bundhelkhand and the Deccan. It is easily raised from seed and coppices well. The pods which ripen before and during the rains contain, when scarcely ripe, a considerable quantity of a sweet farinaceous substance. It is used for human consump tion in the Punjab, Gujrat and the Deccan, in some localities by all classes, in others only by the poor, and in times of scarcity. The pods form a good food for camels, cattle and goats, and the loppings called lanji are said to be specially valued in the trans-Indus regions as a fodder for sheep and goats. The trees are carefully preserved for this purpose and are poilarded during the cold weather when grass and other fodders are scarce. In other localities of the Punjab the leaves are also highly valued as a fodder.

Satix tetrasperma. Vern. *bed leila, bed, bent*

This is a moderate sized, deciduous tree, found throughout India on river banks and in moist places and in the Himalayan valleys up to 6,000 feet. The leaves which are 3 to 6 inches long are four times longer than broad and narrowed at both ends and are lopped and fed to cattle.

Salvadora oleoides. Vern. *jhal, jal, pilu*

A large evergreen shrub or tree of the arid tracts of Sind, the Punjab and Rajputana, often forming the greater part of vegetation of the desert. It is found up to 3,000 feet in the trans-Indus hills and up to 2,400 feet in the Salt Range. The tree flowers in April and its fruit ripens at the beginning of the hot weather. Cattle are fond of the berries which are reputed to increase both the quantity and the sweetness of the milk.

Spondias mangifera. The hog plum. Vern. *amra, amara, ambodha* and *amara*

A small, deciduous tree, found wild or cultivated throughout India, from the Indus eastward, and southward to Ceylon. It is found up to 5,000 feet in the Himalayas, and the leaves which are 12 to 18 inches long are eaten by cattle.

Tamarix dioica. Vern. *jhau, farwan, harwan*

This is a gregarious shrub, found near rivers and on the sea coast throughout India from Sind to Burma and is often planted as an ornamental shrub. It grows freely in soil impregnated with salt and is easily reproduced from coppice shoots. The leaves which are minute and scale-like are smooth and greyish green with a broad white margin. They are sometimes used as fodder for cattle in the Punjab and the United Provinces.

Terminalia belerica. Belleric myrobalan. Vern. *bhaira, bahera*

This is a large deciduous tree, common in the plains and lower hills throughout India, with the exception of the arid tracts to the west, and extends to Ceylon and Malaya. The leaves are 3 to 6 inches long and crowded towards the ends of branches. In Kangra they are considered an excellent fodder for milch cows.

Reed Grasses

Munj. (Saccharum Munj Roxb)

This is a reed grass which is extensively found all over northern India and might be used in times of scarcity. Its content of protein is very low and work carried out at the Imperial Veterinary Research Institute, Izatnagar, by Dr. Kehar (private information) shows that the amount of digestible protein it contains is negligible. On the other hand it contains nearly 50 per cent fibre, and although it might be of some use in very early stages of growth it is not a maintenance ration.

Khans. (Saccharum spontaneum)

This is a common reed usually found on canal banks, and is used for thatching purposes. Like *munj,* it would be of use as a fodder only in the young and succulent condition, but it is inferior to *munj* in its protein content. It is not a maintenance ration, and would ordinarily be used for feeding purposes only as a last resort.

BIBLIOGRAPHY

1. Acharya, C. N. 1933. Ind. Jour. Agri. Sci. Vol. III, 851-863.
2. Charlton, J. 1928. Memoirs of the Dept. Agri. India, Chem. Ser. Vol. IX. p. 1-36.
3. Chopra, R. N. and Badhwar, R. L. 1940. Ind. Jour. of Agri. Sci. Vol. X. p. 1-44.
4. Dunstan. 1906. Bull. Imp. Inst. p. 222-224.
5. Ghosh, M. N. 1919. Agri. Jour. Ind. Vol. XIV, p. 107-115.
6. Lander, P. E. 1942. The Imperial Council of Agri. Research. Misc. Bul. No. 16.

7. Lander, P. E. and Dharmani, L. C. 1924. Mem. Dept. of Agri. India. Chem. Series, Vol. VII, No. 4, p. 77.

8. Lander, P. E. and Dharmani L. C. 1927. Mem. Dept. of Agri. India, Chem. Series, Vol. IX, No. 3. p. 63, and Pusa. Bull. No. 158.

9. Lander, P. E. and Dharmani, L. C. 1929. Mem. Dept. of Agri. in India, 10, 6. Ind. Jour. of Vet. Sci. and Animal Husbandry, 1935, 5, 6 343-349. Ind. Jour. of Vet, Sci. and Animal Husbaudry. VII. No. 1937.

10. Linton, R. G. 1927. Animal Nutrition and Veterinary Dietetics. p. 357.

11. Long H. C. 1917. Plants poisonous to livestock. p. 13. (London: His Majesty's Stationery office).

12. Mellanby, E. 1934. Nutrition and Disease.

13. M'Gowan, J. P. and Crichton, A. 1924. Cottonseed Meal Poisoning. Biochem. Jour. Vol. XVIII.

14. Sastry, M. R. S., Nayar, C. K. V., Ramaswamy, R. and Venkatachalam. K. 1942. Studies on Lupinus angustifolius. Ind. Jour. of Vet. Sci. and Animal Husbandry, Vol. XII, Part III, pp. 219-223.

15. Stockman, R. 1917-1919. Edin. Med. Jour.

16. Taylor, C. S. and Ghosh, M. N. 1916. Agri. Jour. Ind. Vol. XI, p. 377-383.

17. Walker, 1925, Lathyrus Poisoning in the Horse. Vt. Jour. LXXXI. No. 3, p. 134.

18. Watt. 1889. Dictionary of Economic products in India, Vol. 14, p. 594.

19. Withers and Canreeth, 1915. Jour. Agri. Res. 5, pp. 261-288 & 12. 1918, pp. 83-102.

CHAPTER IX

FEEDING OF CATTLE AND BUFFALOES

The preceding chapters have dealt with the constituents of, and different types of feeding stuffs used for farm animals, and the methods by which they are digested and made available for animals. A review has also been given of the general principles involved in feeding for maintenance and production, and of the chief standards by which a reasonable measure of exactitude may be attained in computing rations for particular purposes.

We are now in a position to consider the application of the data thus reviewed to different classes of livestock.

There are in India approximately 97.5 million cows and young stock, 62.9 million working bullocks and bulls, and 44.8 million buffaloes with young stock. One of the main difficulties facing the owners of livestock is to obtain sufficient feeding stuffs to meet what must be acknowledged to be, even for India, a very unsatisfactory average milk yield per cow.

There is a story told of a District Officer touring in the Madras Presidency who was camping near a certain village. He sent his luggage ahead and when he arrived in camp with his wife and children he was astonished to see 21 cows standing in a row. He asked the reason for this and was told that they had been mustered to supply milk for the family while in camp. 'But we do not want so much milk', he said. 'These animals will give you just enough,' was the reply—and it was right.

This represents an extreme picture of uneconomic production, but it is illustrative of the fact that the greater part of India's cattle does not get either the proper amount, or the right type of food.

Grazing conditions are bad over the greater part of the country for most of the year, and the general poverty of the people and the owners of cattle does not permit a standard of feeding such as is found in countries where the dairy industry flourishes. During lactations the owners of cattle certainly follow a certain feeding regime depending on the locality and time of year, but in between lactations, cows are generally turned loose to graze, and get little more food than they can find on totally inadequate grazing lands. Furthermore, the vast majority of cattle in India are of a nondescript type and are not high yielders. Nevertheless the yield of milk could undoubtedly be increased by better feeding.

On the other hand there is a considerable number of dairies where good breeds of Indian cattle are kept, and fed according to scientific rationing regimes, and it is not unusual to find lactation yields up to or exceeding 10,000 lb. (See Appendix 4.)

Some advance has been made in recent years in improving Indian cattle by the distribution of improved types of bull in villages, but little systematic attempt has yet been made to devise better rationing schemes.

It is only by a combination of improved breeds and better feeding that milk yields can be increased, neither of which can be successful without the other.

It is not difficult to say what a cow should be fed for a certain level of milk production, but it is a matter of considerable difficulty to arrange that she gets it.

Any attempt at improvement of the general milk situation in India must accept the two major premises of improved breeds and improved feeding, and there is a task of great magnitude awaiting the administration in enabling owners of cattle to feed up to a better standard, thus implementing the scientific facts concerning feeding which are already available. It is not to be expected that in a country the size of India, with her present poverty, startling results will be rapidly obtained, but as the health of the population could be much improved by a greater consumption of pure milk, it is not unreasonable to suggest that the problem should be treated as one of national concern.

FEEDING OF MILCH COWS

Some of the most important factors essential to efficient milk production are:—

1. To retain only high yielding cows of selected breeds which are more economical than low yielding ones.
2. To feed them throughout the year with the requisite quantities of well balanced rations containing all the essential nutrients.

3. **To ensure that each cow gets a ration which supplies both the total digestible nutrients needed, and the proper amount of protein according to the milk yield.**
4. **To ensure that the rations comprise a variety of palatable feeds, and are not confined to one or two concentrates or roughages.**
5. **To arrange as far as possible that some green fodders are fed throughout the year.**
6. **Not to feed rations that are constipating, but to balance those which tend to be so with some laxative feed, such as bran.**
7. **To ensure that animals do not calve too frequently but are given a reasonable interval to recoup their body strength.**
8. **To take particular care in feeding very high yielders for which it may be found difficult to give sufficient food to comply with the nutritive requirements for the milk yielded.**
9. **To keep the animals in proper surroundings which should be clean, and the stalls light and airy.**
10. **To ensure regularity in times of feeding and milking, and to provide an adequate supply of drinking water.**
11. **To avoid any circumstances, as far as possible, which upset the animal, such as harsh treatment, changes of routine and so forth. The cow is a creature of habit and responds to disturbed conditions by giving a lower milk yield.**

Breed of the cow

The milk producing capacity of a cow is essentially an inherent function of breed, and it may be taken as a general rule that for any particular breed, additional feeding beyond that indicated as necessary by the average yields for that breed will not influence the amount and composition of the milk to any appreciable degree.

Some breeds are naturally higher yielders than others, and cows which produce a large amount of milk and fat must be fed better than those yielding less. It will generally be found, however, that the increased revenue returns from the higher yields will more than offset the additional expenditure on food —and high yielders are consequently more economical to feed than low ones. As a rule, the higher the yields, the greater is the credit balance from the milk over the cost incurred for the additional feed.

Correspondingly, in herds of the same breed it will always

be found, assuming that the proper nutritive standards are being observed, that individual cows differ in the efficiency with which they convert their food into milk—some will always be lower yielders than others.

It has been found [Morrison, 1936] that high producing cows require as much food for maintenance as low producers, and that there is little difference in the efficiency with which they digest their food. High producers and low producers do not differ very much in the amount of milk and fat they can produce from each 100 lb. of food they eat beyond the amounts they need merely for maintenance requirements. High producers have certain inherited capacities for milk production, and as a rule have good constitutions, and consequently secrete an abundance of milk.

Unless high producers are adequately fed they may fall to the level of low producers, hence it is just as important to give the former adequate rations as it is not to feed the latter beyond what is actually needed for their yields. A well fed dairy cow producing one pound of butter fat per day will require about 48 per cent of the food she eats for maintenance. If the ration of such a cow be cut down to about two-thirds, she will still need as much food for maintenance, but will now have for milk production, probably only about 30 per cent of the food.

It is false economy to feed high yielders less food than is necessary for the milk yields as given in the recognised standards. Furthermore, the farmer's overhead charges will remain the same whether his cows are giving 10, 20 or 30 lb. of milk per day.

It is admitted that dairy herds can be built up by careful breeding and feeding, yet once an individual cow's inherent milk capacity is relatively fixed, this cannot be appreciably increased by further feeding. Any additional food given in an endeavour to increase the milk yield will be wasted, as all the cow can do is to store the excess nutrients in the form of body fat and not turn it into milk.

This is important for the dairyman, and he should try as far as possible to differentiate between his high and low yielders and, while not giving every cow in his herd a separately calculated ration, he should at least attempt to sub-divide them into groups so as to ensure economy in feeding.

He should also try as far as possible to replace his low yielders by high ones, because a high yielder, in spite of the increased cost of its food, is a better economic unit than a low one.

Total digestible nutrients needed

The various feeding standards advocated in the past and in vogue to-day have varied to some degree in the amounts of total digestible nutrients which they advocate, in addition to the maintenance rations, for each pound of milk of varying fat percentage produced. For example, it has been reported by Meigs [1925] on experiments carried out by the United States Department of Agriculture, Bureau of Dairy Industry, that the Armsby standards fell short, by about 15%, of the then prevailing standards followed in the Eastern States. A feeding standard, however, is only intended to be an approximate guide for the stock owner who obviously cannot carry out detailed experiments for himself, and where expense is a secondary consideration and the maximum possible production is aimed at, greater quantities of nutrients may be supplied—always with the proviso that every animal has a maximum production capacity, beyond which additional feeding will not produce further yields.

In the case of heifers and cows during pregnancy, an additional amount of food will be needed over and above that given in the standards to provide for the growing foetus, in addition to that needed for maintenance or milk production. It has been concluded by certain American investigators [Morrison, 1936] that for the first two-thirds of the gestation period additional food need not be fed above that required for maintenance and milk production, but that during the last part of gestation the food supplied should contain twice as great a quantity of total digestible nutrients, including an adequate proportion of digestible protein, as is contained in the body of the newly born calf. (See also page 310).

Protein Requirements

The protein needed for milk production is known to vary according to the amount of fat the milk contains. Milk that is rich in fat is usually higher in protein content, and also in lactose, than a milk of low fat content. Consequently a higher proportion of total digestible nutrients and digestible protein is needed, over and above the usual standards, to produce milk of a high fat content than for one containing less fat.

Sanmann and Wright [see reference] have recorded that there is an increase of about 0.42 per cent in the protein content of milk for each one per cent increase in fat, and that this increase is fairly regular. Allowances have therefore been made in the feeding standards for the digestible protein and total digestible nutrients required for the production of milk of varying fat content.

Haecker [see reference] recommends that in addition to the requirements for maintenance for milk of different fat percentages, additional amounts of digestible protein should be given at a rate of 1.7 times the estimated protein contained in the milk. Taking these considerations into account Morrison [1936] recommends the following standards of feeding for milk of varying fat content:—

TABLE I

Showing the additional nutrients which should be added to the maintenance rations of cows for milk of varying fat content.
(Per pound of milk)

Fat percentage	DIGESTIBLE PROTEIN		TOTAL DIGESTIBLE NUTRIENTS	
	Minimum allowance advised.	Recommended for good cows under usual conditions.	Minimum allowance allowed.	Recommended for good cows under usual conditions.
	lb.	lb.	lb.	lb.
For 2.5% fat in milk.	0.034	0.040	0.238	0.251
For 3% —do—	0.036	0.043	0.261	0.276
For 3.5% —do—	0.038	0.046	0.284	0.300
For 4% —do—	0.041	0.049	0.307	0.324
For 4.5% —do—	0.044	0.052	0.330	0.349
For 5% —do—	0.046	0.056	0.353	0.373
For 5.5% —do—	0.049	0.059	0.376	0.397
For 6% —do—	0.052	0.062	0.399	0.422
For 6.5% —do—	0.054	0.065	0.422	0.446
For 7% —do—	0.057	0.068	0.445	0.470

Protein is of special importance as it is the protein supplements of the ration rather than the cereal concentrates that are usually most expensive, and much work has been done in America to determine the minimum protein requirements. According to Perkins and Monroe [see reference] cows in good production should receive, in addition to the protein for maintenance, about 1.25 times as much digestible protein as is contained in the milk produced. Cows of higher productive capacity may be fed a somewhat higher ratio than this, say, up to 1.6 times, but when the protein given is increased beyond this level, the production of milk is not further increased to any extent. In some of the trials recorded by these workers, cows of very high productive capacity yielded extremely large amounts of milk when fed even less than 1.25 times as much digestible protein as was contained in their milk. Working with Holstein cows, Perkins [see reference] found that with a ration having the comparatively wide nutritive ratio of 1 : 11, over 11,000 lb. of milk a year containing 4 per cent fat were produced. The same cows, nevertheless, produced more milk when the protein was increased, thus confirming what was said above, that it is not economical to restrict the ration to a level below the producing capacity of the cow. Other experiments conducted by these workers using nutritive ratios as narrow as 1:2 have shown that an excess of protein fed over milk capacity requirements has tended to delay breeding. Experimental considerations such as these have been taken into account in working out the Morrison feeding standards for dairy cows given in Appendix 2. It will be seen that allowance has been made over a certain range to provide both for moderate, and more liberal milkers, and while the lower figures may be taken as satisfactory for moderate yielders it is generally considered advisable to take the higher one for cows producing one pound or more butter fat per day.

Figures obtained from trials on Sahiwal cows at Lyallpur yielding on an average 20 lb. of milk a day containing from 4.5%—5% fat, were given rations which conformed to a standard slightly below Morrison's lower figure. The data obtained at Lyallpur were as follows (Morrison's figures are also given for comparison) :—

	Fat percentage in the milk.	Digestible protein consumed per lb. of milk produced. lb.	Digestible total nutrients consumed per lb. of milk produced. lb.
Lyallpur.	5%	0·048	0·357
orrison.	5%	0·046-0·056	0·353-0·373

Proportion of concentrate to roughage

The proportion of protein supplied by the concentrates and roughages respectively is an important matter in feeding dairy cows. The problem of computing rations will be gone into in more detail later in this Chapter (see also Chapter V), but it may be stated here that if the roughage fed is berseem, lucerne or other rich leguminous hay, the amount of protein supplied in the roughage will be so large that the balance required may be made up from the less rich protein type of concentrate, such as maize or a cereal mixture. For very high producing cows, even when fed a rich legume roughage, it is nevertheless advisable to add a certain amount of bran to guard against a possible calcium and phosphorus deficiency in the roughage, even if the protein is otherwise sufficient. Maize and cereals are deficient in these minerals. If the roughage fed has a considerably lower protein content than an average good legume hay such as lucerne, then a high producing cow will need a protein supplement to the ration. For example, oat hay grown at Lyallpur was found to have from 1.9—2.2 per cent digestible protein, berseem about 2.2 per cent., and maize grain about 5 per cent.; therefore 25 lb. of average oat hay and 10 lb. of maize would supply only about 1.5 lb. of digestible protein, which is insufficient for a high producing cow. Hence with such a roughage a small quantity of protein supplement is necessary.

In deciding on the relative proportions of roughage and concentrate therefore, the relative proportions of protein they yield must be taken into account, and the mineral needs also must be adequately provided for. A good general guide is that the concentrate must be richer in protein when the roughage is a non-legume such as ordinary hay or silage, or a mixture of these with some legume, in which case a protein rich supplement should be added. With rich protein roughages, the concentrates may consist mainly of cereals or others less rich in protein.

Protein of high and low quality

It may be assumed, that if the ration contains a good legume hay or silage, the quality of the protein—quite apart from considerations of quantity—will be satisfactory. Certain concentrates, however, such as maize, maize meals, dried grains from breweries and distilleries, contain poor quality proteins which are deficient in certain of the essential amino acids. Such concentrates should therefore be fed with others such as linseed meal, gram and other protein rich grains and cakes, all of which supply proteins of high quality. It is not sufficient that a certain quantity of protein be fed. The protein must also supply the right type of amino acids which are contained in all the rich concentrates mentioned.

Feeding on roughage only

Feeding milch cows on roughage only with little or no concentrates is all too common in India, and in the periods between lactations the vast majority of milch cows get little or no concentrates, but are turned loose on the fields to find whatever they can pick up. Even during lactation periods the majority subsist mainly on roughages with a handful of mixed grain thrown in. Such animals cannot be considered to come within the purview of any feeding standards mentioned in this book. They stand so to speak outside the pale of scientific nutrition. Most of them are of non-descript type, their milk yields are notoriously low, and it would not be a paying proposition to feed the majority of them in any scientifically conducted dairy herd. The writer has described in considerable detail in a recent publication [Lander, 1942] the grazing conditions under which the majority of India's approximately 215 million cattle subsist, but it is not possible to state with any degree of accuracy what proportion of them is uneconomic. They no doubt pay for their keep because they are sent out to graze on whatever grazing grounds are available and so their keep costs nothing, but the fact that some of them produce milk is no guide as to whether cows can be made to produce milk economically on roughage alone. Some farmers keep their cattle largely on green fodder supplemented by random grazing, as only a very small proportion can get even grazing which is worth the name.

The question, nevertheless, arises as to what extent milch cows can subsist on roughage alone. A prime essential of good dairy husbandry is that an abundance of good quality roughage should always be available, and in certain localities and under certain conditions it may be necessary to decide whether any concentrates need be fed, and if so, which. The answer will depend in large degree on the milk yielding capacity of the cows, and on the relative costs of the roughage and concentrates available, and the price which can be obtained for the milk.

In some of the cattle districts of the middle west of America, cattle subsist largely on lucerne grazing and roughage, which is cheap in comparison with concentrates, although a certain amount of concentrates is fed. Under some conditions it may pay to feed no concentrates at all.

When cows are fed a liberal amount of hay during the winter months where grazing is scanty, and get plenty of good irrigated pasture during the summer, it may be possible to obtain good milk production with little or no concentrates. Such ideal conditions, however, perhaps exist only on Government Farms, Remount Depots and large estates where grazing and irrigation facilities are ample.

The writer has seen in South Africa first class herds of pedigree Ayrshires, Jerseys and grade Friesians of high milk producing capacity which subsist throughout the year entirely on summer grazing on *kikuya* grass and *Paspalumn dilactatum*, and winter pastures such as rye grass, cocksfoot and tall fescue. Cows of the larger breeds usually do better on such rations than smaller animals owing to their greater capacity to consume roughages. It must be admitted, however, that such pastures received special treatment designed for the purpose of developing the dairy industry in a 35 inch rainfall belt on nothing but pasturage. Such ideal conditions are rare in India, where the yield of milk will usually be decidedly less on nothing but roughage, even of good quality, than when at least a moderate quantity of concentrate is fed. Roughage alone, even good lucerne hay, is usually too bulky to provide the necessary nett energy required.

Where ideal grazing and plenty of rich roughage are not available, cows should be given enough concentrates to ensure the maintenance of a high level of milk production. Even when the price of concentrates is so high that it is advisable to reduce the concentrate allowance below normal, cows should always get all the good roughage that they will consume, otherwise the yield of milk may be so reduced that the balance of income over the cost of the rations will be seriously lowered.

Fat requirements

It is extremely difficult to state with any degree of accuracy, what is the optimum amount of fat which dairy rations should contain. With modern methods of extracting oil from oilseeds in the production of oilseed cakes the tendency is for such concentrates to be somewhat low in fat content. For example, 4F cotton seed contains 20 per cent fat, but the corresponding cake made by expression contains only about 8 per cent while cakes manufactured by the solvent process may contain but little or none.

The animal body is able to convert materials other than the fat of the food into body fat, *e.g.,* the proteins and carbohydrates, and it was formerly thought that, provided the ration contained sufficient total digestible nutrients the percentage of actual fat contained in it was immaterial. It was generally conceded, however, that the animal could more easily form fat for the milk from the fat in the food than from the other ingredients. The earlier experiments which were conducted to throw light on this subject were made with so called low fat rations, which still contained a relatively high percentage of fat, and the results obtained were inconclusive, and did not enable them to be compared in this respect with high fat rations.

Experiments carried out at the Cornell Experimental Station [Maynard, 1929], U.S.A., on cows receiving a roughage of maize silage, mixed clover and timothy grass hays, and fed a concentrate mixture containing less than 4 per cent fat, showed that the milk yield and fat content of the milk were considerably decreased, but that on a ration containing 4 per cent of fat, the fat of the ration supplied about 70 per cent of the fat found in the milk, and that 4 per cent of fat was adequate to maintain the milk yield and fat content.

With a corresponding roughage to which was added a concentrate containing up to 7 per cent of fat, the milk yield and fat content were not found to be appreciably increased, although the fat content of the milk on the higher fat ration was slightly higher. It is generally assumed therefore that a fat content of from 3 to 4 per cent in the ration is satisfactory, and no beneficial results will accrue by increasing it.

Mineral requirements

Minerals have varied and vital functions to perform in the animal economy, and unless the rations supply these in adequate quantities and in the right proportions, untoward results are likely to ensue. With heavy milking cows the possibility of there being a qualitative or quantitative deficiency increases as the milk yield increases. Under natural conditions a cow would only be likely to give sufficient milk for one or two calves, and would get a longer period for recuperation between lactations than is the case in modern dairies. Many of the pastures on which cows graze are deficient in minerals, and so are many of the hays and cereal concentrates. It is therefore often a matter of serious concern as to how to make sure that a high milking cow will get all the minerals she needs in her ration. The average mineral composition of cows' milk is shown below in percentages:—

Calcium	Phosphorus.	Magnesium.	Potassium	Sodium.	Chlorine.	Sulphur.	Iron.
%	%	%	%	%	%	%	%
0·118	0·093	0·012	0·143	0·051	0·106	0·034	0·0002

So that a cow giving 30 lb. of milk per day will secrete the following amounts of minerals daily in the milk, and as a rule she will require about three times this quantity, particularly calcium and phosphorus, in the ration to provide for this outlay:—

Minerals in grammes in 30 lb. of milk.

Calcium.	Phosphorus.	Magnesium.	Potassium.	Sodium.	Chlorine.	Sulphur.	Iron.
16·06	12·66	1·63	19·46	6·94	14·42	4·63	1·39

It is doubtful whether many of the rations fed to high yielding cows are able to provide adequately for this outlay for any length of time.

The evil effects which result from long continued mineral deficiency in rations are not always clearly defined and obvious, and may be characterised by an unthrifty appearance and general malaise, a falling off in the milk yield, and the production of weak and sickly calves. Many cases of abortion are now regarded to have a direct or indirect connection with mineral deficiency.

The minerals which are most likely to be lacking in cows' rations are calcium and phosphorus, and possibly iodine and iron. It would be a matter of great difficulty, on account of the wide variations in the mineral content of fodders, and also on account of the fact that it is not physiologically possible to say with any precision how much minerals an animal needs, to work out feeding standards for minerals on a par with those for digestible protein and total digestible nutrients. A reasonable guide is to allow heifers up to the time of calving at least 10 grammes of phosphorus (P_2O_5) per head per day, and at least twice that amount of calcium (CaO). For milk production, about 0.5 grammes of phosphorus is needed, in addition, for each pound of milk produced, and about 1 gramme of calcium.

The best guard against mineral deficiency is to feed a mixed ration, to avoid an undue preponderance of foods deficient in calcium, such as maize, wheat offals, etc., and to include a good legume roughage and mixed silage. The addition of a small quantity of fish meal and steamed bone meal will furnish a useful mineral adjunct, besides adding to the protein rich part of the ration. For very heavy milkers, one or two ounces of steamed bone meal *per diem* may be given.

The various proprietary mineral mixtures which are so common should be used with caution, as they are not necessarily prepared to deal with a specific mineral deficiency for which a special adjunct is needed, and may possibly do more harm than good. Nevertheless, if this precaution is borne in mind, mineral foods may improve both the health of the animal and the milk yield, although such results are not always immediate.

Orr and co-workers have described trials on 12 pedigree Ayrshire heifers over a period of two lactations which illustrate the importance of a wise use of mineral adjuncts.

The winter ration they employed consisted of oat straw fed *ad lib.*, turnips (6 cows), and silage (6 cows), distillers grains, and a compound cake made up of equal parts of rice meal, palm kernel meal, decorticated ground nut meal, wheat offals and molasses. Half of the cows were fed a mineral adjunct consisting of 56 lb. calcium carbonate, 28 lb. common salt, 6 lb. iron oxide and 2 ozs. potassium oxide, per ton of cake, the consumption of which was adjusted to the milk yield. Three of the cows on silage, and three of those on turnips were given the cake without the added minerals, and the others the cake with the minerals. During summer a less amount of cake was given. The results obtained are shown below:—

Average yield of milk.

	1st lactation.	2nd lactation.	Increase or decrease
	lb.	lb.	lb.
Silage & turnips ration	10,699	9,358	– 1,341
Silage & turnips ration plus minerals	10,029	10,229	+ 200

Showing weight of calves born.

	First calves	Second calves
	lb.	lb.
Non-mineral group	83.2	77.8
Mineral group	81.3	82.6

The general health of the mineral fed animals was good, though Orr states that with four of the six cows there was a marked delay in calving, and in one case the calf was born dead. Apart from this delay, however, all six cows in the mineral group remained in perfect health.

These workers conclude that, although definite conclusions could not be drawn, the results suggested that it is possible to improve the nutritive value of rations by adding minerals, and that the mineral ration showed a beneficial effect both on the health of the cows and on the milk yield.

On the other hand there was also evidence that the addition of a mineral mixture to a ration might have a detrimental effect.

Necessity for common salt

All herbivorous animals need a considerably larger quantity of common salt than their rations usually supply, and they should always be provided with a 'salt lick'. The best way to give this is in the form of lumps of rock salt placed in the mangers.

Both sodium and chlorine are required for a variety of purposes in the body, such as maintaining the acid-base equilibrium, and for the gastric juice. A considerable amount of salt is lost in the sweat and urine, and if a proper supply is not available in the rations, a marked degree of ill health is likely to appear sooner or later. A satisfactory daily allowance of salt for dairy cattle is about half an ounce for an animal of 1,000 lb. weight and an additional 1/3rd of an ounce for every 10 lb. of milk produced.

Calcium and phosphorus

These make up about three quarters of the mineral matter in the entire body of animals, and about 90 per cent of the skeletal structure, and they constitute about half the minerals present in milk. It is obvious, therefore, that growing animals and those in milk must be supplied with sufficient calcium and phosphorus in the rations. In some districts, such as Bihar, there is a deficiency of phosphorus in the soil (and, to some extent, calcium), and in many of the hilly tracts, which get a very heavy rainfall, a deficiency of both. In such areas stock are liable to suffer on account of deficiencies of minerals in the soil which are in turn reflected in the crops. Ordinarily there should be no lack of phosphorus or calcium in crops grown on normal soils, and dairy cattle which are fed good rations including protein rich concentrates and good hay, particularly legume hay, should not suffer on this account.

It has been suggested by Huffman [see reference] and his associates at Michigan, U.S.A., that dairy cows need about 0.35 ounce phosphorus daily per 1,000 lb. live weight for maintenance, and 0.026 ounce for each pound of milk produced, and that not less than 0.60 ounce should be fed during low production and prior to calving.

The daily ration of a 1,000 lb. cow producing 25 lb. of milk a day should therefore contain on a dry basis about 1.0 ounce of phosphorus, and *pro rata* for higher yields. It is possible from the tables of analyses in Appendix I, to compute approximately how much phosphorus any particular ration will furnish, and adjustments may be made accordingly. Even lucerne hay contains only about 0.22 per cent of phosphorus and most other hays much less.

The calcium requirements of cows are generally considered to be somewhat higher than for phosphorus, and there is less likely to be a lack of calcium, in most dairy rations, than of phosphorus. For high producing cows the rations should contain at least 0.2 per cent of calcium per day on a dry basis, and preferably much more, so that when legume hays are not available a calcium mineral supplement should be given.

Dry matter or bulk

Every ration must contain sufficient bulk or dry matter to enable the digestive organs to function to their best ability; on the other hand too bulky a ration should be avoided or cows may not be able to ingest sufficient nutrients to provide for the milk they should produce.

Kellner [1908], as a result of numerous feeding trials, arrived at the conclusion that the dry matter requirements for maintenance should amount to approximately 2.5 per cent of the body weight of the animal. Thus for an 800 lb. heifer, the upper limit would be about 20 lb. of dry matter. For average animals in India this figure might be reduced somewhat.

Dry matter for milk production

Kellner [1908] also ascertained, using his standard based on starch equivalents, that for every gallon of milk produced approximately 2.5 lb. of dry matter were required over and above that needed for maintenance. Therefore, in drawing up a ration for milk cows it is necessary in the first place to make a calculation to find out how much dry matter is needed to meet the needs for milk production. Thus an 800 lb. animal giving 24 lb. of milk per day will require approximately 6 lb. of dry matter per day for the milk, and in addition about 20 lb. of dry matter for maintenance. Therefore the total amount to be supplied should be 26 lb. according to Kellner's stipulations.

Indian cattle appear to be conditioned to a somewhat lower level of dry matter requirements (as well as digestible protein and total digestible nutrients) and it is unusual to find a greater total bulk of dry matter than 20 lb. being fed to an 800 lb. cow giving from 20 to 24 lb. of milk a day.

Naturally the amount of dry matter which cows will need or can take will vary within certain limits for different individuals.

Palatability and variety of rations

The palatability of a ration is perhaps of greater importance than the variety of ingredients it contains, and it is desirable that both the concentrates and roughages should be palatable and attractive, especially for high yielders.

If much roughage of an unpalatable nature is given to cows they may not be able to eat as much as they should, and additional concentrates would then be needed to balance the ration. Furthermore, it is maintained by some authorities that unpalatable feeding stuffs are not so easily digested, and adversely affect the digestibility of the whole ration. Tasteless and unattractive feeding stuffs such as soybeans or groundnut cake may be made more attractive by steaming, or by adding one or two ounces of salt, or molasses in small quantities, as an appetiser. Feeding stuffs which have become in any way tainted must be avoided, and not given with some condiment added to mask their flavour, as they are distasteful to cows and have a harmful effect on the milk.

If the normal feeding stuffs available are fed in reasonable variety there should be no difficulty in computing palatable rations, and once suitable mixtures have been decided on there is no advantage to be gained by changing them merely to give a change.

If a single roughage and concentrate, or, maybe a roughage only, is fed for any length of time this will not only be monotonous and possibly become unpalatable, but it may not supply the necessary variety of nutritive ingredients. As is well known, different proteins vary considerably in their biological values, and a variety of feeding stuffs is more likely to provide the requisite protein both as regards quality and quantity. Furthermore, the mineral balance is likely to be met better with a mixed ration than from a simpler one.

Times of feeding and milking

It is not easy to suggest a time table that will be suitable in all circumstances, but it is essential that feeding and milking regimes once started should be regular. Cows are sensitive animals and often react to an extraordinary degree to alterations in routine and disturbing conditions.

Linton [1927] quotes the following regime as followed in England for a herd of 60 Ayrshire and cross-bred cows:—

4 a.m.	The cows are milked.
6 a.m.	A warm wash is given to all the cows.
6.30 a.m.	The allowance of roots is distributed, usually about 30 lb. per head.
6.45 a.m.	A bundle of straw (14 lb.) is put down between each pair of cows.
10.11 a.m.	The cows are thoroughly groomed.
1 p.m.	The byre is cleaned out.
1.30 p.m.	The feed of dry concentrates is given to those cows producing more than one gallon of milk daily.
2.45-4 p.m.	The cows are milked.
4 p.m.	Ensilage is given to all cows.
4.15 p.m.	Hay allowance is distributed, bedding tidied, dung channels cleaned out and the cows are left for the night, except that at
8 p.m.	A visit of inspection is made by the foreman to see that everything is in order.

The practice followed with the author's experimental herd at Lyallpur is to feed part of the daily roughage allowance in the manger at 3 a.m. or about half an hour before the first milking. Half the day's concentrate ration is then fed at the time of milking. If, however, silage or root crops constitute part of the ration these should be fed after milking, as they are liable to taint the milk. For moderate yielding cows this procedure is repeated in the afternoon at 3.30 p.m. although the roughage may be given after milking so that the cows have something to do during the evening and night. Some dairymen prefer to feed most of the hay or other roughage after the last milking. In the case of cows yielding 25 lb. of milk per day or over it is customary to feed and milk again about noon. One-third of the concentrates may be given then so that the cows are enabled to get sufficient concentrates during the day to meet requirements for the milk.

The exact routine to be followed will however depend on circumstances.

FEEDING STUFFS SUITABLE FOR DAIRY CATTLE

The various feeding stuffs usually fed to different classes of stock have been dealt with in some detail in Chapter VI. Those most commonly available and fed to dairy cattle in India are as follows:—

FEEDING STUFFS FOR DAIRY CATTLE

Dry roughages.	*Green fodders.*	*Concentrates.*
Wheat *bhusa.*	Maize.	Gram.
Oat hay.	Lucerne.	Bran.
Rice straw.	Berseem.	Arhar.
Maize stalks.	Shaftal	Cotton seed.
Juar stalks.	Senji.	Cotton seed cake.
Legume *bhusa.*	Oats.	Sarson cake.
Grass hay.	Juar (cholam).	Toria cake.
Bajra stalks.	Bajra (cumbu).	Linseed cake.
Berseem hay.	Guara.	Til cake.
	Guinea grass.	Ground nut cake.
	Elephant grass.	
	Napier grass.	
	Silage of maize, oats, or mixture of Juar and Guara or Maize and Guara.	

In order to provide an efficient and economical ration, account should be taken of the various advantages and disadvantages which each type of feeding stuff offers. The detailed descriptions given in Chapter VI may be studied in conjunction with the analytical and other data given in Appendices 1 & 2, and used in conformity with the requirements for balanced maintenance and production rations shortly to be described. There is no one particular ration which can be said to be the best ration for dairy cows, but the usual feeding stuffs available should be sufficient from which to make a selection for all ordinary circumstances.

Most of the grains are of equal value to dairy cattle, though oats, wheat and barley are usually too expensive. Good legume hays such as berseem and lucerne are excellent, and if these are replaced by less nutritious hays, the rations must be otherwise adequately balanced. Silage can replace a certain amount of green fodder, and although it is not essential, it is an excellent dairy feed. The food needed for different types of dairy stock should be carefully worked out, and adjusted to their needs according to the milk they are giving, and it will naturally be affected by economic considerations based on the cost of locally available feeding stuffs, and by considerations of the value as indicated in Chapter XVI, on the Economics of Feeding. Some of the special advantages or disadvantages which some of these feeding stuffs offer for dairy cattle may now be considered.

Gram

Gram is one of the staple concentrates for dairy cattle as it is rich in digestible protein, and although low in calcium, it is rich in phosphorus, and is usually available everywhere at a reasonable price.

Bran

Bran has special merits. It is a somewhat laxative and bulky material and its special value is on account of these properties rather than the actual amount of nutrients it yields. It is richer than most grains in protein, and high in phosphorus, although its calcium content is low. It is usual to feed bran in conjunction with other concentrates up to about one third of the whole.

Arhar (tur)

Arhar is of approximately the same feeding value as gram but considerably cheaper and may replace most of the latter, where available, such as in the United Provinces and south India.

Cotton seed cake

Cotton seed cake is one of the most valuable protein feeds for dairy stock; it contains a high percentage of digestible protein of good quality, is comparatively cheap, and can be fed safely to dairy cows for long periods as a sole concentrate provided plenty of good quality roughage is available. In feeding trials conducted at Lyallpur no ill effects have been produced by feeding either decorticated or undecorticated cotton seed cake to fully grown cows for long periods. Not more than 2 lb. should be given per day to heifers, or 6 lb. for medium yielding milch cows. Undecorticated cake should not be given to calves. It is more satisfactory to combine cotton seed cake with other concentrates as there may be a tendency, if fed in too large amounts, for the butter to be somewhat hard.

Cotton seed

Cotton seed as such is usually dearer than cake, and contains much more fat or oil (and the undecorticated cake much more fibre). The seeds should be crushed and soaked or merely soaked before feeding. They are however a rich protein feed and may safely be fed up to 5 or 6 lb. per head per day to 800 lb. cows when combined with bran and gram.

No bad effects have been noticed on the butter or ghee made from buffaloes' milk by feeding as much as 10 lb. of either *desi* or fuzzy American cotton seed daily for several months at a time.

Oats, barley and wheat

Oats, barley and wheat are not usually fed to dairy cattle in India, but when they are used they should be ground to a medium fineness before feeding. All are much relished by cows and may take the place of an equivalent amount of maize.

Linseed cake

Linseed cake (or meal) is very valuable for dairy stock, both on account of its richness in protein and because it is palatable and slightly laxative. It may usefully be employed to make up from 5 per cent to 10 per cent of the concentrate part of the ration. It contains somewhat less protein than cotton seed cake, the oil helps to produce a soft butter, and it is thus useful when fed with rations which tend to have an opposite effect. On account of the high conditioning value of linseed cake and meal, it may pay to use these in moderate amounts, even when cheaper cakes such as cotton seed cake are available.

Ground nut cake

Ground nut cake is a rich and palatable protein food and should be used, to begin with, only in small quantities, but may be gradually increased to not more than three or four pounds per head per day. If given in larger quantities it tends to make a soft butter, if too much fat is left in the cake due to crude methods of extraction. Some people hold that this cake taints the milk and butter.

Til cake and toria cake

Til cake and *toria* cake are cheap and of approximately the same feeding value as cotton seed cake. They may be fed to dairy cattle as substitutes for the latter, but they contain a somewhat lower percentage of digestible protein.

Succulent feeds

Succulent feeds should form part of dairy rations whenever possible. They have a slight laxative effect, are palatable, and induce cows to eat a larger amount of dry roughage than they otherwise would. They are not absolutely necessary, however, provided plenty of rich grass or legume hay is available, but should always be used when high priced concentrates such as bran or linseed cake, valued for their laxative properties, are not available. Whether silage should be fed to dairy cattle will be largely governed by local circumstances and farm economy. Well prepared silage constitutes a very valuable succulent feed, especially for those periods when green fodder is scarce.

All green fodders provide carotene, which is essential for all farm stock, particularly pregnant females.

Green fodders specially suitable for dairy stock in India during the summer are:—maize, a mixture of *jowar* (*cholam*) and *guara*, berseem which is especially rich in calcium, and *senji* during the winter months. *Sarson* and turnip tops are also valuable winter succulents. In circumstances when green cultivated fodders cannot be obtained, good quality *dhub* and *anjan* grass may take their place in proportionate amounts, although these grasses are richer in protein and total dry matter.

If green fodders such as maize, or mixtures of *jowar* and *guara* are fed towards the end of the growing season they become very similar in nutritive value to ordinary dry roughages, especially in fibre content, and at this stage they may replace most of the usual dry roughages of low quality in the rations. Dry roughages should always be fed with green berseem and *senji*, as the latter, if fed as a sole roughage, are too laxative and may cause tympany.

Bajra (*cumbu*) is extensively used for dairy cattle over most of north and south India.

DRY ROUGHAGES

Wheat BHUSA

Wheat *bhusa* is one of the staple roughages for dairy cattle in wheat tracts, but it is low in protein and minerals, and does not by itself constitute a maintenance ration. It may be given as part of the total roughage to make up the bulk of dry matter needed.

Dry maize and JOWAR (CHOLAM *south India,* CHARI *north India) stalks*

These may replace wheat *bhusa,* but should usually be chopped before feeding. Neither constitutes a maintenance ration alone. In Madras *cholam* stalks are fed whole.

Rice straw

Rice straw largely takes the place of wheat *bhusa* in Bengal, south India and east of the Central Provinces. It is also fed in certain other parts of India and forms the staple ration of most cattle in Kangra in the Punjab. It is low in protein and minerals and is not a maintenance ration, particularly for animals which are not accustomed to it.

Silage

Silage is not used to as great an extent in India as it might be, although maize and oats silage either separately, or better still mixed, is a succulent and highly valuable feeding stuff for dairy cows and can be made available throughout the year, especially when other green fodders are scarce.

Henry and Woll [1888-89], have recorded that they obtained a much greater feeding value for dairy cattle from an acre of maize when ensiloed, than when the same crop was dried and fed as hay, and that an average of 7.4 lb. more milk was obtained from each 100 lb. of dry matter in rations containing silage, than in those containing maize hay—even when silage formed only a part of the roughage in the ration.

Maize silage has been found in certain American trials to increase the production of milk by nearly 10 per cent, and the fat nearly 6 per cent, when fed as a part substitute for lucerne hay in a ration of lucerne hay and concentrates. The amount of silage which may be fed to cows varies, but in trials at Lyallpur up to 50 lb. have been fed per day to dairy cattle of 900 lb. weight, and replaced about 18 lb. of good hay. A satisfactory amount is about three pounds of silage and one pound of hay per 100 lb. body weight.

There are many other crops, such as jowar (*chari*, north India, *cholam*, south India) in its early stages, pasture grasses, berseem, etc., which can be used for making silage, and the dairyman will have to decide for himself whether it will pay him to ensile his fodder crops or not. Silage should always be fed after milking as it is inclined to impart an odour and taste to the milk if fed before milking.

Details for making silage have been given in Chapter VII.

FEEDING BEFORE AND AFTER CALVING

Before calving

Cows should, whenever possible, be dried off about from 6 to 8 weeks before calving to enable the body to provide for the developing foetus and to recover losses in fat, protein, and minerals, which may have resulted during the previous lactation, especially if this has been heavy. Proper feeding during the dry period is essential to enable these replenishments to take place, and to get cows in proper condition for calving and to enable them to stand the strain of the next lactation. Some dairymen are in the habit of easing off feeding some weeks before calving on the assumption that this will lessen the tendency to milk fever, but the general consensus of opinion, based on many trials carried out in America and England, is that the incidence of milk fever is not likely to be any greater if cows are specially well fed—or flushed as it is called—prior to calving. Therefore the more common practice is to feed an extra supply of concentrates and high quality roughage, and to include in the former some laxative feeding stuffs, such as bran and linseed cake, for the last week or two before actual calving.

The question of precisely how much food to give during the pre-calving period is somewhat complicated, as a cow has to provide for the developing foetus, and she may also be giving milk almost up to the time the calf is born. While a cow is still giving milk she may be fed according to the standards required for maintenance and the milk produced. The allowance of 3.5 lb. of total digestible nutrients per day for, say, the first gallon of milk is generally considered more than ample, although it should not be lower than this *pro rata* for additional milk.

To be on the safe side, an additional 1 lb. of total digestible nutrients per day may be allowed for the developing foetus.

Thus an 800 lb. cow in calf and giving 2 gallons of milk per day would need:—

	Total digestible nutrients.	Digestible protein.
	lb.	lb.
For maintenance	6.19	0.46
For milk production (5.0 per cent fat).	7.14	0.96
For the calf	1.00	0.20
Total	14.33	1.62

Suitable rations common to north India which would provide these requirements are given below:—

TABLE II

			Total digestible nutrients.	Digestible protein.
			lb.	lb.
(a)	Maize green ..	60 lb.	9.00	0.60
	Wheat bran ..	4 lb.	2.84	0.36
	Gram ..	2 lb.	1.60	0.24
	Toria cake ..	1½ lb.	1.11	0.45
			14.55	1.65

			Total digestible nutrients. lb.	Digestible protein. lb.
(b)	Berseem green	60 lb.	6.0	1.32
	Wheat *bhusa*	6 lb.	2.7	—
	Bran	2 lb.	1.4	0.18
	Maize	6 lb.	4.3	0.30
			14.4	1.80

For south Indian conditions the following rations would provide the same amount of nutrients:—

			lb.	lb.
(c)	*Cumbu* green	60 lb.	7.72	0.55
	Cholam straw	5 lb.	2.20	0.02
	Cocoanut cake	2 lb.	2.04	0.38
	Cotton seed meal	2 lb.	1.88	0.57
	Rice bran	2 lb.	1.09	0.12
			14.93	1.64

			lb.	lb.
(d)	*Cholam* green	50 lb.	9.68	0.50
	Kollukottai grass hay (*anjan*)	5 lb.	1.70	0.25
	Ground nut cake	2 lb.	1.70	0.84
	Gram husk	2 lb.	1.06	—
			14.14	1.59

In the weeks immediately prior to calving, these amounts may be exceeded, and protein rich concentrate with cereals and plenty of good hay, preferably legume hay, should be given.

When good roughages are available the protein in the concentrate part of the ration should be around 10 per cent to 12 per cent but if poor protein roughages are used the concentrates should contain not less than 16 per cent of protein.

A cow has to provide for the skeleton of the developing foetus, and the ration must be sufficiently rich in calcium and phosphorus, and also in vitamins A and D, when the full benefit of the sun cannot be obtained. Well cured legume hays such as berseem and lucerne are excellent, but if they are not obtainable, and the cereal and concentrate mixture does not contain at least one quarter of its bulk made up of phosphorus rich feeds, such as cotton seed cake or bran or linseed meal, one pound of high quality bone meal should be added to each 100 lb. of the concentrates. A suitable mixture for a cow, if dry prior to calving, could be made of the following:—

100 lb. ground maize, 80 lb. crushed gram, 100 lb. of wheat bran and 80 lb. of linseed cake.

From three to four pounds of such a mixture per day, with plenty of lightly cured roughage or green fodder, should be sufficient for a cow in good condition, but this amount may be considerably increased if she needs conditioning prior to calving, when as much as 6 or 8 lb. of concentrates may be fed. If cows are well fed prior to calving they are much more likely to stand the strain and give good yields of milk soon afterwards.

Feeding at, and after calving time

Cows should be kept in the stalls a day or two before calving, and should be given a warm bran mash or saline laxative if there are any signs of constipation. The food may be cut down somewhat at the actual time of calving, and for the first day or two afterwards should be decidedly laxative and moderate in amount, up to about one pound, say, of the concentrate mixture. After the first day the amount of concentrates may be gradually increased, so that after a week or two low producing cows will be getting the maximum they need for their milk yields. High producing cows may take as long as three weeks to attain their maximum concentrate feeding stage, and this should then be continued in accordance with the combined needs of the cow for maintenance and milk production.

Cows should not be kept on short rations longer than necessary, but when they are well under way the total digestible nutrients may be augmented by about 1 lb. per day over and above the stipulated amount to see if they respond by a further yield of milk. If they do, the extra amount should be kept up or even slightly exceeded. When the cows have finally settled down to their maximum yields, they may be fed according to standard requirements. It should be possible to keep heavy yielding cows in good condition by proper feeding, but in the case of some exceptionally heavy milkers it may be difficult to feed the amount of nutrients needed. Such cows are likely to show the strain, any may suffer from abortions and premature death. They should therefore be given as long a dry period as possible between calvings which should not be too frequent. Good balanced feeding in the dry period should help to enable them to stand the strain better than if they are kept on short rations.

MAINTENANCE RATIONS AND RATIONS FOR MILK PRODUCTION

From the digestibility trials which have been conducted at Lyallpur during the last 20 years on heifers and resting bullocks a number of charts, or curves, has been prepared showing the amounts of digestible protein and total digestible nutrients which animals of different weights require for maintenance.

In order to remind the reader what a maintenance ration is, it may be recalled that in the resting condition all the metabolic processes continue to function, heat is evolved, and a certain amount of body tissue broken down as a result of the activities of the various organs. A certain amount of energy is thus required to balance this expenditure of energy. Certain other needs also have to be provided for, such as a moderate amount of exercise and so forth. A maintenance ration therefore, forms the basis upon which additional rations are computed for work and milk production. Figure 10, (page 317) shows the digestible protein and total digestible nutrients required for maintenance for Sahiwal heifers of different weights. For the sake of comparison a corresponding curve evaluated from Morrison's standards (see Appendix I) is shown on the same chart. The data shown in this curve represent, as far as the author knows, the first effort to compute definite feeding standards under Indian conditions. Many years have been taken to collect these data, but the number of animals worked with and the number of trials it has been possible to conduct are small in comparison with the very large number of animals employed and the trials conducted in foreign countries, from which the usual feeding standards have been computed. These foreign standards have hitherto been used in India for such scientific computation of rations as is employed on large Government and private dairy farms.

It is to be hoped that in course of time a more complete set of data will be worked out for Indian cattle on Indian feeding stuffs. It may also be recalled that although very complete data for cattle have been worked out in foreign countries very little, if any, work has been done to obtain corresponding information for other classes of farm stock. The feeding standards given by Morrison for animals other than cattle are to some extent empirical, as they have been worked out in large degree from digestibility data regarding the feeding stuffs from trials on cattle.

Computation of rations for a dairy cow weighing 800 *lb. and giving* 24 *lb. of milk per day*

In computing rations for dairy cows it is first necessary to ascertain the various nutrients required for maintenance, quite apart from the milk yield, and then add the additional nutrients required per pound of milk in accordance with the accepted standard.

For example, the amounts of digestible protein and total digestible nutrients required for an 800 lb. dairy cow per pound of milk produced as determined for Sahiwal cows at Lyallpur are shown below. The amounts required according to the Morrison standard are shown for comparison:—

	Total digestible nutrients. lb.	Digestible protein. lb.
Lyallpur	0.357	0.048
Morrison	0.353 to 0.373	0.046 to 0.056

Therefore the total amount of nutrients needed for the production of 24 lb. of milk will be, on the basis of the Lyallpur figures:—

Total digestible nutrients. lb.	Digestible protein. lb.
8.57	1.152

If, therefore, these respective figures are added to 6.19 lb. of total digestible nutrients, and 0.463 lb. of digestible protein required for maintenance for a cow of this weight it is necessary to find a suitable ration which will provide:—

	Dry matter. lb.	Total digestible nutrients. lb.	Digestible protein. lb.
Lyallpur			
For maintenance ..	10.06	6.19	0.463
For producing 24 lb. of milk ..	10.00	8.57	1.152
Total requirements ..	20.06	14.76	1.615
Morrison			
For maintenance ..	18.00	6.53	0.536
For producing 24 lb. of milk ..	10.00	8.71	1.264
Total requirements ..	28.00	15.24	1.800

The two sets of figures are thus very similar, and the required ration may be computed according to the Lyallpur standard.

Assuming that the rations available are those shown in the table below, in which the quantities of dry matter, total digestible nutrients and digestible protein, per lb. of each are given, it is possible by a little arithmetical jugglery to ascertain how much of each will meet the cow's requirements and provide 20 lb. of dry matter in the ration per day.

TABLE III

	Dry matter per lb of the feed lb.	Total digestible nutrients per lb. of the feed lb.	Digestible protein per lb. of the feed. lb.
Dry roughages			
1. Wheat *bhusa* ..	0.92	0.45	nil
2. Oat hay ..	0.93	0.48	0.022
3. Berseem hay ..	0.87	0.50	0.097
4. *Jowar* ..	0.91	0.43	0.031
5. Rice straw ..	0.94	0·39	0.002
6. *Anjan* or *Kollukottai* grass hay ..	0.95	0.34	0.050

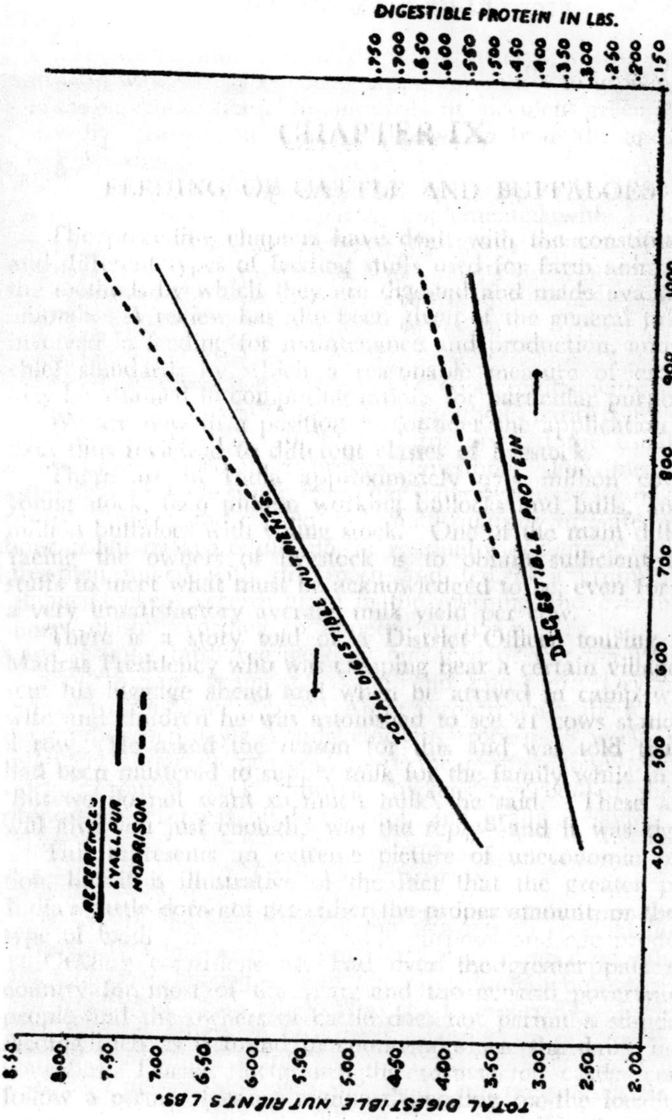

Fig. 10. Maintenance Requirements of Montgomery Dairy Cattle

	Green fodders			
1.	Berseem ..	0.15	0.10	0.022
2.	Oats ..	0.16	0.10	0.012
3.	Maize	0.21	0.15	0.010
4.	*Bajra* ..	0.22	0.13	0.009
5.	*Jowar* ..	0.40	0.21	0.008
	Concentrates			
1.	Gram ..	0.92	0.80	0.120
2.	*Arhar*	0.90	0.67	0.180
3.	*Toria* cake ..	0.92	0.74	0.300
4.	Wheat bran ..	0.93	0.71	0.090
5.	*Sarson* cake ..	0.93	0.82	0.260
6.	Cotton seed cake	0.92	0.57	0.180
7.	Cotton seed ..	0.94	0.70	0.120
8.	Maize grain ..	0.91	0.71	0.050
9.	Gram husk ..	0.93	0.53	—
10.	Rice bran ..	0.93	0.54	0.060
11.	Ground nut cake	0.93	0.85	0.420
12.	Cocoanut cake ..	0.92	1.02	0.190
13.	Rape cake ..	0.93	0.93	0.280
14.	Linseed cake ..	0.94	0.75	0.230

Suppose it is decided to select at random the following ration for north India.

TABLE IV

		Dry matter. lb.	Total digestible nutrients. lb.	Digestible protein. lb.
Wheat *bhusa*	2 lb.	1.84	0.90	—
Maize green	100 lb.	21.00	15.00	1.000
Wheat bran	4 lb.	3.72	2.84	0.360
Gram	2 lb.	1.84	1.60	0.240
Toria cake	2 lb.	1.84	1.48	0.600
		30.24	21.82	2.200

It will be seen that this is not a well balanced ration, as it is too bulky on a dry matter basis, and far richer in total digestible nutrients and digestible protein than is needed according to the standard. Therefore, if the bulk of the ration is reduced by cutting down the green maize from 100 lb. to 60 lb. and eliminating *bhusa* entirely, the following ration is obtained which is quite satisfactory on the whole; the digestible protein is slightly in excess, which is an error on the safe side:—

TABLE V

		Dry matter. lb.	Total digestible nutrients. lb.	Digestible protein. lb.
Maize green	60 lb.	12.60	9.00	0.60
Wheat bran	4 lb.	3.72	2.84	0.36
Gram	2 lb.	1.84	1.60	0.24
Toria cake	2 lb.	1.84	1.48	0.60
		20.00	14.92	1.80

REDUCING THE COST OF THE RATION

The cost of the available rations is always an important point to be taken into account in deciding what rations to feed in order to provide the necessary nutrients. Some of the concentrates such as gram and *toria* cake mentioned above are expensive, and, to obtain a cheaper ration, they may be replaced by less expensive items. The following table shows a computation for a ration containing *arhar* and cotton seed cake giving approximately the same nutrients as shown in table V above:—

TABLE VI

		Dry matter. lb.	Total digestible nutrients. lb.	Digestible protein. lb.
Maize green	60 lb.	12.60	9.00	0.60
Wheat bran	4 lb.	3.72	2.84	0.36
Arhar	2 lb.	1.80	1.34	0.36
Cotton seed cake	2 lb.	1.84	1.14	0.36
		19 96	14.32	1.68

This ration, using cotton seed cake instead of *toria* cake, and *arhar* instead of gram, is almost identical with the ration given in Table V, as far as the various nutrients are concerned, all of which, as in the preceding example, have been worked out from actual digestibility trials at Lyallpur.

Other rations may be computed in exactly the same way.

SOME WINTER RATIONS SUITABLE FOR NORTH INDIA FOR AN 800 LB. DAIRY COW GIVING 24 LB. MILK

Berseem is available in winter and a winter ration selected at random is shown in the table below:—

TABLE VII

		Dry matter. lb.	Total digestible nutrients. lb.	Digestible protein. lb.
Wheat *bhusa*	6 lb.	5.52	2.70	—
Berseem green	60 lb.	9.00	6.00	1.32
Bran	4 lb.	3.72	2.84	0.36
Arhar	2 lb.	1.80	1.34	0.36
Cotton seed cake	2 lb.	1.84	1.14	0.36
		21.88	14.02	2.40

It will at once be seen that this is not a strictly well balanced ration according to the standard, as the digestible protein is higher than necessary, although the total digestible nutrients are approximately correct. The protein rich *arhar* grain may therefore be replaced by maize, when the following ration results:—

TABLE VIII

		Dry matter. lb.	Total digestible nutrients. lb.	Digestible protein. lb.
Wheat *bhusa*	6 lb.	5.52	2.70	—
Green berseem	60 lb.	9.00	6.00	1.32
Bran	4 lb.	3.72	2.84	0.36
Maize	2 lb.	1.82	1.42	0.10
Cotton seed cake	2 lb.	1.84	1.14	0.36
		21.90	14.10	2.14

This table reveals that this ration is normal as regards dry matter slightly lower in total digestible nutrients and still somewhat high in digestible protein compared with the standard but it should be slightly cheaper than the ration given in Table VII. It should prove quite satisfactory but the following is suggested as a further alternative:—

TABLE IX

		Dry matter. lb.	Total digestible nutrients. lb.	Digestible protein. lb.
Wheat *bhusa*	8 lb.	5.66	3.60	—
Berseem green	60 lb.	9.00	6.00	1.32
Maize grain	6 lb.	4.46	4.26	0.30
Wheat bran	1 lb.	0.90	0.71	0.09
		20.02	14.57	1.71

This ration is also well balanced for all nutrients, and is the cheapest of the three winter rations so far described.

The following rations may be suggested for the eastern provinces, where rice straw is the staple dry roughage:—

		Dry matter. lb.	Total digestible nutrients. lb.	Digestible protein. lb.
Rice straw	6 lb.	5.64	2.34	0.012
Jowar green	30 lb.	8.00	6.20	0.016
Rape cake	4 lb.	3.72	3.72	1.420
Rice bran	2 lb.	1.83	1.08	0.120
		19.19	13.34	1.568

Some rations for use in south India, where *cholam, cumbu* and paddy straw are used as cattle fodder, are suggested:—

TABLE X

		Dry matter. lb.	Total digestible nutrients. lb.	Digestible protein. lb.
Cumbu green	60 lb.	13.20	7.72	0.55
Cholam straw	5 lb.	4.00	2.20	0.02
Cocoanut cake	2 lb.	1.80	2.04	0.38
Cotton seed meal	2 lb.	1.80	1.88	0.57
Rice bran	2 lb.	1.86	1.09	0.12
		22.66	14.93	1.64

TABLE XI

		Day matter. lb.	Total digestible nutrients. lb.	Digestible protein. lb.
Cholam green	50 lb.	14.00	9.68	0.50
Kollukottai grass hay	5 lb.	4.75	1.70	0.25
Ground nut cake	2 lb.	1.86	1.79	0.84
Gram husk	2 lb.	1.86	1.06	—
		22.47	14.23	1.59

It may be of interest now to select a few typical rations which are fed in well established dairies in India and compare them with the examples previously given.

Example No. 1

The following represents an average early winter ration for an 800 lb. dairy cow giving 24 lb. milk per day, given in the Agricultural College Dairy at Lyallpur:—

Roughages

Wheat *bhusa*	3.4 lb.
Berseem hay	1.2 lb.
Berseem green	68.3 lb.
Maize green	4.5 lb.

Concentrates

Bran	5.95 lb.
Gram	3.45 lb.
Toria cake	2.50 lb.

This ration may be set out in tabular form thus:—

TABLE XII

		Dry matter. lb.	Total digestible nutrients. lb.	Digestible protein. lb.
Wheat *bhusa*	3.4 lb.	3.13	1.53	—
Berseem hay	1.2 lb.	1.04	0.60	0.116
Berseem green	68.3 lb.	10.25	6.83	1.503
Maize green	4.5 lb.	0.95	0.68	0.045
Bran	5.95 lb.	5.53	4.23	0.536
Gram	5.45 lb.	3.17	2.76	0.414
Toria cake	2.50 lb.	2.30	1.85	0.750
		26.37	18.48	3.360

It is known from the results of experimental trials at Lyallpur that an 800 lb. cow giving 24 lb. of milk needs approximately 14.76 lb. of total digestible nutrients and 1.615 lb. of digestible protein and 19 to 20 lb. of dry matter per day.

The above ration therefore appears to be unnecessarily rich, containing 3.72 lb. and 1.745 lb. respectively of total digestible nutrients and digestible protein, and 6 lb. of dry matter in excess of the standard.

Several alternatives therefore may be suggested for levelling it down to the standard:—

1. Cutting out entirely berseem hay and green maize.
2. Eliminating the berseem hay and reducing the green berseem to 60 lb. per head per day.
3. Cutting out the wheat *bhusa* and berseem hay, and
4. Reducing the concentrates.

The amended ration is then:—

TABLE XIII

		Dry matter.	Total digestible nutrients.	Digestible protein.	Phosphorus as P_2O_5	Calcium as CaO
		lb.	lb.	lb.	gms.	gms.
Wheat *bhusa*	4 lb.	3.68	2.20	—	1.50	4.60
Berseem green	60 lb.	9.00	6.00	1.32	18.27	108.82
Maize green	5 lb.	1.05	0.75	0.05	2.00	0.56
Bran	6 lb.	5.48	4.26	0.54	43.15	7.12
Gram	1 lb.	0.92	0.80	0.12	1.82	0.41
Toria cake	1 lb.	0.92	0.74	0.30	4.62	12.11
		21.05	14.75	2.33	71.36	133.62

This is an improvement, although the digestible protein is still on the high side and the calcium and phosphorus are more than sufficient.

Example No. 2.

A typical ration fed in the Military Dairy Farms to an 800 lb. cow giving 24 lb. milk per day is the following:—

Green maize	63 lb.
Wheat *bhusa*	3 lb.
Wheat bran	5 lb.
Gram	1 lb.
Toria cake	2 lb.

When set out in terms of dry matter, total digestible nutrients and digestible protein this shows:—

TABLE XIV

		Dry matter. lb.	Total digestible nutrients. lb.	Digestible protein. lb.
Green maize	63 lb.	13.23	9.45	0.630
Wheat *bhusa*	3 lb.	2.76	1.35	—
Wheat bran	5 lb.	4.65	3.55	0.450
Gram	1 lb.	0.92	0.80	0.120
Toria cake	2 lb.	1.84	1.48	0.600
		23.40	16.63	1.800

This ration is also somewhat richer in total digestible nutrients and digestible protein than the Lyallpur standard by 1.87 lb. and 0.185 lb. respectively. To bring it nearer this standard the gram might be eliminated, or half the gram and half the *toria* cake, or the bran may be reduced by 2 lb.

The examples given above are intended to illustrate how rations for milch cows may be computed in accordance with standard requirements, but they are not necessarily meant to represent absolute data and may be varied according to the breed of the cow, geographical and climatic factors and practical experience.

The inclusion of Roots and Silage in the Ration

So far, the computation of rations containing green fodders has been described, but roots and silage may be included in the ration, as a part or whole substitute for the green fodder, provided a certain amount of good quality hay is included to cater for the vitamin and mineral requirements. An example will now be given to show how a ration for the dairy cow used in the above illustration may be computed when roots or silage take the place of green fodder.

The requirements will be met by a ration containing turnip roots, made up as follows:—

TABLE XV

		Dry matter. lb.	Total digestible nutrients. lb.	Digestible protein. lb.
Turnip roots	**50 lb.**	**4.75**	**4.25**	**0.65**
Berseem hay	**10 lb.**	**8.62**	**4.41**	**0.62**
Maize grain	**4 lb.**	**3.68**	**2.84**	**0.20**
Wheat bran	**4 lb.**	**3.68**	**2.84**	**0.36**
		20.73	**14.34**	**1.83**

If oat silage is used instead of turnip roots the following ration may be suggested:—

TABLE XVI

		Dry matter. lb.	Total digestible nutrients. lb.	Digestible protein. lb.
Oat silage	**50 lb.**	**11.98**	**7.67**	**0.21**
Berseem hay	**5 lb.**	**4.31**	**2.21**	**0.31**
Bran	**3 lb.**	**2.76**	**2.13**	**0.27**
***Toria* cake**	**3 lb.**	**1.95**	**2.30**	**0.93**
		21.00	**14.31**	**1.72**

FEEDING OF YOUNG STOCK

Various methods are employed in feeding newly born calves, and which of these should be followed will depend upon circumstances and the particular object for which the calf is reared, *i.e.*, whether as a milch cow, for meat production, or work.

In India it is the common practice in villages to allow newly born calves to suckle their mothers during the whole lactation. This is the natural method and the best one for the calves, but it is somewhat expensive, and cannot usually be followed in dairies scientifically run for milk production. Some dairymen prefer to take the calves away from their mothers directly they are born, while others prefer to allow them to stay with the cows for several days before resorting to pail feeding. As far as possible calves should always get the colostrum, or first yielded milk, as this protects them against diseases of the alimentary canal. They should then be continued on whole milk, either by suckling or by pail feeding, for at least three or four weeks, although, if expense is no object, this may be continued for as long as 6 months. After the calves have had a good start on whole milk they may then be raised by several different methods.

Pail-feeding

Arrangements having been made for the calves to have their mothers' colostrum, they may be carefully schooled by hand to take milk from the pail until they have learnt how to feed themselves.

During the first few days of pail-feeding they should receive not less than about 5 lb. of milk per head a day, distributed in three feeds; this may be increased gradually to a gallon or more per head per day till the calves are 3 months old, when not less than $1\frac{1}{2}$ gallons per day should be given. As a rule, it is not an economic proposition to continue whole milk feeding longer than a month, at which age calves will be able to nibble hay and succulent fodders and take small quantities of crushed grain and finely ground cakes.

Feeding on Separated Milk

It may not be convenient to feed calves longer than a few weeks on whole milk, in which case a substitute has to be found. Separated milk may gradually replace part of the whole, three pounds of the former replacing two of the latter, until substitution has been completed. Separated milk, however, has had most of the fat removed and contains less than skim milk, so some appropriate substitute will be needed to replace this. Finely ground meal, which should contain not less than 15 per cent fat may be used and some fish liver oil may also be given. Linseed contains about 34 per cent fat and the seed should be crushed and boiled, or boiled first and then crushed, when an excellent feed is obtained and this may be given at the rate of four ounces per head per day.

An alternative is to give a mixture made up of two parts of crushed gram, one of bran and one of cake, in somewhat larger amounts of from 6 to 9 ounces per head per day to start with, this being gradually increased to two pounds at the end of the fourth month.

The following four regimes for calves show the rations used for dairy calves at the Lyallpur Agricultural College Dairy, those given at the Military Dairy Farms in India, those given in Bulletin No. 10 of the Ministry of Agriculture and Fisheries in England, and those followed at the Imperial Agricultural Research Institute, New Delhi, respectively:—

I. Routine followed in the Lyallpur Agricultural College Dairy

Age	Colostrum. lb.	Whole milk. lb.	Separated milk. lb.	Concentrates. lb.
First 5 days ..	6–8	—	—	—
Upto 4th week ..	—	6–8	—	—
5th to 8th week ..	—	8–10	—	—
9th to 12 week ..	—	8–10	—	—
13th to 16th week ..	—	6	8	—
17th to 20th week ..	—	4	8–12	—
Beyond 20 weeks ..	—	No whole milk.	10 if spare.	—

All young stock under two years are fed in one lot and allowed 0.75 lb. of the following concentrate mixture with half an ounce of common salt, per head per day, and green fodder *ad lib.*

Concentrate Mixture

Bran	..	40 per cent.
Cake	..	30 per cent.
Oats or gram	..	30 per cent.

II. The Military Dairies in India observe the following feeding procedure for young stock

Figures under all heads, except fodder, denote the quantity to be fed for every 10 pounds body weight. (Five pounds and over to be counted as 10 pounds, under 5 pounds to be ignored).

The figures under fodder show the actual proportion of the standard* fodder ration to be fed.

Age	COW				BUFFALO				COW & BUFFALO	
	Dam's Colostrum. lb.	Milk. lb.	Separated milk. lb.	Concentrates. lb.	Dam's Colostrum. lb.	Milk. lb.	Separated Milk. lb.	Concentrates. lb.	Fodder.	No. of times Rations fed daily.
1st-3rd day.	1·0	—	—	—	1·0	—	—	—	—	3
4th-7th day.	—	1·0	—	—	—	1·0	—	—	—	3
2nd week.	—	1·25	—	—	—	0·8	0·1	—	—	2
3rd ,,	—	1·25	—	—	—	0·8	0·2	—	—	2
4th ,,	—	1·0	0·25	1/20	—	0·7	0·2	1/20	—	2
5th ,,	—	1·0	0·25	1/20	—	0·6	0·4	1/20	1/20	2
6th ,,	—	1·0	0·25	1/20	—	0·5	0·4	1/20	1/10	2
7th ,,	—	0·75	0·40	1/12	—	0·4	0·4	1/10	1/8	2
8th ,,	—	0·75	0·40	1/12	—	0·3	0·5	1/10	1/8	2
9th ,,	—	0·75	0·40	1/12	—	0·2	0·6	1/10	1/6	2
10th-12th week.	—	0·50	0·50	1/12	—	0·1	0·5	1/10	1/5	2
13th-16th ,,	—	0·25	0·50	1/12	—	—	0·5	1/10	1/5	2
17th-19th ,,	—	—	0·40	1/12	—	—	0·4	1/10	1/4	2
20th-23rd ,,	—	—	0·25	1/12	—	—	0·3	1/10	1/3	2
24th-26th ,,	—	—	0·10	1/12	—	—	0·2	1/10	1/3	2

*Standard fodder ration is the amount allowed for the maintenance of an 800 lb. cow taken as a standard animal. (See Appendix 2.)

III. The following tabulated dietary is taken from Bulletin No. 10. on Calf Rearing, Published by the Ministry of Agriculture and Fisheries:—

1st week. Its own mother's warm milk three times a day, commencing with about 1 qt. and increasing to 2 qt. at each meal by the 3rd day.

2nd week. 2 qt. of warm new milk (not necessarily its own mother's) three times a day.

3rd week. 2 pints of new and 3 pints of skimmed (or separated) milk with ½ pint of linseed porridge or half a table spoonful of cod liver oil, three times a day.

5th week. 3 qt. of warm skimmed milk, with 1 pint of linseed porridge or one table spoonful of cod liver oil three times a day, and a little sweet meadow hay, increased week by week.

9th week. Mid-day milk and cream substitute omitted. 5 qt. of separated milk are given morning and evening and a handful of broken linseed cake, (6 oz.) at midday, and hay, increasing week by week.

13th week. Milk as before, ¾ lb. mixed linseed cake and crushed oats, ½ gal. pulped swedes gradually increasing, hay *ad lib.*

21st week. Milk as before, 1 lb. of mixed linseed cake and meal, increasing quantities of roots, hay *ad lib.*

24th week. Evening milk discontinued.

27th week. Milk altogether discontinued.

Note :—Pound = 16 ounces Quart = 40 ounces.

Pint = 20 ounces Gallon = 160 ounces.

IV. Details of Calf Feeding at the Imperial Agricultural Research Institute, New Delhi.

Age.	Milk in lb.		Skim milk. lb.	Grain. lb.	Salt. ozs.
1 week.	1/10th of the body weight.		–	–	–
2 weeks.	,,		–	–	–
3 ,,	,,		–	–	–
4 ,,	,,	+ ½	–	–	–
5 ,,	,,	+ ½	–	–	–
6 ,,	,,	+1	–	¼	1
7 ,,	,,	+1	–	¼	1
8 ,,	,,	+1	–	¼	1
9 ,,	,,	+1½	–	½	1
10 ,,	,,	+1½	–	½	1
11 ,,	,,	+2	–	½	1
12 ,,	,,	+2	–	½	1
13 ,,	,,	+3	–	¾	1
14 ,,	,,	+3	–	¾	1
15 ,,	,,	+3	–	¾	1
16 ,,	,,	+3	–	¾	1
17 ,,	,,	+3½	2	1½	1
18 ,,	,,	+3½	2	1½	1
19 ,,	,,	+3½	2	1½	1
20 ,,	,,	+3½	2	1½	1
21 ,,	,,	+2½	4	2	1
22 ,,	,,	+2½	4	2	1
23 ,,	,,	+2½	4	2	1
24 ,,	,,	+2½	4	2	1
25 ,,	,,	+2½	4	2	1
26 ,,	,,	+2½	4	2	1
27 ,,	,,	+3	4	2½	1
28 ,,	,,	+3	4	2½	1
29 ,,	,,	+3	4	2½	1
30 ,,	,,	+3	4	2½	1
31 ,,	–	–	6	2½	1
32 ,,	–	–	6	2½	1
33 ,,	–	–	6	2½	1
34 ,,	–	–	6	2½	1

Remarks: The calf is fed milk at body temperature from its own mother, three times daily, till it is one month old. If this is not possible then milk is given from any other cow in the chain of calvings. The first feed is given two or three hours after birth. After a month it is fed twice daily. Lumps of rock salt are kept in the manger for the calves to lick at will. Immediately after drinking milk some common salt mixed with mineral salt is put into the mouth of each calf to prevent sucking in air.

It is essential that the dry feeding stuffs given during the separated milk period are easily digestible; they should have a total digestible nutrient content of about 80 lb. and a digestible protein content of about 15 lb. per 100 lb.

Thus 100 lb. of the mixture recommended above, *viz.* 2 parts of gram, 1 of bran and 1 of linseed cake would contain:—

Digestible protein	*Total digestible nutrients*
lb.	lb.
14.1	78.4

One pound of such a mixture is approximately equal in feeding value to five pounds of whole milk. Thus if the calf has been receiving from 2 to 2.5 gallons of separated milk, and 1½ pounds of the mixture, it will need about 3 pounds of the dry mixture by the time the separated milk is stopped.

Skim milk, which contains more fat, may be used instead of separated milk, but it should be warmed before feeding and on no account be allowed to become stale. Separated milk is to be preferred, however, as it may be fed at once after separation.

Whey is also used to a considerable extent in India, but it is much lower in nutritive value than either whole milk or separated milk, as all the fat and most of the protein have been removed. Therefore, in order that whey may be equal in value to the original milk, it would be necessary to add some concentrate containing not more dry matter than the calf's digestive tract can cope with. For example, 10 lb. of whole milk contain about 2.1 lb. of total digestible nutrients, and 0.28 lb. of digestible protein, whereas the same amount of whey contains only 0.75 lb. of total digestible nutrients and 0.063 lb. of digestible protein. Therefore the additional concentrate must provide the difference between these, that is 1.4 lb. of total digestible nutrients and 0.27 lb. of digestible protein. The deficiency in 10 lb. of whey could be made good by feeding approximately 1.5 lb. of a mixture containing about 75 lb. of total digestible nutrients and 15 lb. of digestible protein per 100 lb., but as it would be difficult for a calf to eat this amount, only about 1 lb. should be fed with every ten pounds of whey to begin with.

Therefore the whey should be increased somewhat in order to balance the ration.

Stock reared on whey will not maintain more than half the increase in weight of that of a calf fed on whole milk. A suitable mixture which might be fed (1 lb. per day with 10 lb. of whey) to meet requirements for growth is the following:—

1. Five parts of ground gram.
   Four parts of linseed cake meal.

   or

2. Two parts of linseed cake meal.
   Two parts of ground maize.
   One part of oats.
   One part of cotton seed cake meal with about $\frac{1}{4}$ ounce of best quality of bone meal.

Whole milk should preferably be used for the first month or six weeks, and when whey is used the substitution should begin gradually, so that at the end of the first month the calf starts off with one pound of the dry mixture, and 10 lb. of whey. This may be increased to about 2.5 lb. of the dry mixture and 18 lb. of whey at the end of the third month. After the third month the calf must be prepared gradually for the time when the whey and the supplementary concentrate fed with it will be changed to a more normal ration, of composition and nutritive value comparable with whole milk. As whole milk is a natural food, it is better if the food to which the calf is now transferred be somewhat richer in digestible protein than whole milk, containing, say, 20 lb. of digestible protein per 100 lb.

At this stage the following ration is recommended by the National Institute for Research in Dairying, England:—

Linseed cake (broken)	3 parts.
White fish meal	1 part.
Crushed oats	4 parts.
Maize meal	2 parts.

Should fish meal not be available, crushed cottonseed cake and finely ground bone meal should be given to supply the necessary minerals.

When the use of whole milk has to be curtailed at an early stage and calves are fed on milk substitutes such as the above, they will not maintain the same initial rate of growth (2.5 lb. per day) as milk fed calves, but this can be made up to some extent by judicious feeding later on. In feeding for eventual milk production, calves should not be given larger quantities of milk substitutes and finely cut hay than their digestive systems can cope with, as this will tend to make them pot-bellied, and affect their capacity for giving milk.

During the transition period from milk to milk substitutes calves should always get sufficient water, which should be slightly warmed in cold weather. When calves are reared with separated milk and fat substitutes, the average increase in weight should be about 8-10 lb. a week, and with whey and protein supplements, somewhat less, but whichever system is adopted the calves must not receive any set back in early age, or this may permanently affect them later on.

The National Calf Starter used in England during the War

During the War the Ministry of Food in England in collaboration with the Ministry of Agriculture and Fisheries has evolved two special feeding stuffs for calves, known respectively as National Calf Starter Cudlets and National Calf Starter Gruel These feeding stuffs are prepared, used and sold under conditions strictly controlled by the Ministry, which has kindly supplied the following data relating to these feeding stuffs:—

National Calf Starter Cudlets

General instructions for calf raising:—

1. Successful calf rearing depends on constant attention to detail.
2. Feed at regular hours twice daily.
3. Make changes gradually, adjusting to appetite.
4. Every calf must have its proper share of food.
5. Feed milk or gruel at blood heat (about 100°F).
6. Scrupulous cleanliness of calf pens, pails and feeding troughs is essential.
7. Ample clean drinking water must be provided as soon as necessary.

Feeding chart

1st to 4th week. As for gruel feeding, but introduce a little good hay and a small quantity of cudlets during the 3rd week. The cudlets should be placed in the bottom of the pail after each feed of gruel, until the calf will feed from a trough.

5th and 6th weeks. Give 4 to 5 quarts of gruel daily in 2 feeds. Increase the cudlets to 1 lb. per calf daily by the end of the 6th week and continue to offer a little good hay.

7th and 8th weeks. Reduce the quantity of gruel gradually and give clean water to drink, allowing up to 6 quarts of liquid daily. Increase the hay and cudlets as the gruel is decreased. Allow 2 to $2\frac{1}{2}$ lb. of cudlets by the end of the 8th week.

After the 8th week calves can be reared on suitable dry concentrates, good hay, and water. A small allowance of green food or cut swedes or mangolds or turnips can be given from the 3rd month onwards.

Specifications of the Composition of National Calf Starter Cudlets

Ingredients.	Minimum percentage by weight in the completed compound.	Maximum percentage by weight in the completed compound.
Dried whey powder ..	29	31
Dried skim milk powder ..	14	16
Linseed cake meal ..	19	21
Linseed	4	6
Cane molasses.	4	6
Fish meal	4	6
Wheat feed or feeding flour ..	9	11
Oats	9	11
Dried grass meal and/or dried lucerne meal having a carotene content of not less than 250 milligrams per gram	2	3
Lime (calculated as Calcium carbonate)	1	$1\frac{1}{2}$
Salt (calculated as sodium chloride)	$\frac{1}{2}$	1

Provided always that—(a) the percentage by weight in the completed compound of:—

(i) Oil shall not be less than $4\frac{1}{2}\%$ or more than 7%.

(ii) Crude protein shall not be less than 22% and not more than 24%.

(iii) Fibre shall not be more than 5%.

(b) Dried grass meal or dried lucerne meal may be replaced by 1% controlled cod liver oil mixture and $1\frac{1}{2}\%$ oats.

National Calf Starter Gruel

Preparation of gruel

The gruel should be made up in the proportion of 1 lb. of the meal to each gallon of clean water.

Mix the required quantity of meal with sufficient cold water to make a thick paste without lumps, then add more cold water up to 2 quarts of water to each lb. of meal used. Stir well and boil for at least 2 minutes. If all the gruel is to be used at one feed, thin and cool by adding 2 quarts of water for each lb. of meal and stir well; the gruel is now ready for adding to milk, or for feeding without milk to the older calves. If enough gruel for two feeds has been boiled at one time, half of it should be thinned and cooled for the first feed as above. The remaining half should be thinned and warmed by the addition of 2 quarts of hot water for each lb. of meal, when required for the second feed.

Feeding chart

1st week. Colostrum for 3 to 4 days, then 3 quarts whole milk daily in two feeds.

2nd week. Give 3 to 4 quarts whole milk daily in two feeds.

3rd and 4th weeks. Begin by giving 3 quarts whole milk with 1 quart gruel daily in two feeds and gradually increase the proportion of gruel, ending with 1 quart whole milk and 3 quarts gruel daily. Introduce a little good hay and a little dry concentrates, such as National Calf Starter Cudlets. Place a small handful of the concentrates in the bottom of the pail after each feed of gruel, until the calf will feed from a trough.

5th, 6th, and 7th weeks. Give 4 to 5 quarts of gruel daily in two feeds. Increase the allowance of concentrates to 1 lb. per calf daily by the 7th week, and continue to offer a little good hay.

8th, 9th and 10th weeks. Reduce the quantity of gruel gradually, and give clean water to drink, allowing up to 6 quarts of liquid daily. Increase the hay and dry concentrates as the gruel is decreased. Up to 2 lb. dry concentrates can be given daily.

After the 10th week the calves can be fed on suitable dry concentrates, good hay and water. A small allowance of green food or cut swedes or mangolds or turnips can be given from the 3rd month onwards.

Dry feeding

If it is desired to reduce gruel feeding to a minimum, National Calf Starter should be obtained in cudlet form. This should be introduced in the 3rd week and can replace gruel entirely by the end of the 8th week.

Specifications of the Composition of National Calf Starter Gruel

Ingredients.	Minimum percentage by weight in the completed compound.	Maximum percentage by weight in the completed compound.
Dried whey powder ..	29	31
Dried skim milk powder ..	14	16
Linseed cake meal	29	31
Fish meal	4	6
Wheat feed or feeding flour ..	9	11
Oats.	9	11
Dried grass meal and/or dried lucerne meal having a carotene content of not less than 250 milligrams per gram	2	3
Lime (calculated as calcium carbonate) ..	1	$1\frac{1}{2}$
Salt (calculated as sodium chloride)	$\frac{1}{2}$	1

Provided always that—(a) the percentage by weight in the completed compound of:—

(i) Oil shall not be less than $4\frac{1}{2}\%$ or more than 7%.

(ii) Crude protein shall not be less than 22% and not more than 24%

(iii) Fibre shall not be more than 5%.

(b) Dried grass meal or dried lucerne meal may be replaced by 1% controlled cod liver oil mixture and $1\frac{1}{2}\%$ oats.

(c) The completed compound must be capable of passing through a mesh of size equivalent to a B.S.S. No. 25.

Feeding of Weanlings

At approximately the age of 6 months milk and milk substitutes may be changed for more normal growth producing rations, and the problem of selecting those which will provide the necessary nutrients for growth will then arise.

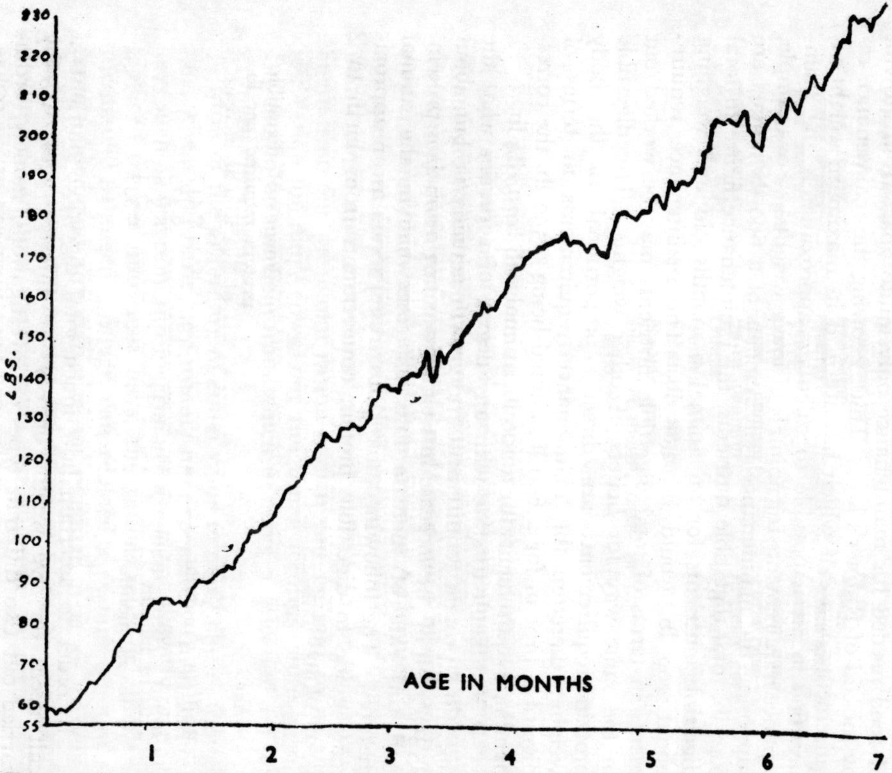

FIG. II. GROWTH CURVE OF A SAHIWAL CALF FROM BIRTH TO SEVEN MONTHS.

Fig 11 shows a typical growth curve for a Sahiwal calf at Lyallpur fed from the first month on whole milk, and then up to the age of 6 months on separated milk and a dry concentrate. The average daily increase was approximately 0.9 lb. over the whole period. It is possible to compute the amount of food needed for maintenance only, quite apart from growth, by the aid of Rubner's law. Thus, knowing the maintenance requirements of a calf which has attained its maximum weight of, say, 600 lb., it is possible to calculate approximately the maintenance requirements for animals of lower weights. For example, supposing the maintenance requirements of a 600 lb. animal are 6.43 lb. total digestible nutrients, the maintenance figure in total digestible nutrients for an animal 6 months old and weighing about 200 lb. would be 2.88 lb. The maintenance requirements in terms of total digestible nutrients may be worked out in the same way for heifers of other weights. The digestible protein requirements are directly proportional to the body weight; therefore the maintenance requirements in terms of digestible protein for a 600 lb. animal being 0.552 lb. the corresponding requirements for a 200 lb. animal will be 0.184 lb.

It is difficult to state with any degree of accuracy what the digestible protein requirement for growth actually is, but about 25 per cent of the increase in weight in young animals is protein, and this rate of increase falls off somewhat as the animal matures. The following table [Ellis, 1937] serves as an approximate guide to digestible protein requirements from birth to 2 years of age:—

Age.	Pounds of digestible protein required per lb. of live weight increase.
Birth to 3 months	0.417
3 to 6 months	0.375
6 to 12 months	0.333
12 to 24 months	0.333

In order to ascertain how much total digestible nutrients a calf needs for live weight increase at different ages, Armsby carried out two different types of investigations, namely, ordinary digestion trials such as have been previously described (Chapter III), and slaughter trials, in which animals fed on different rationing scales were killed at different ages and the composition of their carcases determined. Armsby interpreted his results in terms of Calories of nett energy required per lb. of live weight increase, as shown in the table below:—

Average age. Days.	Composition of Increase. Water. %	Ash. %	Protein. %	Fat. %	Energy content per lb. increase. Calories
8	62.55	3.55	19.24	14.86	1136
15	61.28	3.63	19.15	15.94	1182
21	62.13	3.50	17.15	17.22	1186

If Armsby's Calories are divided by 1360 (one pound of starch yields 1360 Calories) the energy required can be stated in terms of Kellner's starch equivalents, and by a simple calculation these may be interpreted in terms of total digestible nutrients required per pound of live weight. Employing this method the writer has computed the following table which indicates approximately the amount of total digestible nutrients required for Sahiwal calves at different ages and body weights:—

Age.	Approximate weight. lb.	Pounds of total digestible nutrients required per lb. live weight increase.
Birth to 3 months ..	60 - 140	1.56
3 to 6 months ..	140 - 200	1.88
6 to 12 months ..	200 - 325	2.50
12 to 24 months ..	325 - 500	2.81

The above constitutes a working basis on which rations may be computed for growing heifers. Thus, taking a 200 lb. heifer 6 months old, and assuming that the daily increase in weight is approximately 1.5 lb., the total digestible nutrients and digestible protein required for maintenance must be added to the total digestible nutrients and digestible protein required for growth at a specified age. The requirements for maintenance and live weight increase of a 200 lb. calf calculated on the above basis, will be 0.60 lb. digestible protein and 5.76 lb. total digestible nutrients.

The question of bulk also has to be taken into consideration. It is generally accepted that the amount of dry matter which a calf is capable of dealing with without undue gastric distention, is 2.5 per cent of its live weight; therefore a 200 lb. calf would require 5.0 lb. of dry matter, and a calf weighing 300 lb. would need 7.5 lb. of dry matter.

As these respective weights of dry matter should supply 5·76 lb. of total digestible nutrients and 0.6 lb. of digestible protein in the first case, and 6.62 lb. of total digestible nutrients and 0.9 lb of digestible protein in the second, the ration must be reasonably concentrated, and therefore, although the calf will be progressively able to eat more and more roughage such as hay, the nutritive requirements should be kept in mind in order to keep the ration properly balanced. Starchy foods such as the cereals will provide the total digestible nutrients, whereas more concentrated protein rich feed will provide the digestible protein, and the relative amounts of each in the ration will be largely determined by the nature of the more bulky food. If a rich legume hay is fed as part of the ration, less protein rich concentrates will be needed, but if the hay is an ordinary grass hay of inferior quality, more must be given. Keeping in view the above considerations it is possible to compute some representative rations for a 200 lb. heifer at the age of 6 months, on the assumption that it is kept in good growing condition, and the following feeding stuffs are available:—

Roughages	*Concentrates*
Green berseem.	Gram.
Green maize.	Wheat bran.
Green oats.	Linseed cake.
Oat hay.	

The 200 lb. growing heifer should get the following nutrients according to the standard outlined above:—

Total digestible nutrients.	Digestible protein.
5·76 lb.	0.60 lb.

and rations, of which the following is a typical example, may be computed according to the illustrations already given:—

	Total digestible nutrients. lb.	Digestible protein. lb.
Maize green 25 lb.	3.75	0.25
Concentrate mixture 2½ lb.	2.01	0.35
(Gram 2 parts, bran 1 part, linseed cake 1 part).		
	5.76	0.60

Slightly richer rations would be required for calves being reared for beef than for those reared for milk or work.

After the age of 6 months, if good grazing is available, the amount of stall fed concentrates may be diminished somewhat. In localities where rich *dhub* grass or legumes are available for grazing during summer, these may partly replace the roughage and concentrate which would otherwise have to be fed. In some circumstances rich grazing pastures alone may be sufficient to provide for maintenance and growth when the pasture is at its optimum nutritive stage. Later on, when the animals revert to stall feeding during winter, it would be necessary to provide a winter ration consisting of hay, green fodder, grains and cake. Assuming that the heifers are 18 months old, and weighing about 400 lb. each and still growing at the rate of 1.5 lb. a day, their daily requirements in total digestible nutrients and digestible protein would be 8.97 lb. and 1.20 lb. per head respectively. Some typical rations designed to provide these quantities of dry matter, total digestible nutrients and digestible protein are as follows:—

TABLE XVII

		Dry matter. lb.	Total digestible nutrients. lb.	Digestible protein. lb.
(1) Maize green	30 lb.	5.30	4.50	0.30
Wheat *bhusa*	2 lb.	1.84	0.90	—
Sarson cake	3 lb.	2.09	2.46	0.78
Wheat bran	1 lb.	0.77	1.11	0.12
		10.00	8.97	1.20

(2) Berseem green	30 lb.	4.50	3.00	0.46
Wheat *bhusa*	2 lb.	1.84	0.90	—
Cotton seed cake	3 lb.	2.76	2.84	0.54
Maize grain	1 lb.	0.90	1.23	0.20
		10.00	8.97	1.20

The object in feeding growing dairy heifers and young male stock is not to produce fat animals, but thrifty and well-conditioned ones, and the rations should be such as will avoid causing undue distention of the stomach or the laying on of too much fat. Faulty and inadequate feeding in the early stages of an animal's life may impair its subsequent milk producing or working capacity.

Feeding Buffalo Calves

Buffalo calves should be fed on exactly the same lines as cow calves, but as they are, as a rule, of greater weight, their rations will need to be adjusted accordingly, age for age. In other words, the same feeding standards as previously given for cow calves may be employed for buffalo calves according to weight and daily live weight increases

FEEDING OF THE STUD BULL

A young bull calf intended for the stud should be fed according to the principles laid down for female calves, but he should be allowed, if possible, a longer period on milk, which should not be curtailed too early and thus cause retarded growth, otherwise his efficiency as a sire may be impaired. The bull calf will, as a rule, make somewhat greater live weight increases than a corresponding female calf of the same age, and hence should be fed at a somewhat higher nutritive level than the latter. If properly fed, the bull should be ready for service at the age of about 2 years, according to breed, but his activities should be restricted somewhat to begin with. When full grown, he should be fed sufficiently well to keep him in good condition, but not excessively, so as to make him fat. Too great an allowance of dry roughages, or even silage, should be avoided, as these tend to depress his capacity for service. A bull does not need as much concentrates as a cow, and a good average ration to keep a bull in condition would be approximately half the amount a milch cow gets, say 4-6 lb. of concentrates.

It is essential that the bull when in service should be given rations rich in vitamins, particularly vitamin E. A useful adjunct to the the ration is sprouting *dhal* or wheat grains.

FEEDING OF BULLOCKS FOR MAINTENANCE AND WORK

It is unfortunate that but little experimental work has so far been done on bullocks in India and practically none elsewhere in regard to scientific rationing, and exact information on the subject is scanty, as most English and American trials have been carried out on fattening animals for beef production and not for work. Some experimental work, however, has been done at Lyallpur during recent years on working bullocks, from which an approximate estimation of what rations they require for maintenance and work has been made. This work is still in progress (1943) and the conclusions arrived at must, in the meantime, be regarded as tentative.

Table XVIII shows the maintenance requirements for bullocks as determined partly from experimental trials at Lyallpur and partly calculated from body weights in accordance with Rubner's surface law. (Maintenance requirements of animals of various live weights are proportional to their body surfaces. These are proportional to the two-third powers of their live weights, or the cube roots of the squares of the live weights.)

TABLE XVIII

MAINTENANCE REQUIREMENTS OF BULLOCKS (LYALLPUR)

	Body weight.	Dry matter.	Total digestible nutrients.	Digestible protein.
	lb.	lb.	lb.	lb.
Calculated	500	8.66	4.52	0.248
	600	9.90	5.16	0.284
	700	11.07	5.78	0.317
Experimental	800	12.20	6.37	0.350
	900	13.29	6.94	0.381
	1,000	14.35	7.49	0.412

Therefore a 1,000 lb. Hissar bullock requires 14.35 lb. dry matter, 7.49 lb. of total digestible nutrients and 0.412 lb. of digestible protein.

Working bullocks in the Lyallpur trials doing an average day's work of 6 hours ploughing were given the following summer ration which was found to be satisfactory and the bullocks showed a positive daily nitrogen balance.

		Dry matter.	Total digestible nutrients.	Digestible protein.
		lb.	lb.	lb.
Wheat *bhusa*	5 lb.	4.60	2.25	—
Jowar and *guara*	25 lb.	8.00	4.00	0.11
Gram	4 lb.	3.68	3.20	0.48
		16.28	9.45	0.59

Therefore, subtracting from these the amounts needed for maintenance, the amounts of dry matter, total digestible nutrients and digestible protein required for work are 1.93 lb., 1.96 lb. and 0.178 lb. respectively.

Four other experimental rations which were tried were the following:—

		Dry matter.	Total digestible nutrients.	Digestible protein.
		lb.	lb.	lb.
(1) Wheat *bhusa*	10 lb.	9.20	4.50	—
Gram and wheat	4 lb.	3.66	3.28	0.36
		12.86	7.78	0.36
(2) Wheat *bhusa*	8 lb.	7.36	3.60	—
Jowar and *guara*	15 lb.	4.50	2.42	0.21
Gram and wheat grain	4 lb.	3.66	3.28	0.36
		15.52	9.30	0.57

(3) Wheat *bhusa*	8 lb.	7.36	3.60	—
Jowar and *guara*	15 lb.	4.50	2.42	0.21
Gram	1 lb.	0.92	0.80	0.12
Cotton seed	3 lb.	2.76	2.10	0.36
		15.54	8.92	0.69

(4) Wheat *bhusa*	8 lb.	7.36	3.60	—
Jowar and *guara*	15 lb.	4.50	2.42	0.21
Bran	1 lb.	0.93	0.71	0.09
Cotton seed	3 lb.	2.76	2.10	0.36
		15.55	8.83	0.66

and a winter ration was:—

(5) Wheat *bhusa*	7 lb.	6.44	3.15	—
Berseem green	50 lb.	7.50	5.00	1.10
Gram	4 lb.	3.68	3.20	0.48
		17.62	11.35	1.58

The results of the five trials on summer rations are set out below to show the various quantities of dry matter, total digestible nutrients and digestible protein which each provides for work and maintenance:—

	Dry matter.	Total digestible nutrients.	Digestible protein.
	lb.	lb.	lb.
1.	16.28	9.45	0.59
2.	12.86	7.78	0.36
3.	15.52	9.30	0.57
4.	15.54	8.92	0.69
5.	15.55	8.83	0.66
Average	15.15	8.86	0.57

The average figures for dry matter, total digestible nutrients and digestible protein based on the five summer rations are 15.15, 8.86 and 0.57, so if the maintenance requirements of the bullock are subtracted from the requirements for work and maintenance, the figures for work can be obtained thus:—

	Total digestible nutrients. lb.	Digestible protein. lb.
Requirements for work and maintenance	8.86	0.570
Requirements for maintenance	7.49	0.412
Requirements for work	1.37	0.158

The bullocks performed approximately 6 hours continuous work a day, so the actual requirements per hour of work in terms of total digestible nutrients and digestible protein are:—

$\frac{1.37}{6} = 0.228$ lb., and $\frac{0.158}{6} = 0.026$ lb. respectively.

Therefore an 800 lb. animal working only 6 hours daily would require:—

	Total digestible nutrients. lb.	Digestible protein. lb.
For maintenance	6.37	0.350
For work	1.37	0.158
	7.74	0.508

On the basis of these figures it is thus possible to compute rations for a working bullock of any other weight and working for a particular length of time.

The following are examples of how the above principles may be applied:—

Example 1.

If the rations available are wheat *bhusa* and green maize, and if these are fed at the rate of 10.0 lb. of *bhusa* and 30 lb. of maize per day they will provide the nutrients shown below:—

		Dry matter.	Total digestible nutrients.	Digestible protein.
		lb.	lb.	lb.
Wheat *bhusa*	10 lb.	9.20	4.50	—
Maize green	30 lb.	6.30	4.50	0.30
		15.50	9.00	0.30

Such a ration provides sufficient dry matter and total digestible nutrients, but only contains about half the quantity of digestible protein shown above to be necessary for an 800 lb. bullock working 6 hours per day and is consequently not an adequate ration for a working bullock of this type. Hence it is necessary to supply some suitable addition to make up the deficiency in protein. If some of the wheat *bhusa* or the green maize, or both, are cut down and berseem hay, which is a protein rich feed, added, a more suitable ration is likely to result. The following may be suggested:—

		Dry matter.	Total digestible nutrients.	Digestible protein.
		lb.	lb.	lb.
Wheat *bhusa*	5 lb.	4.60	2.25	—
Berseem hay	6 lb.	5.22	3.00	0.360
Maize green	30 lb.	6.30	4.50	0.300
		16.12	9.75	0.660

This ration is an improvement and is slightly higher than the standard indicated previously, but may be considered for all practical purposes to be a reasonably good working ration.

An alternative ration for localities where berseem is not available, is the following:—

		Dry matter.	Total digestible nutrients.	Digestible protein.
		lb.	lb.	lb.
Wheat *bhusa*	4 lb.	3.68	1.80	—
Jowar stalks	7 lb.	6.37	3.01	0.217
Maize green	30 lb.	6.30	4.50	0.300
		16.35	9.31	0.517

The following may be suggested as alternative rations for east and south India:—

Ration for eastern India

		Dry matter.	Total digestible nutrients.	Digestible protein.
		lb.	lb.	lb.
Rice straw	10 lb.	8.00	2.71	0.04
Jowar green	20 lb.	8.00	4.24	0.17
Mustard cake	2 lb.	1.86	1.86	0.56
		17.86	8.81	0.77

Rations for south India

		Dry matter.	Total digestible nutrients.	Digestible protein.
		lb.	lb.	lb.
Cholam straw	15 lb.	13.21	8.34	0.30
Rape cake	1 lb.	0.93	0.93	0.28
		14.14	9.27	0.58
Cumbu green	50 lb.	13.82	8.43	0.47
Ground nut cake	1 lb.	0.93	0.85	0.42
		14.75	9.28	0.89

Following is another suitable summer ration for north India which may be considered to be satisfactorily balanced:—

		Dry matter.	Total digestible nutrients.	Digestible protein.
		lb.	lb.	lb.
Wheat *bhusa*	9 lb.	8.23	4.05	—
Maize green	25 lb.	5.25	3.75	0.25
Cotton seed cake	2 lb.	1.86	1.14	0.36
		15.34	8.94	0.61

Example 2. ***A winter ration***

For winter rationing berseem and green oats are available as succulent fodders, and if 10 lb. of wheat *bhusa* and 40 lb. of berseem are fed they will provide the following nutrients:—

		Dry matter.	Total digestible nutrients.	Digestible protein.
		lb.	lb.	lb.
Wheat *bhusa*	10 lb.	9.20	4.50	—
Berseem	40 lb.	6.00	4.00	0.88
		15.20	8.50	0.88

This is a well balanced ration and is rich in digestible protein.

An alternative ration containing green oats instead of berseem could be made as follows:—

		Dry matter.	Total digestible nutrients.	Digestible protein.
		lb.	lb.	lb.
Wheat *bhusa*	10 lb.	9.20	4.50	—
Green oats	45 lb.	7.20	4.50	0.54
		16.40	9.00	0.54

This is also a well balanced ration, but contains no concentrate. Another ration containing cotton seed cake as a concentrate would be:—

		Dry matter.	Total digestible nutrients.	Digestible protein.
		lb.	lb.	lb.
Wheat *bhusa*	8 lb.	7.32	3.60	—
Green oats	40 lb.	6.40	4.00	0.48
Cotton seed cake	2 lb.	1.86	1.14	0.36
		15.58	8.74	0.84

This also is a well balanced ration although slightly high in digestible protein.

The above illustrations have been devised from the experimental data obtained from the Lyallpur trials, and it may be interesting to see how they compare with some other rations commonly employed in India for working bullocks.

Typical examples of rations given in Punjab villages are:—

			Dry matter.	Total digestible nutrients.	Digestible protein.
			lb.	lb.	lb.
1.	Wheat *bhusa*	18 lb.	16.56	8.10	—
	Toria cake	2 lb.	1.84	1.48	0.60
			18.40	9.58	0.60
2.	Bajra green	80 lb.	16.80	10.10	0.74
	or				
	Jowar green	80 lb.	26.38	11.48	0.33
3.	Wheat *bhusa*	10 lb.	9.20	4.50	—
	Jowar & *guara*	60 lb.	18.00	9.68	0.84
	Gram	4 lb.	3.68	3.20	0.48
			30.88	17.38	1.32
4.	Rice straw	20 lb.	18.40	5.00	—
	Green grass	*ad lib*			
5.	*Jowar* stalks	18 lb.	16.20	7.92	0.014
	Guara grain	2 lb.	1.85	1.45	0.575
			18.05	9.37	0.589

The vast majority of working bullocks in India, as in the case of milch cows, are not fed even approximately adequate rations. It is a common practice to feed working bullocks in north India with *bhusa* and as much green fodder as possible, whilst in the eastern part of the country and rice growing tracts, the staple fodder is generally rice straw, which has been proved to be notoriously deficient. The result of such poor rations is seen in the unthrifty condition of the animals and a diminished capacity for work. The standard given above may be accepted as approximately correct, and may be adjusted according to the amount of work which a bullock is called upon to perform. It is not easy to compute nett returns in terms of work production from definite feeding standards, as was done in the case of milch cows for milk production. A milch cow produces a definite commodity of a certain cash value but the work a bullock does in terms of energy expenditure is not easy to measure and exact data of nutrients required for a given amount of work performed can only be determined by exact and elaborate dynamic trials.

FEEDING OF BUFFALOES

The problem of dealing with the feeding of buffaloes is approached with considerable diffidence, as little or no experimental work has so far been done on buffaloes, either in India or elsewhere, as far as the writer has been able to ascertain, and consequently but few data exist on which rationing standards can be drawn up or rations computed with the same degree of confidence as has been done in the case of cows. The calculations employed in this discussion are derived partly from the results of work recently started at Lyallpur, and partly from recognised custom, and they must therefore be regarded as tentative rather than final.

Buffaloes are ruminants of similar type to cows, and are automatically much the same. It may be assumed therefore that the general principles previously described for cows will apply *mutatis mutandis* to buffaloes.

The buffalo versus the cow as a milk producer

It might be expected from the large size of the buffalo that it would be a much more efficient producer of milk than the cow, but this is only the case when fodder is cheap. Taking cows and buffaloes as a whole, cows always beat buffaloes in total milk yield. The special value of buffaloes' milk lies in the high percentage of fat which it contains, an average figure being round 7 per cent, although 9 per cent may frequently be found.

This factor must therefore be taken into account in computing rations for buffaloes, as they require considerably more total digestible nutrients per pound of milk produced than the average cow. The data given in table No. XX, on page 353, concerning additional nutrients required for different fat contents in cow's milk may be taken as an approximate guide in computing rations for buffaloes, although the digestible protein and total digestible nutrients might be raised slightly.

Rations are computed for cows of a given weight and milk yield, in terms of dry matter, total digestible nutrients and digestible protein by the aid of the data given in appendices 1 and 2. Unfortunately, in the absence of much experimentally ascertained data for buffaloes, it is not possible to compute rations for them on precisely the same basis. Preliminary trials on buffaloes at Lyallpur have indicated that the maintenance requirements of a buffalo of from 1,200 to 1,400 pounds weight, in terms of roughages and total digestible nutrients are approximately 1.5 times those required by an average cow of 800 pounds weight. The Military Dairy Farms in India taking as their standard a cow of 800 lb. and a buffalo of 1,400 lb., allow one and two thirds times the amount of roughage and total digestible nutrients given to the cow, for the buffalo. If the fact previously mentioned that buffalo's milk is richer in fat than cow's milk is taken into consideration, and additional nutrients allowed on a pro rata basis, as was done in the case of cows for milk of different fat content, the following table may serve as a rough working guide for the computation of rations for buffaloes on the same basis as was done in the case of cows:—

TABLE XIX

Maintenance requirements for Buffaloes (Lyallpur)

Body weight.	Dry matter.	Total digestible nutrients.	Digestible protein.
lb.	lb.	lb.	lb.
700	10.94	7.20	0.623
800	12.06	7.93	0.687
900	13.15	8.65	c 749
1,000	14.20	9.34	0.809
1,100	15.29	10.05	0.872
1,200	16.38	10.76	0.935
1,300	17.47	11.47	0.998
1,400	18.56	12.18	1.061

TABLE XX

Additional nutrients required per pound of milk to be added to the above allowance for maintenance.

	Total digestible nutrients. lb.	Digestible protein. lb.
For 5.5 % fat in milk	0.376—0.397	0.049—0.059
For 6.0 % fat in milk	0.399—0.422	0.052—0.062
For 6.5 % fat in milk	0.422—0.446	0.054—0.065
For 7.0 % fat in milk	0.445—0.470	0.057—0.068
For 7.5 % fat in milk	0.468—0.494	0.060—0.071
For 8.0 % fat in milk	0.491—0.518	0.063—0.074
For 8.5 % fat in milk	0.514—0.542	0.066—0.077
For 9.0 % fat in milk	0.537—0.566	0.069—0.080

On the above basis a suitable summer ration for a buffalo of 1,400 lb. body weight, giving 24 lb. of milk a day, containing 7% fat, may be computed thus :—

TABLE XXI

		Dry matter. lb.	Total digestible nutrients. lb.	Digestible protein. lb.
Maize green	105 lb.	20.05	15.75	1.050
Wheat *bhusa*	5 lb.	4.60	2.25	——
Bran	5 lb.	4.65	3.55	0.450
Gram	1 lb.	0.92	0.80	0.120
Toria cake	3 lb.	2.76	2.22	0.900
	....	34.98	24.57	2.520

To enable these figures to be compared with corresponding data for a cow of the same weight and milk yield, the requirements for maintenance and milk production of an 800 lb. cow, according to the Lyallpur standard, are shown below:—

	Dry matter.	Total digestible nutrients.	Digestible protein.
	lb.	lb.	lb.
Maintenance requirements	10.06	6.190	0·463
For 24 lb. of milk with 6% fat	12.00	10.128	1·488
	22.06	16.318	1·951

The following is an example of a summer ration given to buffaloes weighing 1,200 lb. and giving 20 lb. milk per head per day containing 6% fat, in the Lyallpur Agricultural College Dairy:—

TABLE XXII

		Dry matter.	Total digestible nutrients.	Digestible protein.
		lb.	lb.	lb.
Green maize	125 lb.	26.50	18.75	1.250
Wheat *bhusa*	5 lb.	4.60	2.25	—
Bran	6.50 lb.	6.05	4.62	0.585
Gram	2.16 lb.	1.99	1.73	0.259
Cotton seed	2.16 lb.	2.03	1.51	0.259
Toria cake	2.16 lb.	1.99	1.60	0.648
		43.16	30.46	3.001

Prevailing practice in villages in India

It is well known that most village buffaloes in India are fed better than cows, because the zamindar pays more attention to them as they are his main source of milk and ghee. They are nevertheless not as well fed as they might be. A typical example which may be taken as representative of the summer rations fed to buffaloes in villages is as follows:—

TABLE XXIII

		Dry matter.	Total digestible nutrients.	Digestible protein.
		lb.	lb.	lb.
Jowar and *guara*	70 lb.	12.00	10.00	0.70
Wheat *bhusa*	15 lb.	14.70	6.75	—
Cotton seed	2 lb.	1.90	1.40	0.24
Toria cake	1 lb.	0.92	0.74	0.30
Gram	1 lb.	0.92	0.80	0.12
		30.44	19.69	1.36

If this ration is compared with that shown in table No. XXI it will be seen that it is considerably lower, particularly in digestible protein, which is less than half the amount the animal should actually be getting. It could be improved by substituting green maize for *jowar-guara*, diminishing the wheat *bhusa*, and increasing the cotton seed and *toria* cake. The following table represents what the ration improved on these lines would provide:—

TABLE XXIV

		Dry matter.	Total digestible nutrients.	Digestible protein.
		lb.	lb.	lb.
Green maize	100 lb.	21.00	15.00	1.00
Wheat *bhusa*	8 lb.	7.36	3.60	—
Cotton seed	3 lb.	2.82	2.10	0.36
Toria cake	4 lb.	3.68	2.96	1.20
Gram	1 lb.	0.92	0.80	0.12
		35.78	24.46	2.68

This ration provides almost the same quantities of dry matter, total digestible nutrients and digestible protein as indicated above for the standard which may be accepted as a provisional working basis.

Naturally the types of green fodder and concentrates available will vary in different parts of India and in different seasons, but whatever ration is obtainable, buffaloes should be fed rations computed so as to conform approximately to the above standard.

Feeding regime for buffaloes

The times for feeding and milking and all other details of stall management should be precisely the same as those described for cows.

It is a common practice in villages to soak the concentrates in water from 6—10 hours before feeding, mix the *bhusa* in the concentrate mixture, and feed this at the time of milking, the green fodder being given afterwards. The reason for this is that it induces the animal to take a larger amount of bulk in the form of *bhusa,* but, as has already been seen, *bhusa* is extremely unnutritious, and what the village buffalo actually needs is more protein in the form of concentrate. Another difference in the general feeding practice with buffaloes compared with cows, in the cotton growing tracts of India, is that a greater proportion of cottonseed and oilseed cakes is usually fed in the concentrate part of the ration for buffaloes than in rations for cows.

BIBLIOGRAPHY

1. Ellis, J. C. B. 1937. Feeding of Farm Livestock, p. 61.
2. Haecker. Minn. Buls. 71, 79, 140.
3. Henry and Woll. 1888-1889. 6 trials, Wis. Repts. p. 248-49.
4. Huffman, Duncan, Robinson and Lamb. Mich. Tech. Bul. 134.
5. Kellner, O. 1908. The Scientific Feeding of Animals, p. 339.
6. Lander, P. E. 1942. Indian grazing conditions and the mineral content of some Indian fodders. The I.C.A.R. Misc. Bul. No. 16.
7. Linton. R. G. 1927. Animal Nutrition & Veterinary Dietetics. pp. 248-49.
8. Maynard and McCay, N. Y. (Cornell) 1929. Bul. 543.
9. Meigs and Converse. 1925. Jour. Dairy Sci. 8, pp. 177-195.
10. Morrison, F. B. 1936, Feeds and Feeding, p. 480.
11. Morrison, F. B. 1936. Feeds and Feeding, p. 487.
12. Morrison, F. B. 1936. Feeds and Feeding, p. 1004.
13. Orr, J. B., Crichton, A., Crichton, J.A., and Middleton, W. Mineral nutrients in the rations of dairy cows. Jour. Scottish Board of Agric. Vol. VIII, p. 312.
14. Perkins, Ohio Buls., 389, 532.
15. Perkins and Monroe, Ohio Bul. 376.
16. Sanmann and Wright, Ill. Bul. 325.

CHAPTER X

SHEEP

Although practical feeding standards for cattle have been worked out in considerable detail in foreign countries, very little work has been done on these lines in connection with the nutritional requirements of sheep, and those standards which have been put forward from time to time vary very considerably in the scale of rationing which they advocate.

Data in regard to the feeding of sheep in India are even still more scanty and, although for some years sheep breeding has been fostered by certain government farms, very little in the way of actual feeding trials has been carried out, and no standards are available for Indian conditions. Most sheep are not fed at all in the ordinary sense of the word in this country, but they are allowed to roam at large over whatever grazing areas are available, and get little more food than they can find in their wanderings. No data in regard to the feeding of these under-fed masses are therefore of much value.

Where it is intended to keep sheep on a proper commercial basis, such a pastoral industry can only thrive provided proper feeding and management are practised as in other sheep breeding countries. This is particularly the case under Indian conditions, otherwise the price of mutton and the return for wool would not compensate for stall feeding on the lines adopted for dairy cattle. This, however, is a matter of considerable difficulty in India owing to the lack of pasture grazing compared with other sheep breeding countries, and the insufficiency of food resources to meet the requirements of the animal population. Nevertheless, there is room for further research on improved feeding systems to ascertain how far better feeding would be economically useful, as most of the observations so far recorded have been based on grazing alone.

Good grazing conditions on high quality pasture land, especially legume pastures, stimulate growth of wool, but grazing alone does not produce either the best mutton or the best quality wool, and in periods of drought sheep on grazing alone deteriorate considerably and the growth of wool slows down. For the best results to be attained the following therefore appear to be necessary:—

1. The improvement of grazing lands and rotational grazing of flocks.
2. Co-ordinating the breeding and suckling periods to synchronize with the optimum conditions of pasture.
3. Supplementing grazing during the breeding, suckling and fattening periods by suitable added concentrates on an economic basis, and providing fodder reserves to tide over periods of drought, thus avoiding deterioration in the quality of wool, and normal growth.
4. The periodic changing over of grazing land to avoid incidence of disease and digestive disturbances.

Pasture grasses vary in nutritive value within wide limits, and in sheep breeding areas the breeder should ascertain roughly the amount of nutrients which his sheep are likely to get from the pasture at different seasons. Sheep will require much more low quality pasture to meet their needs than a high quality one.

Sheep are bred in India for mutton and wool and to some extent for milk, as in the case of the Lohi and Damani breeds, but there are considerable differences in the food requirements of different breeds and also between different types of the same breed. A particular breed of sheep thrives best in the locality in which it has been reared, and it is notorious that breeds which are suited to one part of the country may not do so well if introduced into another. Changes in soil and climate will inevitably affect the specific nature of a particular breed, and this will be enhanced by a change over to a different feeding system, or on account of variations in the composition of crops in different localities. What may be suitable for one breed therefore, may not suit another. This also applies to different crossbreeds. Sheep which are fed on pastures alone will not thrive as well as when the pasture is supplemented during the breeding and suckling periods by added supplements, unless the grazing is sufficiently rich to meet requirements for maintenance and production.

FIG. 1. A Lohi ewe.

FIG. 2. A flock of Bikaneri ewes on stubbles.

FEEDING STANDARDS FOR SHEEP

Various standards have been proposed by different authori ties in regard to the daily requirements of sheep in terms of dry matter, total digestible nutrients and digestible protein. Ellis, (1937) while discussing the dry matter requirements or appetite, as it is sometimes called mentions the following standards :— Professor T. B. Wood of Cambridge gave 3.4 lb. of dry matter as the daily requirements for a sheep of 100 lb. live weight. Thus sheep feeding on good quality *Dhub* grass in normal conditions would need about 17 lb. of grass per day. Kellner's figure for dry matter requirements is 2.6 lb. Morrison gives 2.9—3.2 lb., whilst Watson gives the dry matter requirements as 2.7—3.0 lb.

Experiments conducted at Cambridge by Woodman, however, have shown that the figure proposed by Wood is somewhat too high, and he has reduced this in his standards for sheep of different live weights by approximately 1/5th. Foreign opinions differ very considerably in regard to the total digestible nutrients and digestible protein required per day, and in the absence of any data whatsoever in India to serve as a guide the standards evaluated by Wood and Woodman and given in Bulletin No. 48 of the Ministry of Agriculture and Fisheries, London, may be accepted as a working basis. These standards give the maintenance and production requirements in terms of starch equivalent, and the table shown below is adapted from these and expressed in terms of total digestible nutrients and digestible protein required for maintenance and live weight increase:—

Live weight	Appetite. Dry matter per week.	Maintenance requirements per week.		Total digestible protein for maintenance & production.	Production requirements per lb. live weight increase. T.D.N.
		T. D. N.	D. P.		
lb.	lb.	lb.	lb.	lb.	lb.
100	20.4	11.25	0.46	2.0	2.50
110	21.7	11.87	0.50	2.0	2.81
120	22.9	12.50	0.55	2.0	3.13
130	24.2	13.13	0.60	2.0	3.44
140	25.5	13.75	0.65	2.0	3.75
150	26.8	14.38	0.70	2.0	4.38
160	28.0	15.00	0.74	2.0	4.69

Thus a sheep of 150 lb. live weight and increasing in weight by 3.5 lb. a week would require for maintenance and growth a total digestible nutrient content in the ration of 29.71 lb. Very little information is available however in regard to the protein requirements of sheep for maintenance and growth, but a safe average ratio between the total digestible nutrients and digestible protein required for sheep is approximately 1:7, which may be narrowed somewhat to say 1:6 or 1:5 during the early stages of fattening, and prior to lambing, or even a little narrower during the weaning period. This ratio may gradually be widened again during the later stages of fattening, and after the weaning period.

Feeding of Breeding Stock

The data which serve as a guide in devising a feeding routine, are the dry matter required per week for different live weights, and the total digestible nutrients and digestible protein required for maintenance and growth. It has been seen in Chapters II and VII that great variations occur in the quality of different pastures, and also at different times of the year for the same pasture, and if the best results are to be obtained from the flock supplementary rations are desirable. The effects of supplementary feeding are seen from a simple feeding trial which was recently carried out at the Government Cattle Farm, Hissar [1939-40]. One group of 20 young ewes produced from mothers of the same age and weight, and of the same wool output in the first year, was kept on grazing conditions only. A second group of 20 similar ewes received one lb. of grain per head daily in addition to grazing, and the results obtained are shown in the following table:—

Group	Average weight on 1.3.39 lb.	Average weight on 1.3.40 lb.	Average wool output per animal lb.	Weight of lambs at weaning. lb.
Grain fed	57.6	90.0	6.1	47.4
Control	57.7	69.0	5.1	43.4

This simple trial indicates the beneficial effects of feeding, on breeding qualities and production, which might otherwise

FIG. 1. A Hissar Dale ram.

FIG. 2. A Bikaneri ram.

remain largely in abeyance. Supplementary feeds for the breeding stock are good berseem and lucerne hays and green fodders, such as maize, jowar etc., etc., but when green fodders alone are fed a small amount of protein rich concentrates should be given. Good silage is an excellent feed for reducing the cost of the ration, and from 1½ to 3 lb. per head daily, or even more, may safely be fed to ewes weighing 150 lb. If no silage is given sheep may be folded after lambing on turnips, where available, as in northern India. Neither silage nor roots need be given if good green pasturage is available.

When the roughage available is poor a grain supplement should be fed, although this will depend on circumstances, and as a rule, not more than ½ lb. of grain per head per day should be fed to ewes weighing about 150 lb. Whole oats are excellent for sheep and need not be crushed, as sheep masticate grain better than any other type of stock.

Maize alone is not suitable for breeding sheep but may be fed mixed with oats or other concentrates. It can, however, be fed as a sole supplement in combination with good legume hay. Should only inferior hay or straw or maize stover be available ½ lb. of cotton seed cake or ground linseed cake per head per day may be fed to pregnant ewes. On the other hand good legume hay alone is sufficient for pregnant ewes until after lambing when a grain supplement is advisable.

The following rations should be found satisfactory for pregnant ewes of 120—150 lb. weight, although many others are equally good when but little natural grazing is available:—

			Total digestible nutrients. lb. per head per day.	Digestible protein. lb. per head per day.
1.	Berseem hay	.. 2 lb.	1.00	0.19
	Maize green	.. 2 lb.	0.30	0.02
	Maize grain	.. ½ lb.	0.36	0.03
2.	Berseem hay	.. 2 lb.	1.00	0.19
	Maize stalks	*ad lib.*		
3.	Oat straw	.. 1 lb.	0.48	0.022
	Maize green	.. 2 lb.	0.30	0.020
	Cottonseed cake	.. ¼ lb.	0.14	0.045
	Oats grain	.. ½ lb.	0.34	0.022

The period of gestation of an ewe is usually about 5 months, and after the young lambs are weaned the ration of the ewe may be reduced considerably and not much more than a maintenance ration such as medium pasture or crop residue from arable lands need be given. This procedure will dry off the milk, and later on, about a week or two before tupping, feeding should be gradually improved so that the ewes are in condition but not too fat nor too lean. From now onwards fresh nutritious food should be available such as good pasture in preparation for pregnancy, and the management of the flock should be arranged so as to synchronize the breeding period with supplies of nutritious pasturage or other fresh fodder. When these are deficient, up to a pound of concentrate such as oats or similar nutritious feed should be fed per head per day. Naturally the feeding regime of the ewe will depend on many conditions, such as the breed, climatic conditions, the feeds available, and the time of lambing, but the supplementary feeds should be enough to augment the pasture sufficiently to provide the ewes with the necessary nutrients.

Morrison's [1936] standards give the food requirements of pregnant ewes at different stages, and these (see appendix 2) may be used as an approximate guide for computing the rations.

Feeding at, and after the Lambing Period

At the time of lambing the food of the ewe should be restricted for a few days and be easily digestible, and plenty of clean water given. Grain should be withheld for the first few days from ewes in good condition, although it may be necessary to give a little grain once if they are not in first class condition. After three or four days the grain may be gradually introduced, so that after a fortnight, ewes of 120-150 lb. weight will be getting about one pound of grain per head daily. An excellent alternative feed at this period is a mixture of bran and linseed meal and lucerne hay, plus grazing, which may be followed by a grain mixture of equal parts by weight of oats and maize, with from 2 to 4 lb. of good legume hay. Where first class pasture is available supplementary green feed should not be required, but the feed should be sufficient to ensure the maximum production of milk for the lamb.

Computing Rations for Ewes with Lamb

The above statement forms an approximate guide to the nature of the rations to be fed to ewes with lamb, but the amounts should be proportionate to the milk yield. An average ewe of 150 lb. live weight may be expected to give three gallons of milk per week. A sheep's milk, however, is much richer than cow's milk and contains about 6 per cent of protein and 7 to 7.5 per cent fat. Therefore, a comparatively narrower nutritive ratio will be required for sheep than for cows.

An ewe of 150 lb. live weight will require 14.38 lb. total digestible nutrients, and 0.7 lb. digestible protein per week for maintenance. The total digestible nutrients and digestible protein required for the production of three gallons of milk a week containing 7 per cent of fat will be 4 lb. and 0.48 lb. respectively. Therefore, the total digestible nutrients and digestible protein required for maintenance and the production of 3 gallons of milk a week will be 18.38 lb. and 1.18 lb. respectively. Corresponding figures for the maintenance requirements for ewes of other live weights can be arrived at by applying the surface law, described in Chapter IV.

Typical rations which will conform to these requirements are as follows: —

	Feed		Amount	Directions
1.	Oats	..	2 lb.	One pound of this mixture per head per day in addition to oat hay or grass hay *ad lib.*
	Bran	..	2 lb.	
	Linseed meal	..	1 lb.	
	Maize grain	..	5 lb.	
2.	Maize grain	..	4 lb.	One lb. of this mixture per head per per day with hay *ad lib.*
	Linseed meal	..	1 lb.	
3.	Maize grain	..	0.5 lb.	One lb. of this mixture per head per day in addition to grazing or hay ration.
	Oats grain	..	0.5 lb.	
	Maize green	..	3.0 lb.	
	Berseem hay	..	1.5 lb.	

Raising and Fattening Lambs

Lambs must be kept in thrifty condition if good mutton is to be obtained and fattening should proceed gradually in order that the fat will be laid on evenly with the lean tissues. If sheep are badly fed when young and hurried attempts then made to fatten them just before marketing, the fat will not be properly distributed with the lean as should be the case with first class mutton. Therefore, in the absence of suitable grazing it would be profitable to feed the lambs up to 1½ lb. of a supplementary mixed concentrate per head per day with hay, the proportion of concentrate depending on the quality of the hay.

At this stage an approximate guide to the total food needed may be found by referring to the table on page 359 which gives the food requirements for maintenance and live weight increases.

A variety of food stuffs may be available for this purpose depending on local conditions. For example, a suitable concentrate mixture could be made up as follows:—

(1) Twenty-five pounds each of bruised oats, linseed cake, crushed gram and bruised maize or barley, with a very small quantity of bran occasionally added, or a small quantity of molasses to make the ration more palatable.

(2) A mixture of 50 per cent bruised oats, 20 per cent linseed cake, 15 per cent crushed beans, 10 per cent dried grains and 5 per cent middlings (weatings).

Up to a pound of either of the above mixtures may be fed, but this regime will be governed by economic considerations and market requirements, and it may not be found economical where good grazing is available to feed more than ½ lb. of concentrate.

Feeding the Breeding Ram

The breeding ram* should be fed so that he will attain the highest degree of potency. As a rule he will need no supplementary feeding if on good pasture, but as the mating season approaches he should be given one of the concentrate mixtures indicated above. During the season he must be kept in good condition, and fed according to the demands made on him, and a fairly rich protein supplement will be needed of from

*The term Teg is applied to a male lamb between weaning and the first shearing period (up to 2 teeth) and before it is used for service. A male lamb from the first to the second shearing (4 teeth) or when it has started service is called a Tup and later on a Ram.

Feeding pen for lambs.

1—1½ lb. a day according to his size. This may be made up from 3 parts of oats and 1 part of wheat bran. The ram must not be allowed to lose condition, nor, on the other hand, should he be allowed to become too fat. If he is thin and needs conditioning a mixture such as:—

10 parts of oats by weight
5 parts of maize by weight
3 parts of bran by weight
2 parts of linseed cake by weight

should prove satisfactory as a supplement to pasture grazing.

A ram should be able to serve from 10 to 15 ewes according to age, although he may serve as many as 50 or more if mature and well fed. During the breeding season the ram should be allowed to run with the ewes for a certain time each day, or continuously if he is of a nervous disposition or inclined to fret. Rams differ in temperament, and their management will depend on circumstances, and when the breeding season is over they must be kept in good condition on good pasturage or fed legume hay and silage if available according to the season, with an occasional small allowance of concentrates. Some succulent food, including roots, should be given whenever possible, but mangel wurzels and sugar beet should be avoided as they tend to produce urinary disturbances. Plenty of exercise is essential and the rams should not be allowed to become fat, and should be kept away from the rest of the flock except during mating time.

Sheep have a very high capacity for good roughages and wherever possible they should be allowed as much as they can get, due regard being paid to the supplements needed for fattening and breeding. Growing lambs invariably grow more rapidly in proportion to their weight than dairy cattle and consequently require more total digestible nutrients than the latter for a given live weight increase. The nutritive ratio of the food for the different types of stock previously mentioned is important, and the food must contain protein of proper quality. This will generally be attained by feeding legume pastures or hays which make up deficiencies in the cereal grain.

Salt

Sheep should always be provided with a salt lick placed in suitable containers or troughs so that they can get it when needed. Sheep require more salt per unit of live weight than cattle, and from 0.20 to 0.25 of an ounce per head per day should be allowed.

Calcium and phosphorus

The general considerations in regard to calcium and phosphorus requirements outlined for dairy cattle apply to sheep, and, as the latter consume as a rule large quantities of roughage, the phosphorus content of these, if they are of good quality, should be sufficient. Where the supplements consist of bran and oilseed meals or cakes these should provide sufficient phosphorus for breeding and fattening, but there is likely to be a deficiency of calcium under poor grazing conditions, or when the supplement feeds consist mainly of cereals, and lack of calcium may be responsible for poor results when the rations include no legume hay.

It is impossible to lay down any definite rules as to whether calcium supplements such as bone meal, ground limestone, etc., should be fed, but the breeder will be guided by local conditions and practical experience. A safe rule to follow whenever pasturage or non-legume hays and roughages, which have been grown on soil of low calcium content are used, is to give one quarter of an ounce of a calcium mineral mixture per head per day. It is quite possible that the peculiar susceptibility which sheep show to changes in environment may be due to the different composition, particularly the mineral composition, of the feeding stuffs grown in different localities. This is a matter which has not been thoroughly investigated in India and may be of considerable importance.

Iodine

In districts where iodine deficiency in crops and consequent goitre prevail, treatment similar to that recommended for pigs (Chapter XIV) may be followed, and a very small addition of iodized salts not exceeding 1/20 of a grain per day given, but it is doubtful whether, under ordinary conditions, these precautions are necessary.

Watering of sheep

Sheep require plenty of fresh, clean water, and even in the cold weather, ewes on dry feeds may need as much as one gallon per head per day, and considerably more when suckling lambs. Lambs when being fed on dry feed only may drink up to $\frac{1}{2}$ gallon a day, which will be correspondingly reduced if they are on fresh pasturage or given succulent feeds. In the hot weather the amount of water sheep need will be considerably greater.

As sheep are very liable to parasitic infection they should not be allowed access to stagnant water or sources which may be polluted.

Feeding for Wool Production

Where sheep are reared for wool production the first essential is to select breeds of wool producing qualities. Most of the breeds producing the finer types of wool are the Merino and its derivatives, and these as a rule mature more slowly, and make smaller daily live weight gain than mutton breeds. Consequently they need more feed for each pound gain in body weight, and the general standard mentioned above must be augmented somewhat for wool production. Whilst fine types of wool are produced from the Merino, much coarse wool is produced from other breeds, and if the maximum amount of wool of the best quality possible according to its type is to be produced, the sheep should be fed so that they are in good condition at shearing time. Wool is very rich in cystine, the chief sulphur containing amino acid in proteins, the chemical constitution of which is given below:—

l-cystine, or dicysteine, or di B-thio alanine.

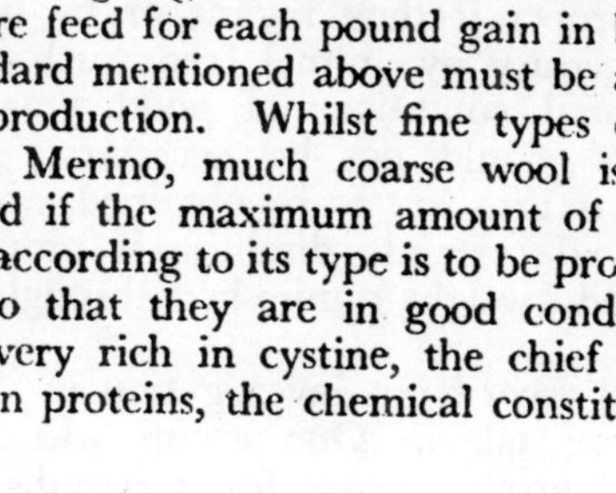

If rations are given which do not contain the right type of protein containing this amino acid the wool will naturally suffer. For example, poor quality grazing either alone or supplemented by cereal mixtures of deficient protein content such as maize will be deficient in this amino acid, but provided the rations are properly balanced as indicated in the general discussion above there should be no deficiency in this respect. The work of Darlow, Heller and Felton [see reference] has shown that there should be ample cystine for normal growth of wool provided ordinary balanced rations are fed.

Certain experiments have also been carried out in Australia on Merino sheep by Marston [1935] who has shown that when sheep are fed only on pasture grass which was very low in cystine containing proteins, great benefits were obtained by feeding blood meal, a feed rich in such proteins. Other animal meals may also be used to supply this amino acid.

For essentially wool producing breeds, therefore, a somewhat higher total level of feed is needed per pound of live weight increase than for mutton types, and the rations must be adequately balanced by a sufficiency of protein of high biological value containing the sulphur amino acid cystine. This can be attained by following a proper feeding régime or by supplementing the rations when necessary by animal feeds such as a blood meal, etc. Under normal conditions of good grazing and feeding such supplements should not be necessary. Good feeding is essential if good returns are to be obtained, and the amount of wool produced will certainly diminish if proper growth is retarded, or the body weight diminished through improper nutrition.

Kellner [1908] records an investigation in which two groups of 12 sheep were taken. One group was fed on ordinary meadow hay and ground beans for 4 months, its total weight at the beginning of the trial being 46.50 kilograms, and at the end 46.55 kilograms. The second group was fed on oat straw and mangels and its total weight decreased from 46.1 kilograms at the beginning of the trial, to 44.1 kilograms at the end. During the 4 months the first group fed the better ration produced 9.12 kilograms of roughly washed wool which contained 5.99 kilograms of pure wool, whereas the second group which was under fed yielded only 7.02 kilograms of roughly washed wool, of which 4.58 kilograms was pure. The difference in feeding therefore resulted in a loss of 1.41 kilograms of wool.

Kellner records another series of four trials in which similar results were obtained, and gives the results shown in the following table as and illustration of the fact that, although the growth of wool does not always suffer when the body weight diminishes, as seen from these figures, yet when the decrease in body weight, or dimunition in increase of body weight, passes a certain limit, then wool production *is* diminished.

	III	I	IV	II
Increase in body weight	0.79 kg.	0.42 kg.	0.17 kg.	1.05 kg.
Growth of wool	0.16 kg.	0.15 kg.	0.15 kg.	0.13 kg.
Growth of wool expressed in percentage of weight of fleece	0.306%	0.292%	0.293%	0.237%

Such trials emphasise the necessity for proper feeding, both as regards quantity and quality. Wool production with sheep is in some respects like milk production with cows, and there is a limit within which feeding can effect yield, that is to say, if more food is given than is necessary it will not necessarily increase the amount of wool obtained.

Kellner [1908] records the average figures obtained in seven other trials with sheep fed a bare maintenance ration, and 14 trials in which a proper ration was given. He quotes the average yield of wool obtained per sheep in the first case as 141.0 grs. representing 0.273% of the weight of the fleece at the end of the investigation, whilst the figures in the second set were 141.0 grs. of wool representing 0.286% of the weight of the fleece. These trials showed only a moderate increase in the weight of wool produced as a result of differential feeding.

In another experiment recorded by Kellner [1908] with lambs 5 months old, one group was given meadow hay and oats, and each lamb weighed 25.4 kilograms at the beginning of the trial of 9 months duration, and 46.25 kiiograms at the end when the animals were fat. The second group was given meadow hay only, and the average weight of each lamb was 25.0 kilograms at the beginning, and 36.15 kilograms at the end. The uncleaned wool from the second group weighed 2.69 kilograms compared with 2.18 kilograms of uncleaned wool from the animals which had received the better ration.

Kellner concluded therefore, that while the growth of wool may sxffer when the live weight of an animal steadily decreases as a result of insufficint food, yet very rich feeding does not cause a greater growth of wool than when the ration fed is sufficient for the maintenance requirements for grown sheep. Too much watery food should not be given, whether for wool or mutton production, as it is not suitable for sheep except in very small quantities ; the main food should be pasture hays with protein and other supplements.

Some General Principles in regard to Feeding and Raising Sheep in India

1. Sheep always prefer short grass and may even starve if pastured on land with long grass; hence it is a good plan in the latter case to graze cattle first.

2. It is desirable that grazing lands should have plenty of bushes to provide reserves of feeding material during the rainy season, and at times when pasture is dry and scanty owing to drought, heat or cold. This is specially desirable when supplements cannot be given.

3. Sheep should as a rule be allowed 12 hours grazing a day which may be reduced to 8 or 9 when there is plenty of pasture, and increased when pasture is very scanty.

4. Sheep should preferably be grazed in small flocks of from 100—150, as those in the front will get the best grazing, and those in the rear will suffer if the flocks are larger than this.

5. Stubbles left over after harvesting crops provide good grazing for sheep. Use should always be made of these but the change over from ordinary grazing to stubbles should be gradual and overcrowding avoided.

6. Pasture land should be grazed in rotation so that flocks do not remain on the same land for lang. This gives new grass an opportunity to grow and helps to prevent parasitic infection.

7. Clean drinking water should always be allowed at least three times a day.

8. It is best to keep lambs separate from adult stock as a prevention against parasitic infection.

Typical sheep grazing in Northern India.

9. Even when grazing facilities are such that concentrate supplements are not needed for ewes, the rams should preferably be fed some concentrates in order to keep them in good condition for breeding. A suitable feed would be 1 lb. per head per day of a mixture of 1 part of bran and 2 parts of crushed gram with 1 per cent of salt added. Some green fodder should be fed to rams when grazing conditions are poor, and when they are in servic about 4 lb. per head per day of any of the following mixtures may be fed:—

1. Oats	..	6 parts	4 lb. per head per day of the grain mixture.
Maize	..	3 parts	
Linseed		1 part	
and Berseem hay		2 lb.	
Maize green		4 lb.	
2. Gram		4 parts	4 lb. per head per day of the grain mixture.
Bran	..	2 parts	
Maize	..	2 parts	
Linseed	..	2 parts	
and Berseem or lucerne hay		2 lb.	
Oats green		4 lb.	

BIBLIOGRAPHY

1. Darlow, Heller and Felton. Okla. Bul. 220.
2. Kellner, O. 1908. Scientific Feeding of Animals, p. 250.
3. Marston, 1935. Jour. Agric. Sci. 25, 103-112.
4. Morrison, F. B. 1936. Feeds and Feeding, p. 1008.
5. Progress Report, 1939-40. Civil Veterinary Department, Punjab. Govt. Cattle Farm, Hissar. Ten year scheme for the investigation of indigenous sheep breeding at the Govt. Cattle Farm, Hissar.

CHAPTER XI

FEEDING OF GOATS

There are thirty million goats in British India, and feeding them constitutes a very difficult problem for the country. As with cattle, the vast majority of these goats belong to peasants who have no knowledge of correct principles of feeding, and the goats get their food from whatever they can find in their wanderings in the countryside, or in their migrations from one place to another. The problem of feeding goats in some parts of India, particularly in hilly tracts, is a matter of the most serious concern on account of the destruction of vegetation which they cause, and the consequent erosion of the soil. Goats will eat almost anything, although they prefer good, clean browsing on young trees and shrubs, and when left uncontrolled they cause great damage.

The herds of goats which roam the countryside subsist almost entirely on what may be called casual feeding stuffs. Pasture grass and the leaves of bushes are supplemented by leaves of trees and shrubs which the graziers lop off by means of a *dhanga,* a long pole with a curved knife at the end. As the owners of most of the goats in India make no special provision for grazing or feeding, it may be said that they are fed at the general expense of the public or by the largesse of nature. The damage which this indiscrimate feeding causes is great and wide-spread, particularly in forest areas, and one of the great problems confronting the country is control in these areas of the numbers of goats which are allowed to graze, and the institution of controlled grazing designed so as to prevent undue destruction of vegetation. This type of feeding, which the nondescript herds of migratory or casually grazing goats obtain, obviously cannot fall within the category of any proper feeding regime, neither can it be said to provide the goats with the feed they need for optimum milk production, or for bringing out the best quantities in any particular type. On the other hand, goats in India are extremely valuable animals because they can be cheaply fed. If properly managed, they are prolific in milk yield according to size, and the milk has specially valuable properties and is a boon to the poor.

The milk of the goat is alkaline, whilst cow's milk is acid in reaction. Also, goats' milk generally contains 4 per cent or more fat which is in the form of a much finer emulsion than is the fat of cows' milk. As a result, the fat forms a very fine curd during the process of digestion which requires on an average only about 30 minutes, whereas cows' milk requires about two hours. This relieves the stomach of a considerable amount of work, and hence goats' milk is exceptionally valuable on account of its high digestibility, for young children and invalids, and those possessing an impaired digestive system. Again, goats' milk has certain special mineral features. The milk of different animals contains a considerable number of different minerals but not all are found in the milk of any one type. Goats' milk, however, contains a greater variety of minerals than any other, and also several times as much iron per unit of volume as cows' milk. It has already been mentioned that iron is one of the minerals in which milk in general is deficient, and goats' milk is the best of all for providing this. Comparing the relative size of cows and goats, goats' milk is also very rich in phosphates and potassium salts.

Goats if properly controlled as regards breeding, management and feeding, should be of even greater value than they are at present. A good, well-fed goat will produce as much milk at less cost for feeding than the ordinary country cow in India. It will eat types of roughages which are useless to cattle, and it can be grazed more cheaply. If the best economic results as a milk yielding animal are to be obtained from goats, however, certain fundamental principles should be kept in mind. Only goats of the best milch breeds should be kept, and only good males of established pedigree and strong vitality used for breeding.

A considerable amount of work has been done in recent years on the development of milch goats at the Government Cattle Farm, Hissar, and at the Goat Breeding Establishment, Etah, U.P.

At Hissar an average milk yield of around 3 lb. milk per day has been obtained from the Beetal breed, or an average of 1.6 lb. for the wet and dry periods. If this milk yield is to be maintained and increased it is necessary to feed the animals in accordance with some nutritive standard. An approximate guide may be obtained if we bear in mind, firstly, that the maintenance requirements for goats per 100 lb. live weight are considerably higher than for cows, on account of their much smaller size, and consequently greater body surface in proportion to live weight, and, secondly, that goats require a somewhat higher amount of total digestible nutrients over and above their maintenance requirements per pound of milk produced, than do cows. Therefore, if the requirements by weight are calculated exactly as was done in the case of cows, and 20 to 25 per cent additional feed for maintenance allowed for the goats and about 10 per cent more total digestible nutrients in addition per pound of milk produced, this will furnish an approximate guide to the amount of feeding stuffs required according to the size of the animal and the milk yield. Goats, however, are able to consume much more food per 100 lb. of live weight than cows, even those of high milk yielding capacity.

Roughage and Concentrate Feeding

The Indian goat is by nature a browsing animal and its natural feed consists of young leaves of trees and plants. It will also eat almost all types of green materials, including coarse grass if not too rank, but prefers short sweet grass or specially grown fodder crops.

In a properly conducted herd, the goats should be allowed from 8 to 9 hours grazing a day, and in addition the milking females should be fed a concentrate mixture in amounts corresponding to their milk production.

Various concentrate mixtures may be used, but gram is always a safe basis. A mixture which has been found to be suitable at Hissar consists of two parts of gram and one part of bran, fed twice daily at the time of milking. A mixture found suitable at the Goat Breeding Farm at Etah is as follows:—

Crushed barley	1 part.
Crushed maize	2 parts.
Linseed cake when available otherwise mustard cake	2¼ parts.
Heavy wheat bran *i.e.*, bran produced by hand milling and not the milled bran	2 parts.

Another suitable mixture employed at Etah for goats consists of :—

Wheat bran	50 parts.
Crushed barley	15 parts.
Crushed maize	15 parts.
Crushed mustard cake	20 parts.

Two per cent of a mineral mixture (Churn brand feeding flour) should be added, or, if this is not available, salt may be substituted, and in all circumstances rock salt should be provided.

The routine followed at the Goat Breeding Farm, Etah, is to give the grain mixture as the first meal in the morning followed by some fodder at 9 a.m. The goats are then allowed to graze from 10.30 to 5 p.m., and at 6 p.m. the second concentrate meal is given, after which the hay racks are filled with whatever fodder is available. From 20 to 24 ounces of concentrates are given during the course of the day to the milking herd, and somewhat less than half this amount to the dry herd, the amount in either case depending on the size of the animal and the amount of milk given.

Fodders which are fed at Etah at different times of the year are the following:—

Napier and guinea grasses, lucerne, berseem, arhar, cabbage or cauliflower leaves, sunhemp pods, leaves of trees, weeds etc.

The feeding of goats in India has not yet reached a stage when such accurate computation of rations can be made as for cows, and it is not usual to weigh the fodder given, but to allow the animals as much as they can eat, an average goat requiring about 8 to 10 lb. of fodder a day.

In general, however, it may be said that milking goats may be given the same feeds as are fed to dairy cows, and a good average guide is that 10 goats will require approximately the same amount of food as one good average milk yielding cow. From 1 to 2 lb. of the concentrate mixtures enumerated above, with plenty of average pasture hay or berseem or lucerne hay, should prove sufficient, and about 1½ lb. of silage or roots may be fed in addition. When out at pasture the amount of concentrates given should be adjusted to the nature of the pasture and the milk yield, but high producing goats yielding three pounds of milk or more per day would need in any case from about 1 to 1.5 lb. of concentrates daily. The concentrate mixture need not be confined to gram or bran but any of the generally available concentrates used for dairy cattle may be substituted. Goats may also be fed clean, fresh garden and kitchen refuse or domestic grass cuttings, which will tend to reduce the cost of the feed.

It must not be assumed that because goats are such ubiquitous feeders that they may be fed on any kind of rubbish. Goats have essentially clean habits and this is why they suffer so little from digestive or other troubles. Their natural requirements are clean and nourishing food and clean fresh drinking water without which they cannot maintain good health or milk yield. Contaminated sources of either food or drinking water will endanger the health of the animals. As for other animals, the goats' ration should be properly balanced and include geen pasturage and concentrates according to milk production, and a suitable amount of green leaves or other succulent feed should always be included in the ration. The propensity of goats to eat the leaves of shrubs and trees gives them an advantage over cattle in the hot weather or in times of drought when green feeds may be scarce or unobtainable, but in a properly conducted herd the acquisition of such supplements should be properly controlled.

Goats will eat many types of leaves such as those of the babal or banyan trees or the shevari. They will also eat with relish the leaves and pods of various acacia and *kikar* trees, the leaves of such trees as the Terminalia, and in fact the leaves of all those trees or plants enumerated in Chapter VIII, dealing with famine fodders.

In winter, goats may be given an additional ration of 2 seers of *pala* (*Ziziphus nummularia*) at night time.

Milking goats should each be provided with a separate small box or container for the feed which should be given at milking time, the average amount generally fed being about ½ lb. or a little more at each milking time. Care should be taken in the feeding regime to ensure sanitary conditions, hence the need for separate boxes, because otherwise goats are extremely liable to suffer from worms and pick up the eggs of worms from their feeding stuff if it is fed on the ground. Worms are one of the most prevalent troubles which the goat in India has to face and the most likely method to avoid this trouble is by separate feeding and strict cleanliness. A useful way to feed green feeds or dry roughages is to have them tied up in bundles and suspended so that the goat can easily get at them, or feed them in a manger in such a way that the goat cannot get into the manger itself or contaminate it by its own droppings. Green feed should be fed fresh, and that left over in the mangers should be removed after feeding and the mangers then carefully cleaned out. An important point to remember in goat feeding is to feed no grass or other fodder when wet, and should such feeding materials become wet, they should be dried in the sun before giving them to the goats.

Goats are liable to suffer from digestive and other troubles if allowed to graze on fresh newly sprouting grass as in the beginning of the monsoon, and care should be taken to see that only small quantities of such material are eaten at a time and at well spaced intervals. Regularity in feeding should always be aimed at, and feeding three times a day, morning, noon and evening, may be found to give the best results, and only that amount of feed should be fed at a time which the goats will clean up completely within a reasonable time. Such a routine will prevent wastage and ensure that the goat obtains just the amount of feed it needs for keeping its digestive system in regular working order and itself in good condition.

Salt and Minerals

The same principles in regard to salt and minerals apply to goats as to other farm stock. Goats have a particular liking for salt and should always be provided with a rock salt lick hung in some accessible place so that they can always get to it when they desire, or it may be placed in the manger, but care must be taken to see that it is kept clean. Under ordinary normal conditions of grazing and feeding, goats should not suffer from mineral deficiencies, but should the exigencies of drought, famine or other factors, indicate the desirability of feeding a mineral adjunct, the same principles may be followed as with dairy cows.

Feeding of Kids

A sound practice is to allow the kids to suckle their mothers three or four times daily for the first week after birth, after which the number should be reduced to twice daily till about 4 months of age. They will thus get on an average from 16—20 ounces of milk a day. After the first four days to a week goats should be milked regularly and completely from one teat, the other being left for the kid until it is about a month or five weeks old. Should, however, a doe be required to rear more than one kid it should not be milked at all. Such a procedure is necessary for the future welfare of the herd. The weaning process should be gradual and by the time the kids are a month or five weeks old they should be able to start nibbling, and may be allowed access to finely cut hay with some lucerne, berseem and green arhar, leading up to some soft feed such as crushed gram or wheat bran, which are much relished by the kids when they are about 6 months of age. By the age of three months the kids should be weaned and then removed entirely from their mothers. After weaning they should be well fed and allowed to graze and browse and given supplementary feeds consisting of bran mash, oats, gram or other cereals. Particular care should be given to the feeding of male progeny intended for breeding purposes, but as with rams they should not be allowed to get out of condition by under-feeding or to get too fat by over-feeding and lack of exercise.

Computing Rations for Goats

Bearing in mind what has been said in regard to the somewhat higher maintenance requirements of goats per 100 lb. of live weight over those required for cows, and their need for a slightly higher total digestible nutrient content in the feed for milk production, a few representative rations which would provide sufficient nutrients for a goat weighing 150 lb. and yielding 5 lb. of milk per day are given on page 379. These have been computed according to the approximate standard indicated. This is to some extent speculative, and is based on experience rather than on data from scientifically conducted trials, as in the case of dairy cows.

		Parts by weight.
I.	Ground barley	1
	Wheat bran	1
	Dried beet pulp	1
	Cocoanut meal	1
II.	Dried beet or turnip pulp	6
	Ground barley	1
	Wheat bran	1
	Cottonseed cake meal	2
III.	Dried beet or turnip pulp	1
	Wheat bran	1
	Oats	1
	Cottonseed cake meal or cocoanut meal	1
IV.	Dried beet or turnip pulp	3
	Ground barley	1
	Wheat bran	1

CHAPTER XII

FEEDING OF HORSES AND MULES

No digestibility trials have been carried out on horses in India, and those which have been conducted in foreign countries such as America, England and Europe are far too few to enable the digestibility data obtained to be used for drawing up standard rations for horses as has been done in the case of cattle. Nevertheless, various authorities have evaluated certain standards which vary somewhat in the amounts of total digestible nutrients etc., which they advocate for different purposes. For example, Linton [1927] has worked out from trials conducted by him, the maintenance requirements of horses, and expressed these in terms of starch equivalents. His figures if expressed in terms of digestible protein and total digestible nutrients indicate that a 1,000 lb. horse requires approximately 6.25 lb. of total digestible nutrients a day, and 0.6 lb. digestible protein. Morrison [1936] adopting a different method for arriving at his data has given the total digestible nutrient requirements for the maintenance of a 1,000 lb. horse as from 7.9 lb. of total digestible nutrients and 0.6—0.8 lb. of digestible protein. These figures have been compared with corresponding figures computed from the rations which were fed to idle horses on a large stud farm in the Punjab, which show that the Punjab horses were being fed slightly higher than the above standard, a 1,000 lb. idle horse getting 9.68 lb. of total digestible nutrients and 0.90 lb. of digestible protein per day.

In the absence of any experimental data on the feeding requirements of Indian horses it may be advisable as a working basis to compute rations for horses according to the Morrison's standard given in Appendix 2. Morrison admits that, owing to the inadequacy of the experimental data relating to horses, his standards have been computed by taking the recognised foods for horses and calculating the digestible nutrients in them from the data of digestible nutrients which similar foods are shown to contain from experimental work conducted on ruminants. This procedure, therefore, assumes that a horse will digest the different ingredients of a particular foodstuff to the same degree as ruminants and *vice versa*.

The standards which have thus been worked out have, nevertheless, been found to constitute a reasonable working basis for feeding horses, and may be utilized for computing rations in India, due allowance being made for the fact that individual horses, as any other type of animal, may differ considerably in requirements. It is impossible to lay down an absolutely hard and fast rule, or to feed according to a strict mathematical formula, but reasonable safety may be assured by using the Morrison standards, if combined with practical experience of local conditions.

Feeds for horses differ from those for ruminants more in degree than in kind owing to the fact that the horse is a single stomach animal, and is consequently unable to digest as much coarse roughage as a ruminant of corresponding weight. Coarse roughages such as good grass hay or legume hay should therefore be fed to horses in less quantity than for ruminants, and the major part given at the end of the day after the concentrate part of the ration has been eaten. The amount given should be not in excess so as to cause undue distension of the stomach, or beyond the animal's actual requirements. When a horse is idle or doing only light work there is little necessity to feed very much high protein feeding stuffs. At the same time even when a horse is maintaining a high production yield in terms of work, the ratio between the protein and the non-protein part of the ration does not need to be so narrow as, for example, in the case of high yielding dairy cows. In the case of the cow the commodity produced is milk whose protein has to be obtained from the food, and allowance must accordingly be made to ensure a fairly narrow nutritive ratio. In the case of a horse, however, doing continuous hard work, or a polo pony doing intense work concentrated within a short time, the commodity produced by the animal is energy, and within certain limits this energy can be provided for just as well, or perhaps better, from the non-protein portion of the ration than from the protein, and therefore there is not the same necessity to feed protein rich foodstuffs to horses at work as there is to a cow producing milk.

A horse even at work will keep in better condition if fed

only that amount of food which he actually requires rather than an excess, particularly of protein, which he does not need. Efforts have beeen made in some highly technical trials carried out in America to compute the food requirements of horses by taking the net energy values of foodstuffs as determined by elaborate calorimetric estimations and determining the amount of energy which a horse expends in terms of foot pounds of work per unit of food. Such data, however, although of great technical interest, are of little use at present for determining rationing standards for horses. It is important, nevertheless, both for the welfare of the horse and from economic considerations to feed rations which are adequate and well balanced but not in excess.

Food Requirements of Horses for Maintenance

Grandeau and Le Clerc [1883] have determined the maintenance requirements of a 1,000 lb. horse from trials conducted over long periods, and have stated these to be from 6.5 to 7.0 lb. total digestible nutrients. Horses are more highly strung than cattle, and as a rule their maintenance requirements are somewhat higher for corresponding body weights. In some of the trials conducted by Grandeau and Le Clerc [1883], the experimental horses remained in good condition on considerably less total digestible nutrients than the figures given. It is no doubt safer, however, to feed slightly more, rather than less, than the minimum needed, and hence Morrison has given the somewhat high figures of 0.6 to 0.8 lb. of digestible protein, and 7 to 9 lb. total digestible nutrients for maintenance for a 1,000 lb. horse.

Idle horses will require about two thirds of the total ration fed to them for the purpose of maintaining body temperature, only one third being required for that part of the maintenance requirements which may be designated general up-keep or general metabolism. Consequently fairly poor roughages such as hay or even straw may serve for the purpose of heat production, but, as has been previously mentioned, straw should not be fed to working horses as the energy expended as heat in mastication, etc. is greater than that which the straw can yield. Horses kept in idleness should not be fed exclusively on straw, although this may form a reasonable part of the ration.

Amount of roughage needed

It is not advisable to feed a horse a larger amount of roughage at one time than is actually needed. Roughages are best fed at intervals during the day, and the major portion at night when the horse can chew it at leisure. The total amount of roughages to be given per day is also of considerable importance and should not as a rule exceed 1.5 per cent of the body weight of the animal, although slightly larger amounts may be fed to horses at work. Owing to the paucity of data available it is extremely difficult to state how far roughages and concentrates can be interchanged for work production, or, in other words, what would be the relative values per pound of roughages and concentrates for this object.

Certain experiments have been conducted in Sweden by Hansson and Meddel [1924], from which it has been concluded that 3 lb. of hay are equivalent to 1 lb. of maize or oats, but only 1.5 to 2.0 lb. are equivalent to 1 lb. of grain for maintaining an idle horse, whereas 2.5 lb. of good average grass hay must replace 1 lb. of barley or maize for work production. In making such substitutions, however, due regard should be paid to the fact that horses cannot utilize more than a certain bulk of roughage and hence this should be kept within the limits necessary according to the animal's weight.

As stated above a horse does not require too high a protein content in the ration even when at hard work, and a reasonable nutritive ratio is 1:10, although a somewhat narrower one may be allowed for pregnant mares, mares with foal at foot and growing colts. Thus growing colts over six months of age would require from 1.6 to 1.8 lb. of digestible protein per day, and from 11 to 13 lb. of total digestible nutrients, giving a nutritive ratio of from 1:6 to 1:7, whereas a mare with foal at foot, but otherwise idle, would require from 1.2 lb. to 1.5 lb. of digestible protein, and from 9 to 12 lb. of total digestible nutrients, and a nutritive ratio of from 1:6.5 to 1:7.5. Idle horses and horses at light work can be fed on a lower scale in conformity with the table below adapted from Morrison [1936] :—

	Dry matter. lb.	Total digestible nutrients. lb.	Digestible protein. lb.
Idle horse	13.0—18.0	7.0— 9.0	0.6—0.8
Horses at light work	15.0—20.0	9.0—11.0	0.8—1.0

As increasingly hard work is demanded from a horse the nutritive ratio may be slightly narrowed, but additional energy required from the feed may be provided by additional cereal feeds and high quality roughages such as legume hay; cereals, however, will provide considerably more energy than the latter.

Mineral needs of horses

The mineral content of the rations fed to horses is of the utmost importance and it is necessary to keep a careful check on the type of ration fed, because a typical standard ration, such as ordinary grass hay with cereal concentrates, is as a rule deficient in both calcium and phosphorus. While such a ration may be adequate from a mineral point of view for fully grown horses, it may prove to be deficient in minerals for growing colts or brood mares. In the latter cases provision has to be made in the ration for body development, and particularly the growing skeletal structure which is largely composed of calcium phosphate. Wherever possible a large part of the roughage fed should be good legume hay which is rich in calcium and phosphorus. Furthermore, a part of the concentrate should be bran which is richer than most other feeding stuffs in phosphorus, but the calcium-phosphorus ratio in bran is poor and needs correction. If rations deficient in calcium and phosphorus are used, bone formation will be defective and osteoporosis which used to be known as Miller's disease may result. Wherever adequate provision cannot be made in the ration for mineral requirements small quantities of high grade steamed bone meal or bone meal flour in amounts from 2 to 4 ounces a day should be given depending on circumstances and the age of the animal. The younger the developing animal the more necessary the mineral adjunct will be. Such mineral preparations should only be obtained from reputable firms which are prepared to state the nature and quantity of the minerals they contain and issue a guarantee of quality. Imperial Chemical Industries, Ltd., (India), supply mineral supplements which have been found to give beneficial results to mares in foal and young growing stock.

A mineral deficiency which is very liable to occur, but much more difficult to detect, in fact impossible without an elaborate chemical analysis, is iodine deficiency. Iodine is found in only infinitesimally small amounts in most feeding stuffs, but plays a most important part in maintaining the animal in good condition. Frequent cases of abortion have been attributed to a deficiency of iodine in the rations. Where such deficiency is suspected a small quantity of an iodine mixture should be incorporated in the rations. A suitable mixture may be made up as follows:—

Four drams of potassium iodide and 4 drams of iodine are dissolved in one pint of methylated spirit and added to 49 ounces of water.

One ounce of this solution may be given every day in the drinking water.

Some years ago the author investigated cases of suspected iodine deficiency in the following rations which were being used for mares with foal at foot:—

Oat hay	12 lb. per day.
Lucerne hay	4 lb. ,,
Green lucerne or green shaftal	20-30 lb. ,,
Grain mixture (Gram, oats, wheat and bran in equal proportion) ..	8 lb. ,,

It was computed that the amount of iodine that a mare was getting from the ration per day was 0.035919 dram. One ounce of the above solution would provide 0.007123 dram of iodine, so that if one ounce per day were fed to a mare, its previous iodine ingestion would be increased by 20 per cent.

It is naturally impossible to make accurate calculations for suspected rations in every case but the above will serve as an indication of a suitable iodine supplement, when this is deemed necessary.

Salt

Horses should always be provided with salt licks so that they can obtain whatever they need, as their requirements for salt are considerable, especially in hot weather.

Vitamins

The only vitamin deficiencies which are likely to occur in the rations of horses are in vitamin A and D. There should be a sufficiency of vitamin A if plenty of green pasture or legume hay is available, but vitamin D may be deficient in the rations at certain times of the year, and specially when they consist chiefly of inferior hay and cereals. The ill effects of lack of vitamin D will be still more pronounced if there is an imbalance in the calcium and phosphorus in the rations and this may lead to rickets in varying degrees of intensity. At the first sign of rickets or osteoporosis an animal should be given from 1 to 2 ounces of fish liver oil a day, and the roughage should consist of well cured legume hay. A calcium-phosphorus mineral mixture may be fed in addition.

Water requirements

Horses should be watered before meals rather than after, but access to water at all times is allowed in some stables. Frequent watering in small quantities, like frequent small meals is better than long draughts at unsuitable times. Opinions differ as to whether horses should be allowed to drink after hard exercise when sweating. There is no particular harm in watering a horse when hot, but it is necessary to guard against chills, and after violent exercise a horse should be kept moving and rubbed down. Cold water should not do a horse any harm if these precautions are taken. Naturally the amount of water a horse needs will be modified by the amount of succulent feed it is getting, the work it is doing and by climatic conditions. At the stipulated watering times horses should be allowed to drink all they need and not be taken away before they have had their fill. The water provided should be clean and of good quality, that is to say, not brackish or otherwise contaminated, and should be supplied in troughs which can be cleaned out regularly. These should be at a sufficient height from the ground to prevent either the horses or other animals from putting their feet in them.

Feeding Stuffs suitable for Horses

Concentrates.	*Dry roughages.*	*Succulent feeds.*
Gram.	Meadow hay.	Pasture.
Oats.	Legume hay.	Lucerne.
Bran.	Oat hay.	Green oats.
Barley.	*Jowar & bajra* stalks.	Green maize.
Linseed meal.	Wheat *bhusa.*	Green *jowar.*
		Green *bajra.*
		Green sudan grass.
		Green wheat.
		Shaftal.

The availability of these feeding stuffs will vary, and the selection made determined, partly by the rationing requirements and partly from considerations of cost and season. When it is desired to ascertain for example, whether it would be more economical, say, to feed oats or gram or barley, the amounts of each necessary will be ascertained in a manner shortly to be described, and their relative values in terms of cash per unit nutrient ingredient ascertained as described in Chapter XVI on the Economics of Feeding. It does not necessarily follow that the most expensive food is the richest in nutrients, nor is a cheaper food necessarily lower in nutritive value.

Gram is one of the standard concentrates used for horses in India and may be fed in amounts from 2.5 to 8 lb. per day according to requirements. The nutritive properties of gram have been described in detail in Chapter VI. As it is very rich in protein there should be no necessity to feed it to idle horses, but in Indian practice gram is probably used more extensively than barley or oats.

Oats contain considerably less protein than gram and the fibre content is considerable, and hence they make a more bulky feed. Horses have a particular liking for oats, which they find very palatable. Barley, wheat or maize have a greater tendency than oats to form a compact mass in the stomach, which may give rise to digestive disturbances and colic. Oats contain about 8 per cent protein, as against 18 per cent in the case of gram, and they may be fed with good pasture hay as a sole concentrate for horses on light work. If gram were fed as a sole concentrate, roughage of a poorer quality might be used, although this would involve a danger of mineral and vitamin deficiency; so it is best not to sacrifice the quality of the roughage unless unavoidable. When horses are being worked hard a small part of the oats may be replaced by richer concentrates such as gram and linseed cake. Oats must be fresh and untainted, as musty oats, or in fact any damaged cereals, are very liable to cause trouble to a horse's digestive tract.

Barley is fed extensively to army horses in India. It should not be fed whole nor as a sole concentrate but given with bran or chopped hay.

Wheat is sometimes fed instead of gram but it is a much more farinaceous food and is inclined to cause colic by forming lumps in the intestines, unless suitably combined with a more bulky feed such as chopped hay. The protein content of wheat is approximately the same as that of barley, but like other cereals it is deficient in minerals and vitamins.

Maize. After gram, barley and oats, maize is probably one of the best concentrate foods for horses if fed in moderation and not exceeding 10 lb. a day. Maize is a highly concentrated food and provided the ration is supplemented by protein rich feeds, such as the legumes etc., to correct the biologically deficient proteins of the maize, the latter may largely take the place of oats, only 85 parts of maize being required as a substitute for 100 parts of oats [Morrison, 1936]. The points to be kept in mind when feeding maize are its highly farinaceous nature and its deficient protein content, and to ensure that these factors are balanced by protein rich supplements. For example, if maize is fed with good quality berseem or lucerne hay this should constitute a satisfactory ration for horses doing moderate work, but the change over from oats to maize should be made gradually. There is little doubt that some of the prejudices against feeding maize are due to neglect of the above mentioned principles, and where the ration is otherwise properly balanced maize can conveniently be used for the cereal part of the ration. Should legume hay not be available in sufficient quantities, good pasture hay mixed with 1/3rd its weight of berseem or lucerne hay should make a well balanced ration for horses taking moderate exercise. The legume hay will also supply the necessary calcium in which maize is deficient. If, however, legume hay is not available and maize has to be given as the main concentrate, with only poor quality pasture grass, such a combination may be satisfactory for a short time but it is advisable to add some other supplement such as bran or linseed meal.

Bran is one of the most valuable protein feeds for all classes of horses and should constitute a part of the daily ration for working horses. It may also be given periodically to idle horses on account of its laxative properties. A bran mash made by pouring boiling water over the bran with a small quantity of boiled linseed makes an excellent feed, preferably at night, as a preventive against constipation. If bran forms any large percentage of the ration, however, a legume hay or bone meal should be included to make up for its low calcium content.

Molasses and Gur

Molasses may be given to horses in small quantities as an appetiser, but not more than one pound a day should be fed. It will also supply about 90 per cent of its bulk weight in total digestible nutrients, mostly in the form of sugar, and is valuable for using with low quality hay which may be unpalatable, as an inducement to horses to eat it. Being a highly concentrated saccharine food, molasses is capable of partly replacing the grain fed, weight by weight, in terms of total digestible nutrients. In hot weather molasses has a tendency to cause unnecessary sweating, particularly in the case of mules, and its use should then be restricted to a minimum. Various molassed foods have been produced in India in recent years, consisting of 50 per cent molasses and 25 per cent bagasse, combined with some bran and oilseed cake. When feeding such composite foods their composition should be known in order to keep the ration properly balanced. When good quality molassed foods are available at cheap rates they form useful, palatable and bulky foods, but they must never be allowed to become mouldy, as they are then likely to do harm.

Peas and beans

Peas and beans are suitable for horses. Peas have approximately the same chemical composition as beans but owing to their smaller size horses are not able to masticate them as well as beans. When they are used they should be crushed and in this condition they may take the place of beans. Beans generally contain about 20 per cent digestible protein and are very useful for inclusion in the ration of horses doing hard work, and 1 to 2 pounds per head per day may replace an equal quantity of oats. It is best not to give horses new beans—those some months old are better. Beans have a very hard seed coat and on this account they should always be gently crushed before being given to horses.

Horse gram or *kulthi* (DOLICHOS BIFLORUS) is largely fed boiled to horses in south India.

Millets

The millets such as *bajra* (*cumbu,* south India), *jowar* (*cholam,* south India), and many of the other common Indian grains may be used for horses. They contain somewhat less digestible protein than peas or beans and as most of the millet grains are both small and hard, they should be soaked before being fed to horses, or even boiled.

Linseed cake and meal

These do not form a prominent part of rations in India but they are invaluable if added at the rate of about ½ to 1 lb. to the rest of the concentrate for convalescent and sick horses. They are also useful for conditioning horses and giving a bloom to their coats for show purposes. Whole linseed may also be fed boiled.

Cottonseed cake and meal

Cottonseed cake and meal may be added as a protein supplement to the cereal rations, but the amount should be restricted to 1 lb. per day or less per 1,000 lb. live weight, as they are somewhat heavy and liable to cause digestive disturbances, especially if introduced suddenly and not gradually in small quantities into the ration. They will supply almost the same amount of total digestible nutrients as linseed cake, but have not the same conditioning value.

Preparation of cereal foods

The cereal grains should be fed crushed but not too finely crushed, as, in the latter form, horses are likely to bolt their food without proper mastication and it will form lumps in the stomach. To ensure proper mastication chaff or chopped hay is often added to the feed. If for any reason the grain cannot be crushed it should be soaked in water or boiled to soften the husk before feeding. Not more grain should be ground at one time than is required for a moderate period owing to its liability to absorb moisture and become musty. It is a usual practice in India to add a small quantity of salt to the concentrate mixture before it is fed.

Roughages

The merits of hay have been exhaustively described in Chapter VI. *Dhub* grass is one of the most valuable pasture grasses in India, and high quality hay made from this grass may

FIG. 1. A stack of lucerne hay being sieved through a wire *'charpoy'* on to a ground sheet.

FIG. 2. Half inch mesh wire *'charpoy'* on end to show construction. Sieved lucerne leaves in bundles on ground and lucerne straw in background. (Coleyana Estate, W. Punjab.)

contain up to 20 per cent protein, as has been found in the case of *dhub* hay at Hissar in the Punjab. If cut at the right time and properly cured for hay, it should constitute a suitable ration in itself for idle horses or even for horses doing light work. *Dhub* grass, however, like all other grasses varies considerably in composition in different parts of India and at different times of the year, and if allowed to become mature loses much of its nutritive value. If fed at this stage it will need some concentrate to balance it.

Oat hay

Oat hay is a useful food but is not as rich as the best *dhub* grass, and will therefore need to be supplemented by legume hay or concentrates in order to balance it. Thus a satisfactory ration for a 1,000 lb. horse doing light work would be:—

		Total Digestible nutrients. lb.	Digestible protein. lb.
Oat hay	12 lb.	5.76	0.264
Lucerne hay	5 lb.	3.00	0.550
		8.76	0.814

Legume hays

Legume hays are particularly fragile and liable to lose much of their value on account of the leaves and fine parts of the stem being broken off during the process of handling and carting. As a result of such losses, if legume hay is fed to horses as a sole roughage, they may have to eat too much dry matter to obtain the necessary nutrients. An excellent system to avoid such loss of legume hays is the following, which is cheap and practicable:—

When the stack is ready to be used, an ordinary *charpoy* covered over by wire netting, ½ inch mesh, may be placed beside the stack, and the hay sieved through it. Broken leaves and the fine parts of the stems will pass through and the coarser stems remain on the wire. The former can then be collected in bulk and carted in sacks direct to the feeding troughs and fed to all classes of stock after the evening grain ration. Later the rougher stalks may be placed in the troughs for subsequent consumption. By this means the horses get the rich leguminous food after the grain and can follow this up by leisurely mastication during the night.

As horses are very fond of well prepared good quality legume hay, it should only be given in the amounts needed, otherwise they will eat too much and may suffer from digestive disturbances. Naturally, less legume hays will be needed, weight for weight, than in the case of hays of lower quality. On no account should legume hay which has become mouldy be given to horses.

Other fodder hays which may be given are *jowar* and *bajra*, but these are much more fibrous than ordinary hays and hence of less nutritive value. They should be finely cut before feeding and form a convenient alternative to straw as an occupational food for idle horses for heat production, but should only be fed in strict moderation to horses at work.

Bhusas of the cereal grains such as oats, wheat and barley will also serve much the same purpose and may be fed in small quantities.

Natural Pastures and Succulent Feeds

Good natural pastures are the exception rather than the rule in India, but where available and properly cared for, are excellent for young horses, brood mares and idle horses. Even when horses are strenuously exercised they may be turned out to graze after they have eaten their evening ration, and during the hot weather it is an advantage to allow them out at night to graze instead of keeping them in stalls. Young pasture grass of good species is very nutritive and just before it has reached the flowering stage constitutes a maintenance ration for idle horses and can replace a considerable part of the roughage. If eaten at too young a stage, however, it may cause diarrhoea.

Where grazing is available pasture legumes should be introduced into the fields where possible. Subterranean and strawberry clover are suitable in areas not subjected to excessive heat or drought, as these require moderate climatic conditions and plenty of moisture. In more rigorous climatic conditions the wild vetch *rewari* (Vicia sativa) should be introduced into grazing areas. This is a valuable wild legume, and it is necessary to soak the seeds in water before sowing, otherwise they are unlikely to germinate. The wild vetch, however, will not stand very heavy grazing.

GREEN FODDERS

Lucerne, oats, shaftal and berseem are the best green cultivated fodders available for horses, and if fed from 20—30 lb. a day, combined with oat hay, to mares with foal at foot but not at work, lucerne hay and a grain mixture will constitute a balanced ration. As examples the following combinations are well balanced rations for a 1,000 lb. mare with foal at foot and doing no work:—

		Dry matter.	Total digestible nutrients.	Digestible protein.
		lb.	lb.	lb.
Oat hay	12 lb.	11.16	5.76	0.264
Lucerne green	20 lb.	5.01	2.94	0.613
		16.17	8.70	0.877
Oat hay	8 lb.	7.44	3.84	0.176
Lucerne green	20 lb.	5.01	2.94	0.613
Grain mixture	2 lb.	1.83	1.51	0.184
		14.28	8.29	0.973

Silage

Silage is an invaluable feed for horses during the winter months in localities where green fodder may be scarce. Maize is suitable for ensilaging, or better still a mixture of maize and *guara,* as the latter is a rich legume. *Jowar* stalks may also be used if mixed with *guara,* but by itself, *jowar,* particularly if ensilaged at the later stages of growth, is inferior. If properly ensilaged (see Chapter VII) these fodder crops form a good adjunct to the ration during the cold weather.

AMOUNT OF FOOD NEEDED BY HORSES

The amount of food that a horse needs will depend on its size, the amount of work and exercise that it is doing, and to some extent on the climate. Temperament will also be a factor, and a highly strung nervous animal will require, other things being equal, a somewhat greater allowance of food than a placid one. According to Morrison [1936] the total amount of concentrates and hay fed should be from 20 to 25 lb. daily for a 1,000 lb. animal and for those of other weights in proportion. Morrison gives the following amounts for different work standards under American conditions for a 1,000 lb. horse:—

		Dry matter.	Total digestible nutrients.	Digestible protein.
		lb.	lb.	lb.
Idle horses,	1,000 lb.	13.0—18.0	7.0— 9.0	0.6—0.8
Horses at light work,	1,000 lb.	15.0—20.0	9.0—11.0	0.8—1.0
Horses at medium work,	1,000 lb.	16.0—21.0	11.0—13.0	1.0—1.2
Horses at hard work,	1,000 lb.	18.0—22.0	13.0—16.0	1.2—1.4

A horse needs a certain amount of bulky food, for although concentrates alone would yield the necessary nutrients such a ration would not be suitable and should not be given. When being fed both concentrates and roughages the concentrates should be fed in three meals, in the morning, at noon and at night, and the more bulky roughage at intervals during the day, although the main hay ration should be fed last thing at night. Where horses are getting plenty of grazing the even distribution of their food will be naturally met.

RATIONING STANDARDS FOR HORSES

In the absence of data for computing rationing standards for horses under Indian conditions, the Morrison [1936] standards may be accepted as a working guide, due allowance being made for the results of practical experience. According to these standards the amounts of dry matter, total digestible nutrients, digestible protein and the nutritive ratios required for 1,000 lb. mares with foal at foot, empty mares and young stock are as follows:—

RATION FOR EMPTY MARES

		Dry matter.	Total digestible nutrients.	Digestible protein.	Nutritive ratio
		lb.	lb.	lb.	
Green *jowar*	12 lb.	3.0	2.16	0.12	
Bhusa	4 lb.	3.7	2.24	0.04	
Lucerne hay	4 lb.	3.7	1.92	0.36	
Maize grain	2 lb.	1.8	1.68	0.12	
Pasture	12 lb.	2.4	1.68	0.30	
Total:	34 lb.	14.6	9.68	0.94	1:9.4

RATION FOR MARES WITH FOAL AT FOOT

Green *jowar*	12 lb.	3.0	2.16	0.12	
Bhusa	4 lb.	3.7	2.24	0.04	
Lucerne hay	8 lb.	7.4	3.84	0.72	
Barley	2 lb.	1.9	1.58	0.18	
Bran	2 lb.	1.8	1.38	0.20	
Total:	28 lb.	17.8	11.20	1.26	1:8.0

RATION FOR YEARLINGS

Oat hay	2 lb.	1.9	1.03	0.06	
Lucerne hay	4 lb.	3.7	1.92	0.36	
Green *jowar*	10 lb.	2.5	1.80	0.10	
Gram	2 lb.	1.8	1.38	0.34	
Pasture	12 lb.	2.4	1.68	0.30	
Total:	30 lb.	12.3	7.81	1.16	1:5.8

The grain fed will depend to some extent on what is available. Suitable combinations would be:—

A. 2/3 by weight of oats
   1/3 by weight of bran

B. 1/3 of oats, 1/3 of gram and 1/3 of bran.

Should gram or bran not be available, maize may be used as a part alternative, but this should be supplemented by legume sievings as described above.

These alternative rations will supply the necessary nutrients and will, to a certain extent, make up for the laxative effect of the bran, which is largely due to its fibrous nature.

In drawing up standards for rationing horses it is necessary first of all to consider the maintenance requirements for a horse doing no work. According to Linton [1927] the maintenance ration for a horse weighing 1,000 lb. should supply not less than 5 lb. of starch equivalents, which, converted into terms of total digestible nutrients are approximately 6.25 lb., and 0.3 lb. of total digestible nutrients should be added or deducted per 100 lb. of live weight over or under 1,000 lb.. Linton [1927] also gives the digestible protein required for a 1,000 lb. horse as approximately 0.60 lb.. Morrison [1936] however, gives a somewhat larger figure, 0.80 lb.. After the maintenance ration has been provided for it is necessary to give an additional ration according to the amount of work being done. The figures advocated for work production are somewhat empirical owing to the difficulty in computing work in terms of energy expended. Various standards have been suggested by different workers, and Linton recommends the following additional nutrients over and above the maintenance requirements for every hour of hard work for a 1,000 lb. live weight animal:—

Total digestible nutrients. lb.	Digestible protein. lb.
1.25	0.15

For horses of other weights the amounts required will be in proportion. On this basis the following would be the requirements expressed in pounds of total digestible nutrients and digestible protein for a 1,000 lb. horse doing 4 hours hard work a day:—

	Total digestible nutrients. lb.	Digestible protein. lb.
Requirements for maintenance ..	8.00	0.70
Requirements for 4 hours' hard work	5.00	0.60
	13.00	1.30

One cannot go very far wrong in regard to the dry matter requirements, if the rations for horses at work are so arranged that the total dry matter contained in them amounts to 1.5 to 2 per cent of the animal's live weight, and for idle horses somewhat less, *viz*:—from 1—1.5 per cent of the live weight. The tendency in India on some stud farms and in the Army is to feed somewhat above these standards, but no hard and fast rule can be laid down and the standards given must be used in conjunction with practical experience.

According to these standards a 1,000 lb. barren mare doing little or no work requires the following nutrients:—

Dry matter. lb.	Total digestible nutrients. lb.	Digestible protein. lb.
13.0—18.0	7.0—9.0	0.6—0.8

Some years ago the author had occasion to investigate the rations fed to empty mares and mares with foal at foot of 1,000 lb. live weight on a stud farm in the Punjab. The rations being fed to the mares, and the nutrients these contained were:—

		Dry matter. lb.	Total digestible nutrients. lb.	Digestible protein. lb.
Oat hay ..	8 lb.	7.44	3.84	0.176
Lucerne hay ..	4 lb.	3.48	2.40	0.440
Lucerne green ..	30 lb.	7.52	4.41	1.020
Grain mixture .. (Equal portions of gram, oats, bran and wheat).	4 lb.	3.66	3.02	0.367
		22.10	13.67	2.003

This represents a somewhat higher nutritive level than the Morrison standard, to comply with which the following rations may be suggested as alternatives:—

		Dry matter. lb.	Total digestible nutrients. lb.	Digestible protein. lb.
(1) Oat hay	12 lb.	11.16	5.76	0.264
Lucerne green	20 lb.	5.01	2.94	0.613
		16.17	8.70	0.877

Or

(2) Oat hay	8 lb.	7.44	3.84	0.176
Lucerne green	20 lb.	5.01	2.94	0.613
Grain mixture	2 lb.	1.83	1.51	0.184
		14.28	8.29	0.973

The 1,000 lb. mares with foals at foot were being fed the following ration 'A', which is again considerably higher in all respects than the Morrison standard. Ration 'B' may be suggested as an alternative more in conformity with this standard.

'A'

		Dry matter. lb.	Total digestible nutrients. lb.	Digestible protein. lb.
Oat hay	12 lb.	11.16	5.76	0.264
Lucerne hay	4 lb.	3.48	2.40	0.440
Lucerne green	30 lb.	7.52	4.41	1.020
Grain mixture (Equal portions of gram, oats, bran and wheat).	8 lb.	7.32	6.04	0.734
		29.48	18.61	2.458

'B'

Oat hay	12 lb.	11.16	5.76	0.264
Lucerne green	20 lb.	5.01	2.94	0.613
Grain mixture	4 lb.	3.66	3.02	0.367
		19.83	11.72	1.244

The lucerne hay has been eliminated, the green lucerne reduced from 30 to 20 lb. and the grain mixture also considerably reduced. Ration 'A' was exceptionally rich and it is questionable whether this was desirable, either from a feeding or an economic stand-point. Such readjustments are matters which the horse breeder must decide for himself from his knowledge of the quality of the feed which he is giving and the practical results he is getting.

The same stud was feeding the following rations to colts of about 500 lb. weight:—

RATIONS FOR COLTS OF ABOUT 500 LB.

		Dry matter	Total digestible nutrients.	Digestible protein.
		lb.	lb.	lb.
Oat hay	8 lb.	7.44	3.84	0.176
Lucerne hay	4 lb.	3.48	2.40	0.440
Lucerne green	20 lb.	5.01	2.94	0.613
Grain mixture	6 lb.	5.49	4.53	0.551
		21.42	13.71	1.780

This again is considerably richer than the Morrison standard for such an animal, which is as follows:—

Dry matter.	Total digestible nutrients.	Digestible protein.
lb.	lb.	lb.
10.9—13.3	6.6—8.4	0.9—1.0

The following rations were accordingly suggested as alternatives:—

		Dry matter.	Total digestible nutrients.	Digestible protein.
		lb.	lb.	lb.
Oat hay	4 lb.	3.72	1.92	0.088
Lucerne hay	2 lb.	1.74	1.20	0.220
Lucerne green	20 lb.	5.01	2.94	0.613
Grain mixture	2 lb.	1.83	1.51	0.184
		12.30	7.57	1.105
Or				
Oat hay	4 lb.	3.72	1.92	0.088
Lucerne hay	2 lb.	1.74	1.20	0.220
Lucerne green	15 lb.	3.76	2.20	0.510
Grain mixture	4 lb.	3.66	3.02	0.367
		12.88	8.34	1.185

The colts were still growing, so both the alternative rations were kept somewhat on the high side. It is always a safe plan to feed somewhat above requirements rather than below.

It is interesting to note that the somewhat high level of feeding followed at the stud farm in question approximates very closely to the rationing standards laid down for Army horses in India. Many of the horses on this farm have been bred from imported stock, and are either of pure imported blood or crosses with country stock. It may be that such stock need a somewhat higher rationing standard in India, whereas a somewhat lower one is suitable for the pure country bred horse.

On this stud farm the above mentioned rations were being fed quite independently of free grazing in the paddocks, which was additional.

The object of the horse breeder is to bring his horses up to a certain weight and in good condition within a certain length of time, and experience will enable him to say whether this object can best be achieved by feeding his animals more, less, or in conformity with a given standard.

In comparison with the above the amounts and kinds of rations fed to Army horses and the nutrients they contain are:—

WHEN NO GREEN FODDER IS AVAILABLE

(A) *Rides, Class I, light draught horses and officers' chargers including Government chargers,* 15.0 *hands and over.*

		Dry matter. lb.	Total digestible nutrients. lb.	Digestible protein. lb.	
Fodder hay	20 lb.	18.20	10.20	1.100	Plus one
Gram	3 lb.	2.76	2.40	0.360	ounce of
Barley	4.5 lb.	4.14	3.20	0.302	salt per
Bran	2.5 lb.	2.33	1.78	0.225	day.
		27.43	17.58	1.987	

(B) *Rides, Class I, riding ponies, Class I, and officers' chargers including Government chargers under 15.0 hands.*

Fodder hay	18 lb	16.38	9.18	0.990	Plus one
Gram	3 lb.	2.76	2.40	0.360	ounce of
Barley	4 lb.	3.68	2.84	0.268	salt per
Bran	2 lb.	1.86	1.42	0.180	day.
		24.68	15.84	1.798	

(C) *Riding ponies, Class II, 14.2 hands and over.*

Fodder hay	18 lb.	16.38	9.18	0.990	Plus one
Gram	2.5 lb.	2.30	2.00	0.300	ounce of
Barley	3.5 lb.	3.22	2.49	0.235	salt per
Bran	2.0 lb.	1.86	1.42	0.180	day.
		23.76	15.09	1.705	

WHEN GREEN FODDER IS AVAILABLE

			Dry matter	Total digestible nutrients.	Digestible protein.
			lb.	lb.	lb.
(A)	Green lucerne	30 lb.	7.52	4.41	1.020
	Gram	3 lb.	2.76	2.40	0.360
	Barley	1 lb.	0.92	0.71	0.067
	Bran	2 lb.	1.86	1.42	0.180
			13.06	8.94	1.627
(B)	Green lucerne	30 lb.	7.52	4.41	1.020
	Gram	2 lb.	1.84	1.60	0.240
	Barley	2 lb.	1.84	1.42	0.134
	Bran	2 lb.	1.86	1.42	0.180
			13.06	8.85	1.574
(C)	Green lucerne	30 lb.	7.52	4.41	1.020
	Gram	1.5 lb.	1.38	1.20	0.180
	Barley	1.5 lb.	1.38	1.07	0.101
	Bran	2.0 lb.	1.86	1.42	0.180
			12.14	8.10	1.481

The rations given when green fodder is available are distinctly higher than the Morrison standard, but approximate more to it in the case of the rations given when no green fodder is available. In the former case although the weights of the animals are not specified the green lucerne provides the necessary bulk. If horses are doing hard work the total amount of dry matter available in the ration is somewhat below requirements, but if the animals are at rest, or only taking light exercise the dry matter appears to be sufficient.

Mares with Foal at Foot

The Morrison standard for a 1,000 lb. mare with foal at foot is:—

Dry matter lb.	Total digestible nutrients. lb.	Digestible protein. lb.
15.0—22.0	9.0—12.0	1.2—1.5

The rations fed at the stud previously mentioned for mares with foal at foot in summer and winter were:—

			Dry matter. lb.	Total digestible nutrients. lb.	Digestible protein. lb.
Summer.	Oats	4 lb.	3.64	2.68	0.176
	Bran	4 lb.	3.72	2.84	0.360
	Maize green	80 lb.	16.80	12.00	0.800
			24.16	17.52	1.336
Winter.	Oats	2.7 lb.	2.46	1.81	0.120
	Bran	2.7 lb.	2.51	1.92	0.240
	Gram	2.7 lb.	2.48	2.16	0.320
	Berseem green	30 lb.	4.50	3.00	0.660
	Bhusa	8 lb.	7.36	3.60	—
	Lucerne hay	8 lb.	6.96	4.00	0.620
			26.27	16.49	1.960

An examination of these rations shows that they contain larger amounts of dry matter and total digestible nutrients than given by the standard for mares with foal at foot. The Army rations are, on the whole satisfactory, but it might be possible by readjustment to effect a slight reduction in the dry matter and total digestible nutrients.

Alternate rations more nearly approximating the Morrison standard are:—

			Dry matter	Total digestible nutrients.	Digestible protein.
			lb.	lb.	lb.
Summer.	Oats	4 lb.	3.64	2.68	0.176
	Bran	4 lb.	3.72	2.84	0.360
	Green maize	60 lb.	12.60	9.00	0.600
			19.96	14.52	1.136
Winter.	Oats	1.4 lb.	1.27	0.94	0.616
	Bran	1.4 lb.	1.30	0.99	0.126
	Gram	1.4 lb.	1.29	1.12	0.168
	Berseem green	20 lb.	3.00	2.00	0.440
	Bhusa	8 lb.	7.36	3.60	—
	Lucerne hay	8 lb.	6.96	4.00	0.776
			21.18	12.65	2.126

The Punjab winters are very cold, and horses take more exercise than in the hot weather and need more net energy from the food for heat production.

An examination of the ration given at the above mentioned stud shows that the dry matter, total digestible nutrients and digestible protein are satisfactory. The summer rations, however, contain no dry roughage and a further alternative might be to replace a small portion of the maize, by some legume hay, so as to keep the dry matter, total digestible nutrients and digestible protein approximately the same. The object of the suggestion is on account of the fact that the summer ration is inclined to be deficient in calcium and phosphorus. This, however, has been provided for by the addition to the rations of all the animals, of one chattak per day of Churn brand flour mineral supplement, supplied by Imperial Chemical Industries, India. The winter ration is well balanced and comprises a considerable variety of feeding stuffs.

From 1—2 ounces per day of the iodine mixture previously mentioned are also given.

Rations for tonga ponies

The average weight of a Punjab tonga pony is approximately 600 lb. and whilst some are well fed and cared for, many do not get the necessary nutrients required for maintenance and work. On the assumption that a tonga pony is doing four hours hard work a day it would require the following nutrients for maintenance and the work performed:—

'A'

	Total digestible nutrients. lb.	Digestible protein. lb.
Requirements for maintenance	4.50	0.320
Requirements for work (4 hours)	5.00	0.800
	9.50	1.120

The ration which a moderately well cared for tonga pony actually gets in the summer is the following, which is well balanced and adequate:—

'B'

		Dry matter. lb.	Total digestible nutrients. lb.	Digestible protein. lb.
Dhub grass green	30 lb.	6.00	4.20	0.75
Chaffed *bhusa*	2 lb.	1.84	0.90	—
Gram	6 lb.	5.40	4.80	0.72
Bran	2 lb.	1.86	1.42	0.18
		15.10	11.32	1.65

A representative winter ration for this type of pony, which is shown below, appears to be somewhat richer in nutritive content than required by the standard 'A' above:—

'C'

		Dry matter.	Total digestible nutrients.	Digestible protein.
		lb.	lb.	lb.
Lucerne green	30 lb.	7.60	4.40	0.90
Chaffed *bhusa*	4 lb.	3.68	2.20	—
Gram	6 lb.	5.40	4.80	0.72
Bran	2 lb.	1.86	1.42	0.18
		18.54	12.82	1.80

This ration could be readjusted thus:—

'D'

		Dry matter.	Total digestible nutrients.	Digestible protein.
		lb.	lb.	lb.
Lucerne green	20 lb.	5.07	2.93	0.60
Chaffed *bhusa*	2 lb.	1.84	0.90	—
Gram	6 lb.	5.40	4.80	0.72
Bran	2 lb.	1.86	1.42	0.18
		14.17	10.05	1.50

Ration 'C' is richer in total digestible nutrients, digestible protein and dry matter than the standard, therefore the quantity of wheat *bhusa* has been reduced to 2 lb. and that of green lucerne to 20 lb.

It is well known however, that there are many tonga ponies which do not get anything approaching the rations shown above.

Riding horses

Oats are generally considered to form the ration *par excellence* for this type of horse, and may form the staple concentrate. Although oats probably hold prior place in the feeding regime, gram or other carefully selected concentrates may be given up to 8 lb. a day in three or four meals, the largest being in the evening. The concentrate mixture should contain a certain proportion of bran, especially if there is liability to constipation. Sufficient hay, depending on the animal's weight, and averaging 12 lb. a day, may be given but this amount should not be exceeded, as hay is a bulky feed and must not be allowed to cause the horse's body to get out of trim and proportion, as style and proper movement are essential for the riding horse.

The bulk of the hay should be fed long and the major portion given at night, the rest being fed at intervals between meals when the horse has leisure to eat it.

The race-horse

Whilst all the general principles of feeding apply to the race-horse, speed is the primary object needed, and consequently the ration must be such that no unnecessary fat is carried by the horse, as every ounce of weight beyond what is necessary will reduce speed. The Arab horse is noted for its speed and capacity for endurance and in Arabia is generally fed barley rather than oats. In India, practical experience and the availability of these two grains will determine to what extent oats should be fed to the partial or total exclusion of barley.

No better advice could perhaps be given in regard to the general feeding regime of race-horses than that of Woodruff, reproduced below:—

"After weaning, trotting-bred colts should be fed about 2 lb. of oats per day, with an unlimited allowance of hay. As the colt grows older the amount of oats should be increased to 4 lb. for the yearling, 6 lb. for the 2-year-old before training and 8-12 lb. for the colt 2 to 3 years old in training, an unlimited allowance of hay being given all this time. When going into winter quarters, the feed of the trotter should be reduced fully one-half in order to prevent fattening. A few carrots may be given and a bran mash occasionally, with good clean sweet hay. Horses whose legs must undergo blistering or firing should have more cooling feed, as mashes and carrots, with less oats, in order to reduce the tendency to feverish, inflammatory symptoms. Care must be taken not to permit the animal to get flabby or washy by too much soft food while undergoing treatment. Horses turned out in the field should be fed oats twice a day, for the exposure increases the need of heat-giving food.* In the spring when animals are shedding

**In north Indian winter.*

their coat, bran mashes are given in order to keep the bowels open. With the beginning of the training season the feed should be increased to 8 or 10 lb. of oats daily, in which case the horse wants less hay, but may still have all he will clean up unless he is a glutton. It is necessary to muzzle some greedy horses to prevent their eating the bedding long before the time for the race. No carrots or corn should now be given, unless it is necessary to induce a light feeder to eat his oats by mixing a handful of corn with them. During the jogging and after preparation, a bran mash about once a week, depending on the condition of the horse's bowels, will be proper. The trainer must never relax his vigilant observations or let his judgment sleep. During fast work, preparatory to the coming trial, the horse will be put upon his largest allowance of strong food. Some will not eat more than 8 or 10 lb. of oats a day, and it is necessary that such light feeders be not over worked. A good feeder ought to have about 12 to 13 lb. of oats with a fair amount, say 6 to 8 lb. of hay. Some will eat 16 lb. of oats a day."

Feeding Brood Mares and Foals

Great losses may be incurred in breeding due to neglect or improper feeding conditions and care of the mare. A cardinal point to observe is that exercise and light work combined with good feeding are essential for the production of good foals, but excessive work or strain should be avoided, and as foaling time approaches work should be reduced to a minimum and the mare allowed plenty of exercise out in the open. Special care is needed for brood mares to ensure an adequate supply of suitable food rich in protein, calcium and phosphorus, particularly if the mare is not fully mature. During winter a considerable part of the roughage should be well cured legume hay, and bran should form a part of the concentrate. The bran may be kept at a minimum if good hay is supplied and the mares are idle. A modification of the feeding regime is necessary at the time of foaling when the feed should be temporarily reduced. Just after foaling the mare may be given a light bran mash, and succeeding meals may consist of ground oats or a mixture of oats and bran. For the next few days the ration should remain moderate and not too rich in protein and may be gradually worked up to normal after about a week, when the mare may be allowed to graze on any pasture available.

Data on the weights of foals at birth and their rates of growth are extremely scanty, and in the absence of such it is not easy to give any reliable figures for Indian conditions. Morrison, [1936], however, has recorded data obtained at the Macdonald Agricultural College in Canada on over 400 foals regarding their weights at birth and rates of growth. The average weight of the sires of these foals was 2,050 lb. and of the dams 1,760 lb. The average weights were as follows:—

WEIGHT AND GAINS OF DRAUGHT FOALS

		Weight at the given age. lb.	Daily gain during previous periods. lb.
At birth	..	120	—
6 months old	..	730	3·4
1 year old	..	1,020	1.6
18 months old		1,350	1.8
2 years old	..	1,480	0.7
3 years old	..	1,790	0.9
4 years old	..	1,980	0.5

During the first six months, therefore, these colts increased in weight, on an average, 3.4 lb. a day; for the second six months the increase per day was 1.6 lb.; for the third six months 1.8 lb. a day, and during the fourth six months 0.7 lb. a day. During the third year the average increase in weight per day was 0.9 lb. and during the fourth year 0.5 lb. a day.

Where a suitable machine for weighing animals is available it would be interesting to record corresponding data for different types of animals and plot these in growth curves. This would also facilitate the compilation of rations where the weight of the animals has to be taken into account.

Useful data might also be obtained by grouping colts, under different rationing systems based on price and nutrient content to ascertain the optimum feed required to bring a colt to maturity within a specified time at a particular cost.

FIG. 1. Foals at foot feeding by themselves at troughs.

FIG. 2. Weanlings at feeding.

Young colts should be encouraged as early as possible to begin eating from their mothers' food, which they are usually able to do when about a month old. An excellent feed may be made from ground oats and bran, or from bran, linseed meal and ground maize in proportion by weight of 3, 1, and 4. Young colts require plenty of protein and minerals, and at the time of weaning when the mothers' milk supply of these nutrients is shut off, they should be getting from 2 to 2.5 lb. of protein rich concentrates per day, with sufficient good quality hay, preferably legume hay, but not in excess such as to cause distension of the stomach. Young colts may be given daily a small quantity of good quality ground bone-meal, from 1/4th of an ounce per head upwards according to their weight and the nature of the roughage available. An excellent preparation is Churn brand bone meal flour supplied by Imperial Chemical Industries, Ltd., India.

Part of the necessary protein may be supplied by feeding the colts with carefully pasteurised separated cow's milk, and when sufficient milk is available they may be allowed to drink as much as they can. The supplementary concentrate needed will thus be considerably reduced or may not be required at all for some months, especially if pasture of high nutritive quality is available. Whenever possible young colts should be given access to good grazing, which is ideal both for them and the mares, and they will be initiated into the art of grazing by imitating their mothers and nibbling the grass. It is only under exceptionally good pasture conditions, however, that colts can be maintained without any concentrate supplement. Should there be any tendency to rickets, fish liver oil may be administered daily in small quantities.

After the age of six months when foals are usually weaned, they may be fed equal parts of oats and barley, or 3 parts oats and 1 bran, or ground maize may be fed instead of oats, or a mixture of 2 parts maize, 2 oats and 1 bran. As they grow up a certain amount of crushed roots mixed with small quantities of bean or pea meal or some other protein rich food may be fed The amounts of nutrients in the feeding stuffs available may be computed from the tables of their digestible nutrients given in Appendix 1, and from the standard requirements shown in Appendix 2. Wherever the feeding stuffs fall short of the standard requirements, particularly as regards protein, the ration must be supplemented by other protein rich foods, such as linseed meal, cottonseed cake meal or other oilseed cakes, including soybean. In north India crushed gram soaked in cow's milk is often given to high priced weanlings.

The computation of such rations forms an important part of the duties of every breeder and will constitute an interesting and profitable pastime.

Feeding the Stallion

The most important points to be kept in mind in feeding and caring for the stallion are, to give him sufficient palatable and nutritive rations as required by the standards, and in conformity with the general principles enumerated above, and not to feed such an undue proportion of roughage as may cause him to become pot-bellied. The rations should be properly balanced in protein, minerals and vitamins and an integral part of the roughage wherever possible should be good quality legume hay. The latter, with the usual grain mixture will form a satisfactory basal ration for the stallion. When, however, only inferior hay can be obtained, a certain quantity of protein rich concentrates will be needed. For example, if only pasture hay is available and no legume, a suitable concentrate ration would be as follows :—

1. Four parts oats, one part bran; *or,*
2. Four parts oats, six parts maize and three parts bran; *or,*
3. Four parts oats, four parts maize and one part linseed meal.

The actual amounts of the various mixtures and the roughages fed cannot be definitely specified, but should conform approximately to the standard requirements for idle horses, or horses at work, according to the amount of work or exercise which the stallion is getting.

Examples of the rations fed to stallions by the Army Remount Department in India are:—

WINTER RATION FOR A THOROUGH-BRED STALLION

		Dry matter.	Total digestible nutrients.	Digestible protein.
		lb.	lb.	lb.
Oats	6 lb.	5.46	4.02	0.264
Gram	2 lb.	1.84	1.60	0.240
Bran	2 lb.	1.86	1.42	0.180
Lucerne	10 lb.	8.70	6.00	1.100
Green fodder (Maize or *jowar*)	40 lb.	7.96	5.48	0.360
		25.82	18.52	2.144

SUMMER RATION FOR A THOROUGH-BRED STALLION

		Dry matter.	Total digestible nutrients.	Digestible protein.
		lb.	lb.	lb.
Oats	5 lb.	4.55	3.35	0.220
Gram	1 lb.	0.92	0.80	0.120
Bran	2 lb.	1.86	1.42	0.180
Lucerne	10 lb.	8.70	6.00	1.100
Green grass (*Dhub* or *anjan*)	30 lb.	7.70	5.15	0.330
		23.73	16.72	1.950

DAILY FEEDING PROGRAMME

Horses may be fed according to the following daily programmes:—

1. The concentrate mixture may be given three or four times a day, the largest meal being reserved for the evening. The hay may be fed periodically in small quantities during the day after the concentrate feed has been taken, but, as with the latter, the major part should be given in the evening after the last meal of concentrate.

2. An alternative method is to cut up the roughage and mix it with the concentrate and allow the horses to have access to

it continuously through the day when in the stable, or from the nose-bag whenever leisure is available during work. This method may be preferred for hard-worked horses, but for lightly worked or idle animals the first method has much to recommend it.

Whenever it is necessary to feed larger rations than normal it is far better to give an additional meal or two rather than increase the size of the usual ones.

As an example of a feeding time-table the following is taken from Animal Management, 1933, page 131, compiled by the Veterinary Department of the War Office:—

	Heavy draught.		Other Horses over 15 hands.		Horses under 15 hands.	
	Oats	Hay	Oats	Hay	Oats	Hay
	lb.	lb.	lb.	lb.	lb.	lb.
Reveille	–	2	–	1	–	1
Morning	3	2(C)	2	1¼(C)	1½	1¼(C)
Mid-day	4	2(C)	2½	1¼(C)	2	1¼(C)
Afternoon	–	1½	–	1½	–	1½
5 p.m.	4	2(C)	2½	1¼(C)	2	1¼(C)
8 p.m.	4	2(C)	3	1¼(C)	2½	1¼(C)
Hay up	–	3½	–	2½	–	2½
Totals	15	15	10	10	8	10

Note: (C) = chop.

If it is not possible to feed the long hay as above, then give half at mid-day and half at evening stables. It is better to feed four times instead of three.

It is indisputable, whichever method is adopted, that the horse is best fed in small quantities at a time, and as often as possible, as this is its natural method of feeding and suited to its small stomach, which cannot, like that of ruminants, store large quantities of food. Furthermore, if the horse is fed less frequently but in greater amounts, he is inclined to bolt his food and the stomach which performs its work most efficiently when only partially full, will become overloaded, and indigestion may result due to insufficient salivation and insufficient room in the

Young mules feeding (Brucepur Estate, W. Punjab.)

stomach. If such unsatisfactory feeding is carried to excess, especially if a large quantity of succulent feed is given at one time, the food may ferment with the evolution of gas and possibly cause rupture of the stomach.

Large quantities of green berseem or other succulent fodder should not in any case be given at one time to horses, owing to the liability of distension of the stomach due to the evolution of gas. *Senji* is even more unsatisfactory in this respect and should be withheld altogether.

Feeding of Mules

The general principles of feeding, the types of feed fed, and the amounts given, as outlined for horses, apply in every respect to mules, with the proviso that, weight for weight, mules can get along satisfactorily on somewhat smaller rations than are needed for horses. Mules, however, are not so particular as horses about the quality of their rations, nor are they as a rule such greedy animals as horses, and they are less likely to bolt their food or take more than they need. In this respect they resemble cattle and may be fed roughage in the stalls and allowed to eat all they want. Mules are somewhat more particular than horses in regard to their drinking water and will often refuse to drink water which does not appeal to them, but which a horse will readily take. Mules, as is well known, are very temperamental animals, and changes either in the feeding or drinking regimes will be distasteful to them. For example, if they have been accustomed to be fed from troughs and watered from buckets they may refuse for a time to take their food from nose-bags, or their water from a stream. They require therefore particularly gentle and knowledgeable care in handling.

Scale of Mule Rations in India

The following table shows the standard Army ration fed to mules in India as given in Animal Management, 1933, prepared by the Veterinary Department of the War Office:—

	Gram crushed.	Barley crushed. (a)	Bran.	Salt.	Fodder hay, other than oat hay or lucerne dry. (b)
	lb.	lb.	lb.	oz.	lb.
Light draught mules, *viz.*:— draught mules in artillery, engineer, signal units and cavalry brigade transport companies other than those working in army transport carts.	2½	3½	2	1	18(c)
Pack artillery mules	2½	3	–	2/3	20
Sapper and miner and Class I equipment mules of signal service.	–	5½(d)	–	2/3	20
Class II equipment machine gun and Lewis gun mules of pioneer and infantry battalions.	–	5½(d)	–	½	15
Army transport draught and 1st class pack mules.	–	5½(d)	–	½	14
Army transport, 2nd class pack mules.	–	4½(d)	–	½	12

(a) In Burma paddy will be issued in lieu of barley.
(b) In stations where hay is not available, *bhusa* will be supplied in lieu. The ration for mules in all stations will be hay or *bhusa*, whichever is cheaper.
(c) Also 3 lb. bedding.
(d) Or gram, whichever is the cheaper.

In the U.S.A. Army, mules from 900—1,000 lb. weight are allowed 9 lb. of a mixture of oats, maize, barley and bran, and 14 lb. of hay per head per day. A light meal of the grain mixture is given in the morning, and the balance of the grain and all the hay in the evening.

In the Abyssinian campaign 8 lb. of beans and 15 lb. of hay were allowed daily to mules.

The actual amounts of concentrates fed will, as in the case of horses under service conditions, depend on what is available locally, but no great error will be made if their rations are the same as those given to horses but in slightly lower proportionate amounts. The same general principles as outlined for colts will apply *mutatis mutandis* to young mules.

The mule is a more hardy animal than the horse and can stand more rigorous conditions, or even neglect, without serious results ; the horse can only be maintained in good condition by proper care and suitable feeding.

BIBLIOGRAPHY

1. Grandeau and Le Clerc. 1883. L'alimentation du Cheval de Trait. III.

2. Hansson & Meddel. 1924. Centralanst. Forsoksv. Jordbruk (Sweden) No. 253.

3. Linton, R. G. 1927. Animal Nutrition and Veterinary Dietetics. p. 280.

4. Morrison, F. B. 1936. Feeds and Feeding. p. (i) 431. (ii) 431. (iii) 1006. (iv) 336. (v) 461. (vi) 1006. (vii) 1006. (viii) 468.

5 Woodruff. The Trotting Horse in America, pp. 90-105.

CHAPTER XIII

CAMELS

There are two classes of camel, the double humped or Bactrian camel which is found in Turkestan and Central Asia, and the single humped camel found in Arabia, Northern Africa, Egypt, the Sudan, Abyssinia, Somaliland and India, but it is chiefly the Indian camel which will be dealt with here.

Camels are very much creatures of habit and thrive best in the particular region to which they are accustomed, and those which are, so to speak, indigenous to the plains of India require considerable acclimatisation both in regard to feeding and treatment if transported to hilly or mountainous country.

The too prevalent opinion in uninformed quarters that camels will stand any amount of rough treatment, and can go for days on end without water or food, is a bad mistake to make if camels are to be kept in good condition. They need satisfactory and regular treatment and feeding just as much as horses or working bullocks, and their feeding regime should be regulated accordingly, keeping in mind that they are ruminants, that their stomachs consist practically of only two cavities, and that their capacity to eat and digest very large quantities of roughage at a time is not commensurate with that of true ruminants. Camels, being by nature herbivorous animals, get their food in normal conditions by grazing on whatever fodder or leaves of shrubs or trees they can find in the locality and when being worked normally in these conditions may, if the grazing is adequate, be able to do without any supplementary ration. Such camels are well supervised by their owners who take care that they are not turned out to graze in the heat of the day which the animals will generally refuse to do. If grazing is scanty, as it is in many parts of India, especially in the cold weather, camels will need a certain amount of grain ration, whilst natural grazing and other types of fodders constitute their natural food. They cannot digest large quantities of concentrates and therefore these must be regarded essentially as supplementary rations under normal conditions.

PLATE XXXVI

A Jacobabad camel type.

When grazing, camels keep continously on the move, and will eat a little here and a little there during their wanderings.

Under service conditions camels are unlikely to get the same facilities for regular grazing, and may frequently find themselves in localities where no grazing at all is available. While it is admittedly true that camels can, if driven to it, go for several days without food, they should nevertheless, be fed a proper quantity of dry fodder in the form of good *bhusa* perferably *missa bhusa* (legume *bhusa*), rather than white *bhusa* (wheat *bhusa*), or good quality hay wherever natural grazing is lacking. Camels which are fed a regular supply of bulky forage are likely to keep their condition better than when grazing is the chief source of roughage. It may not always be possible to adopt this regime, but it should be adopted as far as possible when hard work and long marches are called for. Green fodders are not suitable for camels doing hard work as they do not provide sufficient nourishment for the animals' nutritive requirements.

The regular feeding of camels with bulky forage ensures that they get sufficient food and obviates the danger of poisoning by poisonous plants. Green food of any description is relished by camels and they will eat almost any green crop, particularly legumes. When however they have been accustomed to dry roughages they should not be given too much succulent feed to start with, as they are inclined to overeat in such conditions and may suffer from colic as a result.

A safe general rule to follow, whether camels are living under normal conditions or more strenuous service conditions, is to treat them like any other herbivorous animal and ensure:—

1. That a sufficient amount of dry fodder be fed as an alternative to, or in combination with natural grazing, *and*
2. That they have sufficient time to chew the cud to ensure proper digestion.

When grazing is available and circumstances permit, they should be allowed at least six hours grazing a day, and when hard worked and this is difficult to obtain, dry fodder must be supplied.

Camels when free to graze will eat almost any green stuff they can find. Watt gives a long list of plants which are eaten by camels:—"Acacia arabica (babul), A. Farnesiana, Aegiceras majus, Albizzia Lebbek (Siris), Alhagi maurorum (camel thorn or Shutar Khar), which is collected in the Pishin valley in October and November, beaten up into Bhusa, and stored; Amarantus polygamus, Anthrocnemum indicum (Machur), Atriplex Stocksii, Avicennia officinalis, Bauhinia racemosa, Berberis (several species), Calligonum polygonoides, Carduus nutans, Carchorus antichorus, Cressa cretica, Crotalaria Burhia (Sis), Dalbergia Sissoo, Dodonaea viscosa (aliar) said not to have suited camels at Thul, Eclipta alba, Haloxylon multiflorum and Hal. recurvum (the salt plant most relished by the camel in India, from it Kharsajji is chiefly made), Haloc haris violaceae, Indigofera pauciflora, Kochia Indica, Lippia nodiflora, Leptadenia Spartium, Lycium europaeum, Melia azadirachta (Nim), Mimosa rubicaulis, Mollugo hirta, Phoenix dactylifera (Date palm), the ground kernels of the fruit and the leaves are given to camels, Pistacia integerrima and P. mutica, Prosopis spicigera, Psora lea plicata, Quercus ilex (the Holly oak), Rubia tinctorum (Madder), Salicornia brachiata, Salsola foetida et kali, Salvadora oleoides et persica, Suaeda fruticosa et maritima et nudiflora, Tamarix gallica, Tria nthema crystallina et monogyna et pentandra, Vitis carnosa, Zizyphus mummularia (Jhari) the staple camel fodder in Rajputana, Zygophillum simplex."

In Australia it has been found that camels are very fond of Sterculia or native poplar (Brachtchitob Gregorii), also of Swainsonias and Psoraleas. They do not care for oily scented Myrtaceae, such as Eucalyptus, Melaleuca, etc."

Feeding of Grains

Camels will eat almost any kind of grain once they become accustomed to it. Gram and barley are equally good and are usually fed in India. When camels are being hard worked, not less than five or six pounds of gram should be fed per day. Part of this may be given in the morning and the rest in the evening. The amounts fed at these two meals, however, should be adjusted according to the length of working hours, and under conditions of light work the major portion or even the whole of the grain may be fed in the evening. The gram should always be fed crushed as the teeth of camels are not suitably formed for crushing grain, and if the whole grain is fed a considerable amount may be passed whole

Camels feeding.

with the faeces without being digested. This is uneconomical and may also lead to digestive troubles. It is advisable that crushed grain should be fed with a liberal amount of *bhusa,* otherwise the camel, which is a voracious eater is likely to swallow the grain without proper mastication.

It is not a sound policy to feed camels an excessive amount of grain at the expense of fodder, otherwise the necessary bulk will not be obtained.

Salt

It is necessary to give camels a salt ration every day, and a good way to feed this is to dissolve the salt in water and damp the food with it. This makes the food more palatable and prevents loss by scattering and being blown away by the wind. It is customary in the army in India to give up to five ounces of salt a day and this has been found to produce a very beneficial effect on the health of the animals. If salt is not given there is great danger of a disease called Jhooling, a form of necrosis of the skin. When plenty of salt is provided in the ration, however, this disease is unlikely to occur.

Rations for Camels in India

In India it is customary in the Army to feed five pounds of gram or grain mixture per day under normal conditions, and seven pounds on active service conditions or when the camels are hard worked. The amount of dry fodder should be 18 lb. of *bhusa*, or 25 lb. of dry fodder, or 40 lb. of green fodder. Forty pounds of green fodder, however, although useful as a succulent vitamin yielding feed, which provides considerable bulk, will not yield the same amount of dry matter or digestible nutrients as twenty five pounds of hay. These standards have been found from experience to be suitable, and provided the camels keep in good condition and can perform their work when called for, there is little reason to unduly increase this allowance. The *bhusa* usually given is *missa bhusa,* and the grain a mixture of 2.5 lb. each of barley and gram. Other grains, such as rice, bajra and other millets may be used if judiciously fed. Salt is fed in addition as noted above.

The following table shows the standard rations for camels in India as laid down by the Veterinary Department of the War Office :—

		Peace.	War.
Ground barley or gram if barley is not obtainable		5 lb.	6 lb. (If inferior quality, 7½ lb.)
Fodder	green	40 lb.	40 lb.
	or		
	dry	25 lb.	25 lb.
or			
Bhusa		16 lb.	16 lb.
Salt		1 oz.	1½ oz.

Under strenuous conditions 1 lb. of gur may be fed per animal per day. It is useful also to feed Cumin seed (*Zira*) up to 1 lb. per month as an appetiser and gastric stimulant. The millets are difficult to crush and this cannot always be performed, but crushing is not so necessary as in the case of grains.

Other suitable alternative feeds which may be given to camels under different conditions are:—

1. 4 lb. grain and millets.
   20 lb. *bhusa* or wheat straw (without grazing).
2. 4 lb. grain.
   8 lb. roughage (when good grazing is available).
3. 5 lb. grain.
   12 lb. hay (without grazing).
4. 4 lb. grain.
   30 lb. dry fodder (without grazing).

Steel (1929) states that fresh beans and chopped straw may be employed as rations for camels, but that beans are found to be too heating under the hot arid conditions of the desert and should be replaced by barley and cotton seeds. Jowar however is considered to be the best grain to feed under such conditions. The same writer quotes the following as also being suitable rations under Indian conditions according to the amount of work being performed:—

(a) 20--25 seers *missa bhusa* daily, no gram.

(b) 6 seers *missa bhusa* plus 6 seers white or straw *bhusa,* plus 2 seers crushed and soaked gram or *mote*.

(c) 8 to 10 seers white or straw *bhusa,* plus 3 seers crushed and soaked gram or *mote*.

Camels and drivers at feeding time.

Camels watering.

In cold climates *mote* is better then barley, and in the cold weather in India, one pound of *ata* mixed with ghee may replace twice its weight of gram, the *ata* and ghee being useful for heat production.

Watering the Camel

It is as important that camels should be as well watered as well fed, and although they can undoubtedly subsist for long periods without water, even as much as ten days or longer in the case of Somali camels, they should be watered regularly according to their needs whenever opportunity permits. Under strenuous conditions of hard work or active service they should not be deprived of water longer than necessary, for the harder the work or the heavier the load camels have to carry, the more water will they need. Different breeds of camel vary much in their capacity to do without water, but the Indian camel will do well if watered every two days. When a camel has been deprived of water beyond its customary period for any length of time, it should be judiciously watered, and given only a small quantity to begin with and then after a short time allowed to drink as much as it can.

Under strenuous conditions during the hot weather a camel may drink over twenty gallons, a considerable proportion of which it can store in the form of a special physiological subcutaneous oedema. In addition, a considerable reserve can be held in the muscles and hump, but the actual amount needed will depend on the type of animal, the climate and working conditions. If they are working hard in high temperatures it is best to water camels every day, preferably in the afternoon or evening, on their way back from grazing. The large type of camel such as the Egyptian Delta camel should in any case be watered daily, whilst the Somali camel can go up to five days without water without loss of condition.

A sound general rule on service would be to water the animals every day if water is available. Camels, however, can be accustomed by careful training to take water less frequently than is their usual habit if necessity arises. They soon learn when watering may be expected, and so the watering regime should be made as regular as possible.

Camels are particularly fastidious animals in regard to their water and much prefer still to running water. The water should not to be too cold, and watering should usually take place before they receive the evening meal of grain. When they are allowed to drink water from a stream they should be given plenty of room, and they should not be taken away before they have had their fill. Camels like to be leisurely in drinking, as in all their other activities, and will usually take a long drought to begin with, and then stand about for some time before taking their full requirements, and if very thirsty they may take a considerable time.

Feeding in Stalls

Camels are unsociable animals when feeding, and under normal conditions each animal should be allowed to have its own separate trough or stall, these being arranged in lines. If they are allowed to feed communally from the same though they are inclined to become restive and kick or bite their neighbours, and young or weaker animals may be prevented from eating their proper allowance. Waste will also be avoided by employing separate troughs. Convenient troughs (Fig. 12) or mangers may be made by constructing carefully plastered mud circular depressions about six inches deep below ground level, with a circular rim of similar height above ground level, and so arranged that the surrounding earth slopes gently away from the rim. Permanent troughs may be constructed with cement.

Experimental Rations with Camels

Little experimental work has been done in India on the rationing of camels, but Cross [1917] has recorded the results of various experimental rations tried on medium sized camels which were doing no work and had poor grazing. He summarises his findings as follows: —

"The average amount eaten per animal was ;—

1. Gram 12 lb., plus *missa bhusa* 20.6 lb.
2. Gram 6 lb., plus *missa bhusa* 23.1 lb.

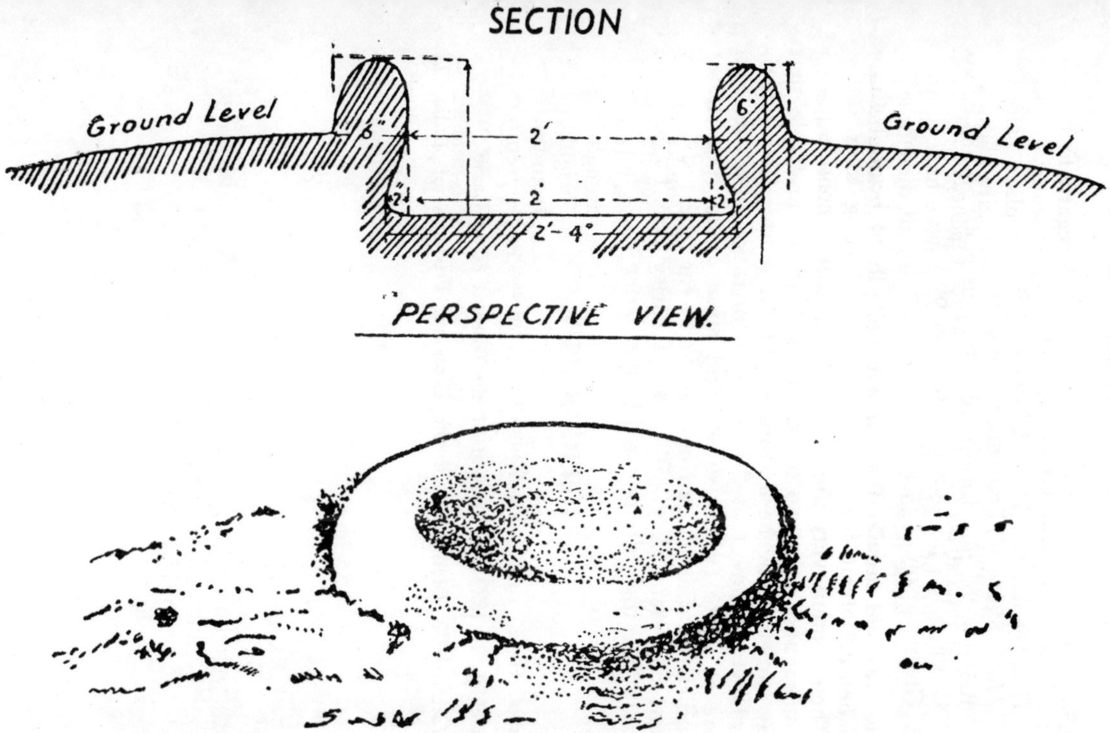

Fig. 12. MANGER FOR CAMELS.

SPECIFICATION—The mangers are circular, dimensions as above. They are made entirely of mud the surface being multi-worked. The idea of the outward slope is to prevent food blowing out as it does with saucer shaped mangers. Mangers should be placed 6 to 8 feet apart in lines. The Camel standings where possible being sloped.

3. *Missa bhusa* without grain, 24.9 lb.
4. Gram 6 lb., plus *moth bhusa,* 26.7 lb.
5. *Moth bhusa* without grain, 29.9 lb.
6. Barley 5 lb., plus turnips 26.6 lb., plus *moth bhusa* 26.4 lb.
7. Gram 6 lb., plus green taramira, 66.1 lb.
8. Green sarson, 135.4 lb.

Two year old camels ate on an average 4 lb. of gram plus 14.7 lb. of *bhusa*."

"These experiments were carried out with camels suffering from surra and doing no work. It is therefore probable that healthy camels doing hard work would eat more. The experiments however show that the Government ration of 16 lb. of a mixture of white (wheat straw) and *missa bhusa* (pea straw) is insufficient. Camels can be trained to eat wheat straw, hay and oat hay etc. They will also eat barley and Indian corn (maize), but they do not relish the latter".

BIBLIOGRAPHY

1. Cross, H. E. 1917. The Camel and its Diseases p. 24.

2. Steel, J. H. 1929. A Manual of the Diseases of the Camel.

3. Watt, 1889 Dictionary of the Economic Products of India. II. p. 58-59.

CHAPTER XIV

PIGS

The popular idea that it is the easiest possible thing to feed pigs, and that they may be allowed to eat anything, and hence do not need the same carefully planned rations as in the case of other farm stock is fallacious. The object of breeding pigs is to bring them to a certain live weight in the shortest possible time, and to ensure that the meat yielded by their carcases is of good quality. Good breeding and judicious feeding will attain these ends, but breeding will not produce a good carcase if feeding has been conducted on wrong lines.

The pig is one of the most efficient of all farm animals for converting the nutrients contained in the food into live body weight, and if properly fed, the rate of growth of the pig is greater than that of other animals. In feeding, certain basal principles must be kept in mind. The pig is essentially an herbivorous animal, and its natural method of obtaining its food is by grazing and feeding on roots and other ground crops which it can find. Nevertheless, although grazing is natural to the pig the pig's stomach, unlike that of the ruminant is unable to contain and digest large quantities of bulky foodstuffs. On the contrary, in order to obtain optimum growth, the major part of the food should be in the form of concentrates. Furthermore, the concentrates best suited for the pig are cereal grains, millets, gram and so forth, most of which are essentially carbohydrate foods rather than protein. Under natural conditions the pig is able to look after itself and chooses by instinct those foods which suit it best, but under controlled conditions of feeding in piggeries, the pig's activities in this direction are limited, and, consequently, feeding stuffs must be selected which are consonant with good growth. As the pig is largely fed on cereals, sufficient minerals must be supplied in the ration to counterbalance the mineral deficiencies of the former. During the early stages of the pig's life, when growth is greatest, it is necessary that sufficient protein be fed to provide for growth; hence the ration must have a narrow nutritive ratio, which may be widened considerably as the fattening period approaches. At this stage the primary object is to bring the pig into good selling condition, with well-conditioned fat distributed throughout the lean part of the carcase.

Bearing in mind that the prime object of pig production is to produce the maximum weight of good quality pork or bacon with the minimum expenditure on feeding stuffs, the feeding regime must be adjusted accordingly. It is desirable, therefore, to review the live weight increases which pigs may be expected to make under good feeding conditions, and the feeding stuffs and quantities needed to make these increases. It is also important to keep in mind the fact that, as for other animals, the ration must provide for both maintenance and production, *i.e.*, the ration must first of all cater for and maintain the pig's general bodily functions as described in previous chapters, and after the maintenance ration is provided, an additional amount of food must be fed for production in the form of growth or fattening. The pig's maintenance requirements at any particular age or weight must always be provided for, and therefore, when feeding for meat production the full production ration should be fed. Unless this is done the rate of growth will be retarded, and consequently the difference between the food required for production increase over that required for maintenance will be greater ; or expressed differently, food will be wasted if full production is not maintained and the pig brought to the required condition as early as possible. The principle is precisely the same as in the case of a cow producing milk. The cow needs its basal maintenance ration whatever the yield of milk; so will the pig. Hence, whether the pig is intended for pork or bacon, it should be ready for the slaughter house and of a certain body weight as soon as possible. Prolonged feeding beyond this stage will not produce proportionate increases in pork or bacon and the food then fed will be largely wasted.

Denmark was until recently the country most noted for its skill in regard to breeding and feeding for pork and bacon for the export market and the national industry had attained a high level of perfection. In recent years the pig industry has also reached considerable dimensions in England and the U.S.A., but in England, as the recent activities of the Pig Marketing Board illustrate, it has not yet reached the degree of perfection

FIG. 1. Large white Yorkshire sow.

FIG. 2. Middle white Yorkshire sow.

FIG. 3. Large white Yorkshire boar.

found in Denmark. In India the pig industry is on a much smaller scale and confined to certain well established piggeries situated in or near a few of the large towns. Keventer's piggeries at Aligarh, Calcutta and Darjeeling, and those of the British India Steamship Coy. at Calcutta are the best examples.

These piggeries are run on approved principles but the general standard of bacon, pork and ham production in India is still on a somewhat different level from that in the above mentioned countries. Furthermore, while in the latter the pig in-industry serves the requirements of the nation as a whole, in India it caters for only an extremely small part of the population, chiefly the European.

Choosing the Breed

The main breeds which are found in India are the Large White, the Middle White and the black Berkshire. In India the best quality pork is said to be produced from the pure Middle Whites, or by crossing the Berkshire boar with Middle White sows. The latter combination is ideal whether for pork or bacon. Excellent quality lean pork and bacon may also be produced by crossing the fine boned type of Large White boar with the Middle White sow, or the Berkshire sow may be crossed with the Large White boar for bacon, and with the Middle White for pork. In breeding from whatever type, however, the pig breeder should look carefully into the antecedents of the animals he selects in regard to their milk production, and the size and number of the progeny they produce.

Sows and boars of established milk yield and progeny production should be preferred. The breeding sow should have not less than twelve teats evenly spaced, and positioned as far forward as possible.

Once the breed has been decided on, and whether required for pork, bacon or dual purpose, the next object will be to adjust the feeding and arrange for a proper balance between stall feeding and pasturage where facilities for the latter exist.

Feeding Weanlings and Young Pigs

The weight of an average piglet at birth is from two to three pounds, and if the sow is in good condition and well fed, the piglet should put on weight rapidly in the first few weeks after it is born, and continue to do so if properly fed after weaning, the best time for which is at the age of six weeks to two months.

The efficiency of the future pig both from a production and an economic point of view will largely depend on the feeding in the transition period after weaning to full production feeding. The weaning process should be gradual and supplementary feeds introduced in very small quantities from about the age of one month, and gradually increased, till at the time of weaning the young piglets should be about 15 lb. in weight. They should then be fed on sloppy food and butter milk, and skim milk whenever available, until they are 6 months old, when their weights should be from 60 to 80 pounds. As soon as the piglets start to eat solid food when about a month old, they should be given the above mentioned food in gradually increasing quantities placed in shallow troughs so constructed that the little pigs can easily get at their food, but not the sows. Too much food must not be given, or allowed to become stale, and trouble is frequently experienced by not taking this simple precaution. From the age of one to two months onwards, the ration of the little pigs must be adjusted to meet a satisfactory live weight increase, and with proper feeding a certain live weight may be expected at any particular age. The following table [Ellis, 1937] shows approximately the weights which may be expected for different types at different ages:—

Age in months.	Pork type.	Bacon type.	Young breeding stock.
	lb.	lb.	lb.
0	3	3	3
1 ..	14	14	14
2	32	32	32
3 ..	55	50	42
4 ..	85	80	67
5 ..	120	115	97
6	—	155	130
7	—	200	165
8	—	245	200
9	—	—	230
12	—	—	280

Porkers feeding.

According to Morrison [1936] the amount of food which should be fed, and the live weights which may be expected at different periods of growth are shown below:—

Economy of gain of pigs at different stages of growth

Weight of pigs.	No. of Pigs.	Feed per head daily.	Feed daily per 100 lb. live weight.	Daily gain.	Feed per 100 lb. gain.*	Dressing percentage.**
		lb.	lb.	lb.	lb.	
Birth to 100 lb.	37	2.2	4.2	0.81	304	77.7
100 to 200 lb.	30	6.1	4.0	1.70	359	83.4
200 to 300 lb.	23	7.6	3.0	1.83	415	86.4
300 to 400 lb.	16	7.8	2.8	1.71	470	88.1
400 to 500 lb.	7	8.0	1.6	1.58	510	88.2

* 6 lb. of skim milk, fed to some of the pigs in the 1st period, considered equal to 1 lb. of concentrates.

** Average for 2 experiments.

The best economic live weight increases are attained somewhat sooner with porkers than with bacon pigs, and it is usual to allow the latter a longer time before reaching maturity. With pigs designed for breeding purposes the age of maturity can be somewhat delayed and the feeding regulated accordingly.

In the early stages of growth when the pig is putting on considerable weight in the form of body tissues the ration should provide for a narrow nutritive ratio of not more than 1:3.5.

It will be seen from the table on page 428 that baconers and porkers put on approximately equal increases in weight for the first five or six months, and at about eight months of age the porker should be marketed if first class pork is desired.

It will also be seen from the above table that the greatest average daily live weight increases occur with bacon pigs at from six to eight months, so that at this age it is necessary to feed more total digestible nutrients per lb. increase in live weight. Furthermore, the average requirements for maintenance are greater as the weight of the pig increases.

Rations for Young Pigs

From the time the young pig is fully weaned there is a considerable variety of rations which may be used, but it is advisable for the succeeding four months to feed swill (the refuse from hotel and domestic kitchens), butter and skim milk, after which about 3 lb. of swill per day together with green material such as cabbages and other green fodders may be given. An alternative to swill is up to 3 lb. of gram per day, one lb. of gram being approximately equal to 2 lb. of average swill.

While a considerable variety of rations may be employed for young pigs, gram usually forms the basis of these in India when it is obtainable. Gram may be combined with a mixture of barley meal, flaked maize, and middlings (see Ch. VI), with an addition of white fish meal when this can be obtained, so as to get a nutritive ratio of approximately 1 : 3.5. A mixture of gram and wheat, gram and barley or peas and barley is frequently referred to as *bejhar,* and about 2 chataks may be fed per day towards the end of the weaning period, and increased to from 5 to 8 chataks per day gradually from the age of 5 to 6 months. The following table shows the feeding routine practised at the Keventer Piggeries, Aligarh, U. P., and is reproduced by permission:—

Daily ration list for growing pigs

1.	5 weeks to 7 weeks	2 chataks	*bejhar*—half in the morning and half in the evening.				
2.	3 to 4 months (separated from mother)	5	do	do	do	do	do
3.	4 to 5 months	6½	do	do	do	do	do
4.	5 to 6 months	8	do	do	do	do	do

With as much grass and lucerne, preferably lucerne, as they can eat, and as much skim milk or butter milk as is available.

The diet should not be arranged too exclusively from one or two feeding stuffs only, such as wheat offals or other grains, as these are notoriously deficient in calcium. Various legume seed meals such as pea and bean meals may be fed as supplements, or moderate quantities of oil seed cakes, although cottonseed cake should not be given to young pigs. The best results whether for bacon or pork are obtained when dairy produce such as skim milk, butter milk or whey are fed. In the early stages of growth the food may be given in a fairly sloppy condition, but this should gradually be reduced and some of the food fed dry. If available, a certain amount of green food should also be given, but not in excessive quantities, and it should not be allowed to accumulate and become stale.

Young pigs should be allowed good grass or lucerne grazing whenever possible, the latter being particularly valuable on account of its richness in calcium and phosphorus, both of which are essential for growth of bone. Where sufficient quantities of legume feed are not available young pigs should be given a substitute such as fish meal or carefully steamed bone meal or commercial calcium phosphate. Oats should be crushed and sieved to remove most of the husk and fed only in moderation. As young pigs require chiefly concentrates, too much dry bulky material should be avoided. On this account bran is not very suitable as it is too fibrous. Potatoes should also be avoided at too early an age as they are too carbonaceous, are inclined to produce bloating and do not provide the requisite protein. They should be steamed or lightly boiled before feeding. Young pigs should be fed three or four times daily and only given as much at a time as they will finish.

Feeding Pigs for Market Purposes

After the young pigs have arrived at the stage when they can feed themselves they should be fed according to whether they are required for pork, bacon or breeding purposes. If intended for pork they must be fattened and ready for the market at from 7—8 months of age, when the live weight for a Middle White porker should be approximately 150 lb., and 120 lb. when cleaned. Cleaning will thus reduce the gross weight by about 20 per cent with the head still on, and by 27 per cent to 30 per cent when the head is removed. The intensity of feeding will be greater than for bacon pigs (which will not be ready till one year old), and almost the whole of the food will consist of meal. It is necessary in feeding pigs, while keeping these separate objectives in mind, to have a rough working guide as to the total digestible nutrients and digestible protein which the food should contain. But few digestibility trials have been

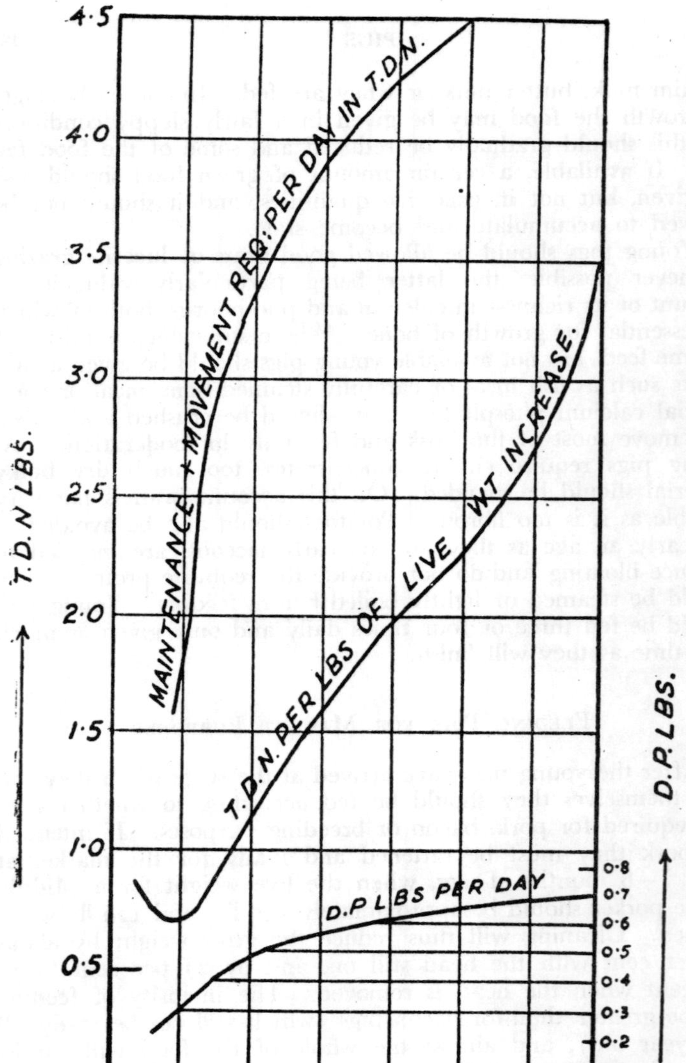

Fig. 13.—Chart showing requirements of total digestible nutrients and digestible proteins

carried out on pigs in comparison with the number on cattle, the most famous being those of Crowther at Newport, England, those of the late Prof. T. B. Wood at Cambridge, and the earlier trials of Kellner. Wood [1937] prepared a chart showing the calories needed per lb. of live weight increase, and the calories needed for maintenance plus ordinary movements per day, and also the protein requirements per day for pigs of different live weights. To facilitate computation from the analytical data in appendix 1, the author has converted this chart as nearly as possible, for sake of simplicity and ease in calculation, into requirements expressed as total digestible nutrients and digestible protein (Fig. 13). This chart may be used for computing rations for pigs exactly as described in the case of cows.

Another method sometimes used is to employ what is called a "unit of pig meal", with a total digestible nutrient content of above 80 which represents the average pig meal on the market. This means that 100 lb. of average pig meal contains 80 lb. of total digestible units.

Ellis [1937] gives the following daily requirements for porkers in terms of such a standard meal:—

Live weight lb.	Average daily requirements lb. meal equivalent.
Between 30 and 70.	$2\frac{1}{2}$ (5% of live weight).
Between 70 and 130.	$4\frac{1}{4}$ ($4\frac{1}{4}$% of live weight).
Between 130 and 170.	$5\frac{5}{8}$ ($3\frac{3}{4}$% of live weight).
Between 170 and 230.	$6\frac{3}{4}$ ($3\frac{3}{8}$% of live weight).
Between 230 and 270.	$7\frac{1}{2}$ (3% of live weight).

If reference is made to the table on page 428 it will be seen that from the second month to the fifth month a porker should put on 90 lb. in weight in 90 days, or one pound a day; hence from the above table the daily requirements in terms of standard pig meal would be:—

$$\frac{2\frac{1}{2}+4\frac{1}{4}}{2}=3\frac{3}{8}$$

Therefore, on an average, one pound live weight increase will be produced by feeding about $3\frac{1}{2}$ lb. of meal equivalent containing the following:—

			T.D.N.
Gram	.. ..	2½ lb.	1.50
Maize	..	¾ lb.	0.32
Barley	.. ..	⅜ lb.	0.33
			3.15 lb.

This mixture will provide 3.15 lb. of total digestible nutrients. Reference to the charts shows that the average total digestible nutrients required for maintenance and movement between these ages and weights is 2.8 lb., and the average daily requirements for live weight increase is approximately 0.8 lb., or 3.6 lb. in all (2.8 plus 0.8).

Feeding, however, cannot be conducted according to a strict mathematical formula, but these figures provide an approximate guide to requirements for the particular purpose in view. When pigs are being fed for bacon the intensity of feeding in the early stages should be less than for porkers, consequently a suitable working guide would be to make a ten per cent reduction from the meal requirements described above in calculating the food needed for baconers.

From the age of 7 to 8 months (when the porker wou'd usually be killed) pigs which are being fed for bacon have arrived at an age when a greater amount of food is required to produce a pound of live weight increase than was the case at earlier stages. By the use of the table it is possible to calculate the daily requirements in meal equivalents to produce a pound of live eight increase in the baconer. Thus a baconer 30 weeks old and weighing 200 lb. will have gained, according to the table, 170 lb. in 22 weeks, assuming it was weaned at the age of 8 weeks. Hence its average daily increase in weight would be $\frac{170}{22\times7}$, which is equal to 1.1 lb. On the basis of feeding for porkers the average daily meal equivalents would be

$$\frac{2\tfrac{1}{2}+4\tfrac{1}{4}+5\tfrac{5}{8}+6\tfrac{3}{4}}{4}=4.78 \text{ lb.}$$

If a deduction of' 10 per cent from this is allowed a figure of 4.3 lb. is obtained or half a pound of meal equivalent more per day than is required for the porker. This supports the contention that it is more expensive to feed for bacon than for pork. When feeding for bacon after the porker stage pigs should not be allowed to become too fat. Corresponding calculations for other weights may be made from the chart. The chart may also be used for computations for general production purposes.

The standards for pig feeding may be summarised by saying that the pig meal unit in terms of mixed meal represents 80 lb. of total digestible nutrients, and that the chart may be used as a guide to ascertain the needs for different weights and ages. Young porkers should be fed almost the whole of their food in the form of meal and dairy products such as milk, with a minimum of roughage apart from the green feed. The baconer, however, may be given from 15 to 20 per cent of the ration in the form of succulent feeds and roughages such as roots and green fodders.

The breeding pig will require a somewhat less percentage, say 5 to 10 per cent, of the weight of the total nutrients required in the form of concentrates. Speaking generally, for a given live weight the bacon pig will be somewhat older than the porker, and the breeding pig than the baconer.

Fig. 13 also shows the digestible protein requirements for different live weights, and it will be seen that these do not rise in the same proportion to weight as do the total digestible nutrients and maintenance requirements. This means that in the early stages of growth up to a live weight of about 100 to 120 lb., the protein requirements are considerable and must be met by rations with a nutritive ratio of not less than 1:3.5 or 1:4. The ratio should be progressively widened as growth proceeds, until for bacon pigs of from 180 to 200 lb. live weight it may be approximately 1:6. It is not advisable to widen the ratio beyond this at too early a stage by giving rations too rich in fats and carbohydrates in proportion to protein, otherwise the carcase will have too great a proportion of fat to lean. After this stage when the pigs are being fattened for market the ratio may be widened to about 1:10. If much green food is consumed some adjustment will be necessary in the nutritive ratio of the meal fed depending on the nature of the green feeds. Thus, if a rich leguminous fodder such as berseem or lucerne is used, the nutritive ratio of the meal may be widened, but with non-legume greenstuffs it will require to be narrowed somewhat. If roots and potatoes are fed in any bulk the nutritive ratio of the meal must be narrowed much more. It is a wise plan therefore when feeding pigs to make a rough practical computation of the proportion between the digestible protein and the total digestible nutrients being fed, in order to maintain the nutritive ratio at a proper level according to the stage of development and the purpose for which the pig is intended. This naturally would be more complicated where tankage or kitchen refuse is employed to any extent, as tankage varies considerably in composition, but practical experience should enable the pig breeder to form an approximate estimation of how to compute the whole ration if the general principles are kept in mind. The nutritive ratio of the concentrate meal part of the ration for nursing sows, weaners and breeding boars should be kept up to approximately 1 : 5, and for pigs a month or so before slaughter 1 : 7 or even wider. Breeding pigs may be fed concentrates with a ratio of 1 : 6, whilst, as previously stated, the ratio for young pigs should be around 1 : 3.5. This may be widened to 1 : 6 for young porkers after three months of age.

Precautions against Soft Fat

One of the most important things to guard against in feeding pigs are rations which are likely to produce soft or tainted fat. Reference has been made to some of these in Chapter VI. The best quality bacon should have fat of a clean white colour and be firm in consistency.

Certain feeding stuffs have a tendency to produce fat which is soft and greasy and the bacon will be of inferior quality. It is characteristic of the pig, more so perhaps than of other animals, for the body fat to take on the characteristics of the fat in the food, and when fats of low boiling points are used, such as the fat of ground-nut cake, soybean meal or linseed or rice meal, the bacon fat becomes soft and oily and has a tendency to become rancid. Maize which is fed with the germ intact has a similar effect if fed in too large quantities. Such material should be fed in small quantities, and in combination with other foods, the fat of which has the opposite effect. The most valuable of all foods in this respect is milk; others are beans, barley, peas and gram. Fish meal should also be used with precau-

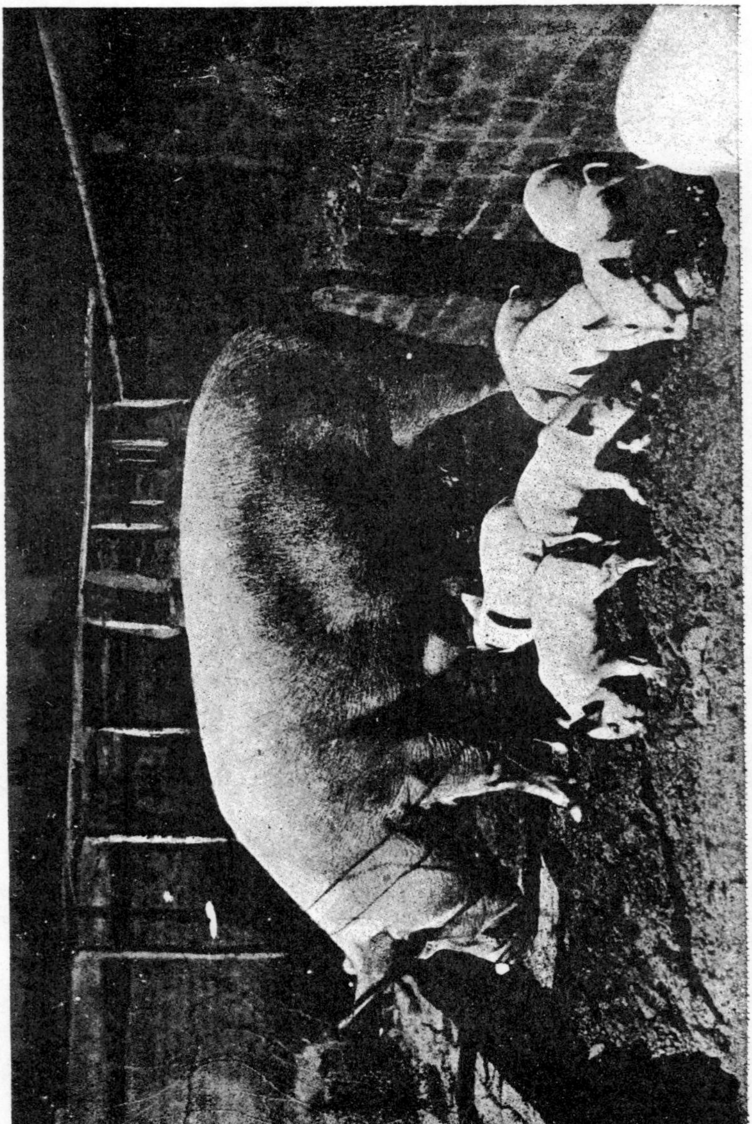

Large white Yorkshire sow with litter.

tion, and only the best quality white fish meal employed on a scale of not more than a quarter to one fifth of the total concentrate ration. Whilst the experiments of Orr [1922] mentioned in Chapter VI have shown that the best quality fish meal can be fed up to the time of slaughter without tainting the flesh or causing oiliness, it might be advisable to discontinue its use a few weeks before slaughter time.

Feeding Breeding Stock

The average number of pigs which a sow will produce will depend largely on the breed and feeding, but a good strain should be able to produce up to twelve piglets twice a year. A considerable strain is imposed on young gilts in producing and feeding large litters and it is advisable to defer breeding until they are two years old. Good feeding is as essential for in-pig sows as for any other type of pig, and they should only be fed with good well balanced rations. Sows in-pig should be given plenty of exercise in the open air until a week or two before farrowing otherwise they will tend to become too fat. About one half of their rations may consist of pasture or other succulent or root feeds, and the other half mixed meal.

The whole ration should thus be approximately divided between the mixed meal and a succulent feed. It is possible however, to effect certain interchanges or replacements with other feeding stuffs provided these supply the nutrients necessary for the growing foetus. It is not easy to give precise figures as to the amounts of other feeds which may be used as supplements, but a comparative guide may be obtained by ascertaining the total digestible nutrients and digestible proteins which other feeds will supply and the nutritive ratio of the resultant ration. For example, three typical meal mixtures shown in table I are expressed in terms of the total digestible nutrients and the digestible protein they yield and the nutritive ratios. Taking into account the nutritive factors thus shown it is possible to compute tables showing how much of other feeding stuffs will yield the same amount of total digestible nutrients as one pound of the meal mixture.

TABLE I

A.

		Total digestible nutrients. lb.	Digestible protein. lb.	Nutritive ratio. 1 :
1.	40 lb. maize meal	28.20	2.15	
	35 lb. ground gram	28.00	4.35	
	20 lb. barley	14.15	1.34	
	5 lb. linseed cake meal	4.13	1.17	
	100 lb. mixture	74.48	9.01	1:8.2
2.	25 lb. maize meal	17.63	1.34	
	25 lb. ground gram	20.00	3.11	
	25 lb. barley	17.69	1.67	
	25 lb. linseed cake meal	20.64	5.85	
	100 lb. mixture	75.96	11.97	1:6.4
3.	60 lb. barley meal	53.24	3.99	
	20 lb. weatings	15.75	2.90	
	10 lb. palm kernel cake	9.37	2.20	
	10 lb. white fish meal	6.62	4.10	
	100 lb. mixture	84.98	13.19	1:6.4

B.

	In terms of D.P.I.	In terms of D.P.I.	
One lb. of mixed feed No. 1 =	0.90	0.75	lb. of gram.
=	1.00	1.80	lb. of maize.
=	1.00	1.40	lb. of barley.
=	0.90	2.27	lb. of oats.
=	1.10	1.56	lb. of wheat.
=	0.90	0.69	lb. of arhar.
=	1.10	0.26	lb. of soybean.
=	1.10	0.48	lb. of linseedcake.
=	5.00	0.90	lb. of berseem.
=	5.30	0.69	lb. of senji.

Naturally, such tables are for guidance only, as it would never do for example, to replace the major part of the meal by its equivalent in terms of total digestible nutrients of, say, green berseem or senji, because though the total digestible nutrients required would be met, the digestible protein needed would not.

The total digestible nutrients required for maintenance for a sow of ordinary size will be provided by about 1.2 lb. of mixed meal containing 80 lb. of total digestible nutrients, per 100 lb. live weight of the sow. This should be increased by 5 per cent to cater for the growing embryoes and keep the sow in condition. Thus a sow of 350 lb. live weight would need about 5 lb. meal in the early stages of pregnancy. At the beginning of pregnancy half the ration should be in the form of meal, and the other half calculated to yield a similar amount of total digestible nutrients from some of the bulky feeding stuffs enumerated in table 1. It is thus possible to calculate the sow's requirements in terms of feeding stuffs. For example, if the sow is allowed 12 lb. of green lucerne or berseem daily she would obtain about 6 lb. of total digestible nutrients, equivalent to about 4.8 lb. of meal. Hence 4.2 lb. of meal would have to be fed as a supplement to grazing. If on the other hand the sow is fed on ordinary pasture grass, a somewhat higher proportion of meal will be needed. As the farrowing period approaches the amount of meal should be increased gradually from the above figure up to about 5 to 6 lb. at the time of confinement, with a somewhat larger allowance during winter. It is not easy to state with any degree of precision how much protein the in-pig sow requires but this must be adjusted so as to provide a proper balance between the total digestible nutrients and the digestible protein of the meal, with the protein content of the bulky feed. For example, with berseem or lucerne as the bulky feed, the nutritive ratio of the meal fed could be 1 : 7, whereas if ordinary pasture grass is fed it should be narrowed to 1 : 6 or 1 : 5.

Typical pig meal mixtures which will supply respectively nutritive ratios of 1:8 and 1:6 are as follows:—

	Total digestible nutrients.	Digestible protein	Nutritive ratio.
	lb.	lb.	1:
1. 40 lb. maize meal	28.20	2.15	
35 lb. ground gram	28.00	4.35	
20 lb. barley	14.15	1.34	
5 lb. linseed cake meal	4.13	1.17	
100 lb. mixture	74.48	9.01	1:8.2

2. 25 lb. maize meal	17.63	1.34	
25 lb. ground gram	20.00	3.11	
25 lb. barley	17.69	1.67	
25 lb. linseed cake meal	20.64	5.85	
100 lb. mixture	75.96	11.97	1:6.4

Corresponding rations can easily be computed according to the feeding stuffs available from the analytical data given in Appendix 1.

Shortly before farrowing a somewhat laxative and more easily digestible food should be given, and one pound of bran or about 5 per cent of linseed cake might replace a part of the previously fed meal. On an average a sow of 350 lb. weight may be fed from 6.5—7.5 lb. of meal with a reasonable amount of green fodder in addition. At the time of farrowing the food should be cut down and fed as slop with milk added. Green stuff should be temporarily withheld immediately after farrowing as it may cause digestive disturbances in the young pigs. The ration should then be progressively increased and the total digestible nutrients per 100 lb. of meal fed raised to something over 80 and should include gram meal, and barley meal in small quantities only, as the latter is inclined to depress the milk yield. At the same time the nutritive ratio should be raised to approximately 1:5 by including richer protein concentrates such as legume grains or white fish meal. The sow should not be overfed too early, but as the size of the litter increases the meal allowance may be gradually increased from 8 lb. to about 12 lb. per day. The food should preferably be given three times a day, and after about three or four weeks may be fed dry or slightly damp, but should not be too sloppy. At this time the sow should be allowed to get whatever exercise she needs and have access to grazing.

Feeding the boar

The boar may be fed a ration similar to that of the fattening pig or the sow; it should be nutritious and properly balanced, but it may contain a considerably higher percentage of bulky food than in the case of the sow's ration. Plenty of exercise is essential and the boar should on no account be allowed to grow fat or lazy, but on the contrary, he must not be underfed, as either of these extremes will prejudice his breeding capacity.

During the off season when the mature boar has access to plenty of good pasture, about one pound of concentrates per day per 100 lb. live weight should be sufficient, and slightly more in cold weather. Young boars should be given proportionately more while still growing and their food should approximate to the requirements indicated in Figure 13.

Shortly before the boar's service are required his rations should be increased so that he may put on weight, and subsequent rationing will be adjusted to the number of services demanded, but in no case should he be allowed to lose weight. It may then be desirable to feed but little maize, or to replace this entirely by other concentrates such as ordinary cereal grains, with a considerable percentage of the total ration made up of richer protein supplements such as bran, middlings, linseed meal and chopped legume hay. The reason for this is, that the proteins of maize are biologically deficient in certain amino acids as well as in minerals and these are as much needed by the boar in service as for the sow after service.

What is the Best Diet for Pigs ?

The question may be asked—what is the best diet for pigs ? There is a considerable range of feeding stuffs to choose from and selections may be made from these in accordance with the specific object of feeding, whether for young stock, sows in-pig, or fattening. Whilst it is unwise to restrict the range of feeding stuffs to only one or two, it is also sound policy on the other hand not to feed too complicated a mixture unless there are special reasons for doing so.

Certain fundamental principles for feeding pigs of various classes are summarised below:—

1. The proper breed must be selected for the purpose required.
2. The pig is a greedy feeder but utilises the food given for meat production much more efficiently than any other animal.
3. The basal concentrates for pigs are usually deficient in calcium, phosphorus and Vitamins A and D. These deficiencies must be made good by supplementing the rations, especially in the case of sows in-pig, nursing sows and young stock.
4. Young pigs make a greater increase in weight per pound of food fed than older ones; hence they require rations with a wider nutritive ratio.

5. Fattening pigs, after having made full growth, require rations with a much wider nutritive ratio than for young pigs.
6. After a certain weight and age the increase in weight per pound of food fed diminishes, and the breeder should adjust the time of slaughter to coincide with the period when further increases in body weight do not compensate for the cost of food and other charges.
7. The maintenance requirements increase until maximum growth is attained, and still have to be met at later stages. It does not pay, therefore, to over fatten pigs for commercial purposes. The additional weight laid on may be uneconomically obtained.
8. Plenty of exercise, reasonable grazing and grubbing for roots, and a correct though small percentage of bulky food in the ration is desirable for all classes of pigs, although weaners and young piglets should be given a minimum of bulky food.
9. Where pigs are confined chiefly to sties, special attention must be given to the Vitamin A and D requirements, and green fodder or legume hay and calcium-phosphorus supplements included in the rations.

Keeping the above in mind a breeder is justified in considering that the best food for his pigs is that which he has found by experience under his own particular conditions to produce the maximum quantity of high class carcases at the minimum cost.

Advisability of using Cooked Food

Opinions differ considerably as to whether food for pigs is better cooked or uncooked. Some are definitely against cooked food and state that the advantage gained, if any, does not pay for the cost of cooking. Others state that all pigs pay better on cooked food which is more digestible. For Indian conditions it may be advisable to give one cooked meal during the day in winter, though it is questionable whether the general practise of cooking food should be advocated.

Value of Mangolds, Turnips and Potatoes for Pigs

The general consensus of opinion is that the potato, either steamed or boiled, is the best root to feed. Turnips and mangolds are not much in favour, and in no case should the latter be given to in-pig sows. Turnips are too acid for young pigs. Some breeders recommend moderate quantities of mangolds as a substitute for grain, but not turnips, whereas potatoes boiled with maize and ground oats are recommended as a good fattening food.

Mineral Requirements

Common salt

Pigs require less common salt than most other farm stock, and when they are given suitable rations and good grazing is obtainable they can, as a rule get most of the salt they need from these sources. Nevertheless, pigs will usually eat up to 1/10th of an ounce of salt a day, depending on the rations they are getting, and should be provided with free access to salt in a suitable trough so that they can take what they need.

An alternative method of feeding salt would be to mix it in with the food, allowing a quarter of a pound of salt to 100 lb. of the combined grain and protein supplement.

Salt is more necessary for pigs when the concentrate part of the ration is largely of vegetable origin.

Calcium and Phosphorus

Calcium and phosphorus are specially liable to be deficient in pig feeds which are made up from grains and their by-products, or protein feeds of vegetable origin. The deficiency can largely be made up by including skim milk, butter milk, swill, fish meal, and particularly legume pasture or hay in the rations. The desirability of adding a mineral supplement to the diet therefore depends on the constitution of the ration and, generally speaking, if this is well balanced and includes the above mentioned supplements, pigs should receive sufficient calcium and phosphorus even when not grazing. It is well in cases of doubt to add a high quality calcium-phosphorus supplement to in-pig sows and sows with litters, and to the food of young pigs if sufficient milk is not obtainable. There is not so great a likelihood of phosphorus deficiency as there is of calcium, as most of the protein rich concentrates are rich in phosphorus. The cereal grains are low in phosphorus and still lower in calcium, so special care must be taken with rations which are chiefly cereal in nature. Where both calcium and phosphorus are needed as supplements, finely ground bone meal or calcium phosphate may be used, but when only calcium is deficient ground limestone may be fed. If pigs on pasture are fed only cereal concentrates, both calcium and phosphorus are likely to be needed.

In cases of doubt, the table of analyses of feeding stuffs will provide a rough guide as to whether mineral supplements are necessary or not. Proper bone structure can only be built up, even when calcium and phosphorus in the rations are sufficient, if the Vitamin D supply is adequate, which can best be assured by plenty of sunlight, green pasturage or well cured hay. Calcium and phosphorus deficiencies will soon be reflected by the appearance of rickets in the young piglets, or by their being born dead, and whenever such events occur, fish liver oil in moderate quantities and appropriate mineral supplements may be fed to the sow and also given in milk to the young pigs. Mineral supplements may also be given in the form of ground egg or oyster shells or ground chalk, wood ashes or any clean refuse of a mineral nature rich in calcium. It is not easy to say precisely how much calcium and phosphorus are required by pigs, but according to the trials conducted by Carroll, Hunt and Mitchell, [1930], they need approximately five grammes of calcium and as much phosphorus per head per day. Thus a 100 lb. pig being properly fed would need about 0.3 per cent of calcium and the same amount of phosphorus in the entire ration. The cost of commercial pig meals does not necessarily bear any relationship to the efficacy of their mineral content. Thus it is recorded in "The pig and pig products", Jan. 1936, that a fine meal of a well known brand sold in 5 lb. bags for the equivalent of eleven annas, contained 0.2 per cent of total minerals, whilst another meal of coarser type costing only one anna per pound contained 1.44 per cent of minerals, or more than five times as much as the more expensive meal. It was also found that the cheap and coarser meal contained twice as much iron, ten times as much phosphorus, five times as much calcium, eleven times as much magnesium, and nine times as much potassium as the finer meal. The conclusion therefore is that the finer types of commercial meal, which may command a higher price, are by no means necessarily as suitable from a nutritive and mineral point of view as the coarser types. Sufficient quantities of iron, phosphorus and manganese are present in 1½ lb. of coarse maize meal to provide for the needs of an average growing pig for one day, but it would contain only 1/10th of the amount of calcium required, which must therefore be supplied from the other parts of the ration.

Iodine deficiency

Goitre is apt to occur in certain districts where iodine is deficient in the soil and crops. Reference has been made to the use of iodine in Chapter XII on Horses. Where pigs are fed iodine deficient rations young pigs when born suffer from goitre, a disease characterised in pigs by lack of hair on their bodies and other symptoms. Wherever goitre occurs or is likely to occur, it is advisable to supply the sows with a mineral mixture including common salt to which one third of an ounce of potassium iodide has been added per 100 lb. of the mixture. This is known as iodised salt and should be made accessible to the sows for the last three months of pregnancy. In districts where goitre happens to be endemic no harm can be done by giving iodised salt to brood sows and growing and fattening pigs throughout the year instead of common salt.

Anaemia in young pigs

In-pig sows or suckling pigs which are kept in sties and do not have access to the soil, and are thus unable to obtain natural minerals in their wanderings may suffer from anæmia, and the young piglets may be born dead or also suffer from anaemia. Under normal conditions the sow obtains enough iron and copper from her rations, and the young will be born with sufficient reserves of these minerals in their bodies to carry them up to the weaning stage when they can begin to feed themselves. In unnatural conditions of confinement in feeding houses, the essential iron and copper (the latter in extremely small quantities) may be lacking and anaemia result. This condition is characterised by weakness and lack of appetite, and even if the young pigs recover, they will take considerable time to make up for the throw back they have suffered, with obvious economic losses to the breeder. Wherever there is a tendency therefore, for anaemia to occur, preventive measures must be taken. One method is to paint the udder of the sow daily with a mixture made up by dissolving 135 grammes of ferric sulphate and 22.5 grammes of copper sulphate in 600 ounces of water. This should be done daily until the pigs are weaned, and at the same time they should be allowed access to a palatable concentrate feed to every 100 lb. of which 1/10th of a pound of the mixture of ferric sulphate and copper sulphate has been added. Correspondingly larger amounts might be given to the sows for their own benefit. The evidence available goes to show that anaemia in piglets cannot be prevented by internal treatment of the sows, as no method of introducing iron and copper into the ration artificially has been found to influence the iron and copper content of the sow's milk. Another precaution which may be taken is to put into the sties fresh earth or wood ashes, in which the piglets can rummage, but direct treatment with a copper-iron solution is advisable. It is doubtful whether copper or iron adjuncts are necessary for other classes of swine.

Vitamins

The only vitamins likely to be deficient in the usually fed rations are 'A' and 'D', and these can generally be adequately supplied by providing all classes of swine with reasonable amounts of green fodder and suitable types of well cured hay, especially legume hay. Plenty of sunlight is also desirable for the provision of vitamin D. In special cases where vitamin A and D supplements are needed, the administration of fish liver oil in amounts depending on the size of the pig can be given. Milk is the food *par excellence* for all classes of swine to guard against vitamin A deficiency.

Water Requirements

Water is much more important than is generally recognised, and in India it is essential that pigs should have continuous access to drinking water in troughs or containers so arranged that they do not become contaminated. The amount of water a pig needs will naturally depend on the weight of the pig and climatic conditions, but may vary from 1 to 1½ gallons per 100 lb. of the weight of the animal when still growing to 1/3rd of this amount per 100 lb. live weight when ready for slaughter. The more succulent food a pig gets and the cooler the weather the less water will be needed. In hot weather a sufficient and continuous water supply is absolutely essential.

Feeding Stuffs for Pigs

A detailed description of the feeding stuffs available in India has been given at length in Chapter VI, and a few of the more important points concerning some of those specially suitable for pigs may now be considered.

Cereal grain

The cereal grains form the basis of pig feeds, but they are all low in protein content and in calcium, and are thus unbalanced and need to be properly supplemented by protein and mineral rich feeding stuffs.

Maize

Maize is extensively used in western countries, but may be largely replaced by gram in India. Maize if fed alone will produce very poor results for growing and fattening pigs owing to its biologically defective proteins, but it is valuable when constituting part of the ration. Maize is low in calcium content and poorer in phosphorus than most other grains and therefore it needs protein supplements and minerals to balance these defects. Yellow maize is better than white owing to its rich vitamin 'A' content in which the white variety is defective. Pigs fed chiefly on maize and not getting sufficient sunlight are likely to develop rickets or produce rachitic offspring.

Barley

One of the particular virtues of barley for pigs is that it produces a hard firm fat, and pork and bacon of excellent quality. Barley should always be moderately ground for pigs as this will give better results than if it is fed whole. If there is difficulty in grinding, it should be soaked in water. Pigs have a particular liking for barley which should, whenever possible, comprise part of the meal mixture, but should not be fed as the sole cereal. In spite of these good qualities, barley shares the defects of all other cereals, its protein being of poor quality, calcium content very low and vitamins 'A' and 'D' almost negligible.

Wheat

Wheat is expensive to feed to pigs but may form a proportion of the cereal ration; as a flesh producer it is slightly more valuable than ground maize. It should, however, be fed as a part only of the cereal mixture in order to obtain the best results. Wheat is richer than maize in both protein and minerals, but, nevertheless, requires efficient protein supplements, unless the pigs are on rich pasture with plenty of legumes. If whole wheat is self-fed, pigs will masticate the grain thoroughly, but when hand fed they are likely to bolt it, and it is safer therefore always to feed it crushed. Soaking wheat, if grinding is not possible, is not satisfactory, and it is better in such cases to feed it whole because the soaked grain appears not to have the same food value as the whole dry grain.

Oats

Oats are excellent when used as a limited part of the pig meal up to about 1/4th of the ration, but they are more fibrous than other cereals or legume seeds and are thus unsuitable in large quantities. If fed in excessive amounts to young growing pigs, growth will be reduced, and correspondingly, a proportionate value of the oats as a feed also reduced. Brood sows however may be fed slightly more oats before parturition, but the amount should again be reduced when the maximum milk yield is needed. The value of grinding oats is probably greater than for any other grain.

Gram

Gram when available, constitutes perhaps the main basis of pig feeds, in India. It contains more than twice the protein content of cereals, and forms an excellent part feed for the growing period, but its proportion in the ration should be moderated during fattening so as to keep the nutritive ratio within the limits given above. The protein of gram is of excellent quality but it is deficient in calcium and rich in phosphorus, and, like other grains, it is deficient in vitamins.

The millet grains

Millets when available are valuable part feeds in the early stages of growth and may be used as substitutes for maize for all classes of pigs. They should preferably be fed threshed.

Weatings and other cereal products

Weatings may constitute a part of the pig food but they are more fibrous than whole wheat and contain more protein, bulk for bulk, as a part of the endosperm of the wheat grain passes into the weatings. Weatings are comparatively rich in vitamin 'E'.

Pollards are mainly composed of carbohydrates with a small quantity of bran, whereas shorts contain somewhat more bran.

Middlings are an excellent feed for all classes of swine, but share the mineral and protein deficiencies of the cereals.

If they are fed with protein supplements such as dairy by-products, tankage or white fish meal they give good results as the latter furnish the proteins which the middlings lack. (See also Chapter VI.)

All wheat milling products may be used for pigs in appropriate quantities if due regard is paid to the particular stage of development of the pig and its nutritive requirements.

Protein supplements

The best all round protein supplements are milk, skim milk and butter milk, and these, or tankage or swill are safe sole protein supplements to feed to pigs on pasture, when the latter will provide the necessary vitamins.

Fish meal

Fish meal is one of the most valuable supplementary feeds for pigs as it is rich in good quality protein and vitamins 'A' and whenever possible should form part of the protein supplement when good pasture or green fodder are not available. Fish meal however, should not be fed in greater quantities than are needed to balance the ration, and on no account should inferior meals be used as they will adversely affect the flesh at slaughter time.

Milk

Whilst milk, skim milk, butter milk and whey are all valuable feeds for pigs, the latter three are deficient in vitamins 'A' and 'D' and therefore, although they are excellent supplements when fed with cereal meals up to a month after weaning (as a young pig is born with certain vitamin reserves in his body) they will need to be supplemented by vitamin containing feeds after this time. Hence if young pigs are fed only butter milk, skim milk or whey with cereal meals from this stage onwards but with no pasture, some fine legume hay should be fed to correct these deficiencies.

Skim milk, and butter milk are rich in protein, and of course watery, and therefore should not be fed alone but in sufficient amounts to balance the cereal mixture. Immediately after weaning (when the young pigs need a feed with a nutritive ratio of 1:3.5, as much as from 3—6 lb. of skim milk or butter milk may be fed per pound of meal, the amount of milk in proportion to meal being gradually increased as the pig grows older.

Whey

Whey, like milk, contains proteins of excellent quality, and a pig over 100 lb. in weight will progress satisfactorily if fed on a ration of whey and cereal meals, provided the necessary vitamins 'A' and 'D' supplements are provided.

Tankage and meat scraps

It is doubtful if tankage (by-products from butcheries) commands a very high market value for the reasons given earlier concerning the pig industry in India *vis a vis* its counterpart in western countries. When obtainable however, tankage, swill and kitchen refuse are excellent supplements of high protein content for all classes of pigs. For purposes of growth good tankage is considered to be worth more than six times as much as an equivalent quantity of maize, due regard being paid to correct balance and nutritive ratios.

Linseed meal

Linseed meal is a good protein supplement for pigs if fed in combination with skim milk, fish meal or tankage, and has been extensively used in American trials conducted by Morrison [1936], in which a mixture consisting of 50 lb. of tankage or fish meal, 25 lb. of linseed meal and 25 lb. of ground legume hay was fed to growing and fattening pigs when pasture was not available. Such a mixture forms an excellent supplement to cereal meals, but linseed meal should not be fed as a sole protein supplement to pigs not on pasture, as its proteins like those of maize are biologically deficient. Linseed is also low in calcium content and contains little or no vitamins 'A' or 'D'.

Cottonseed cake meal

If cottonseed cake meal is fed to pigs it should be of the decorticated variety, and the ration should not contain more than 8 to 10 per cent of the meal, and must be properly balanced with other feeds of good quality protein, vitamin and calcium content. Cottonseed meal is likely to cause too hard a body fat if fed in large amounts and should not be given as a sole protein supplement. A further objection to too much cottonseed meal is the danger of gossypol poisoning, even when vitamin 'A' is adequate in the ration. Cottonseed meal should not be fed to young pigs as they are particularly liable to suffer from this defect.

Soybeans

Soybeans form a valuable protein supplement but should not constitute more than 10 per cent of the total ration on account of their tendency to produce soft pork. They should preferably be cooked before feeding. Soybean meal is very low in calcium and contains much less phosphorus than dairy produce, fish meal, milk or tankage, and therefore, if fed to young pigs not on pasture, a mineral supplement of legume hay should be given to supply vitamins 'A' and 'D'. It is better to feed the meal rather than the whole beans as the meal has been partially cooked in the process of manufacture. Soybean meal may be fed as a sole protein supplement to pigs after 6 months of age when they weigh from 60 to 80 lb., provided the necessary minerals and vitamins are supplied from pasturage or hay. If fed whole, soybeans have a greater tendency than the meal to form soft body fat, and owing to their high oil content they should be fed in conjunction with a hard fat producing supplement such as fish meal or tankage.

Groundnuts

Groundnuts form an excellent feed for young pigs, but like soybeans they should be given in moderation as they also tend to form soft body fat. They should not as a rule be fed after the pig has attained a weight of about 100 lb., when the ration should be confined to hard fat-producing feeds, or contain a sufficiency of the latter to prevent the formation of soft fat.

Pasturage and succulent feeds

In conclusion emphasis must be laid on the value, especially for growing pigs, of good pasturage which supplies most of the food ingredients lacking in meal mixtures. As, however, pigs cannot make satisfactory growth on too bulky feeds, they should always be given from 1 to 1½ lb. of concentrates daily per 100 lb. of live weight even when on pasture.

Good pasture and forage crops help to effect economy in the grain necessary for each 100 lb. of gain in live weight, as pigs on pasture will usually need only about half as much protein supplements as those not on pasture.

Whenever grazing is not available it is very important that pigs should be given plenty of green fodder, preferably lucerne, when this is obtainable.

Rationing Table

The following table is intended to serve only as a guide to quantities which should be fed under average conditions.

Age in weeks.	Approximate live weight. lb.	Food allowance per head per day. lb.	Approximate Nutritive Ratio. 1 :
3 — 8	10 — 25	½ — 1½	4
8 — 12	25 — 50	1½ — 2½	5
12 — 16	50 — 80	2½ — 4	5
16 — 20	80 — 115	4 — 5	6
20 — 24	115 — 115	5 — 6	7
24 — 30	555 — 210	6 — 7	8
In-pig sows		3 — 6	5
Suckling sows		8 —14	5
Stock boars		3 — 7	5

Some Typical Rations

1. *For suckling sows and weaners to* 12 *weeks.*

(a)	Middlings	60 parts
	Barley meal	30 parts
	Fish meal	10 parts
(b)	Middlings	50 parts
	Bran	10 parts
	Maize	15 parts
	Barley meal	15 parts
	Meat meal	10 parts
(c)	Middlings	40 parts
	Cocoanut cake	20 parts
	Barley meal	10 parts
	Wheat meal	10 parts
	Maize meal	10 parts
	Fish meal	10 parts

2. *For young pigs.* 12—20 *weeks of age. Gilts for breeding, boars, and in-pig sows.*

(a)	Middlings	45 parts
	Barley meal	45 parts
	Fish meal	10 parts
(b)	Middlings	30 parts
	Cocoanut cake	15 parts
	Barley meal	20 parts
	Maize meal	15 parts
	Wheat	10 parts
	Meat meal	10 parts

(c)	Middlings	30 parts
	Palm Kernel meal	15 parts
	Barley meal	15 parts
	Wheat	15 parts
	Oats	15 parts
	Fish meal	10 parts

3. *For fattening pigs over 20 weeks.*

(a)	Middlings	35 parts
	Barley meal	60 parts
	Fish meal	5 parts
(b)	Middlings	25 parts
	Cocoanut cake	8 parts
	Barley meal	40 parts
	Maize meal	10 parts
	Wheat or Oats	10 parts
	Soybean meal	7 parts
(c)	Middlings	25 parts
	Palm Kernel meal	10 parts
	Barley meal	20 parts
	Wheat meal	20 parts
	Maize meal	20 parts
	Meat meal	5 parts

BIBLIOGRAPHY

1. Carroll, Hunt and Mitchell. 1930. Illinois Report.
2. Ellis J. C. B. 1937. The Feeding of Farm Live-stock, p. (i) 253. (ii) 255.
3. Morrison F.B. 1936. Feeds and Feeding, p. (i) 808. (ii) 882.
4. Orr, Crichton and Green. 1922. The Value of Fish Meal as a Feeding Stuff, Scot. J. Agric. 5. 2.
5. Wood T. B. 1937. Animal Nutrition, p. 163.

CHAPTER XV

FEEDING OF POULTRY

The innumerable small flocks of poultry which may be seen throughout the length and breadth of India are not fed according to any recognised system. The vast majority get their food at random round the villages and in adjacent fields and pay their owners for their keep, because this costs the latter little or nothing. Most of these birds are of nondescript type, and both the eggs and the poultry fetch very low prices, which would not pay if feeding stuffs had to be bought and fed on any proper system.

True poultry keeping, however, when poultry are bred and fed as a paying proposition is another matter, and if it is to be a success certain definite principles of breeding, feeding and general management must be followed.

Very little research has so far been done in India on poultry feeding but a considerable amount of information is available from foreign countries which can be applied here for feeding well balanced rations.

Although there is still much to be learnt it is possible by utilising this knowledge to compile satisfactory rations for different parts of the country and to use it as a starting point for local research.

Poultry and pigs show some similarity in the nature of their basal foodstuffs, these being more of a carbohydrate than of a protein nature, and they require a considerably greater proportion of concentrates than other farm animals.

The first requisite for poultry production, whether on a large or small scale, is to start with good stock of one or more of the well established breeds, according to whether the object in view is egg production, table poultry or dual purpose birds.

Most of the poultry in India are of a somewhat nondescript type, but considerable improvement may be effected in these by careful selection and breeding from the best and most prolific birds. More rapid progress will perhaps be made, however, if the prospective poultry breeder uses good quality cocks of one

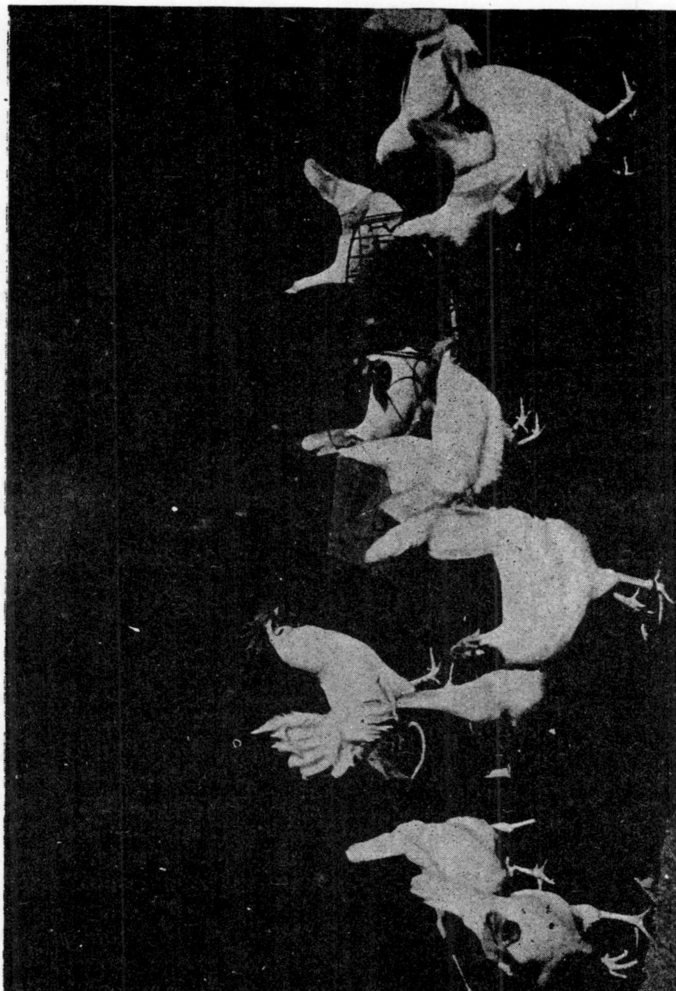

A pen of White Leghorns feeding.

of the standard breeds to mate with *desi* hens, and thus builds up a flock of improved characteristics. Work is progressing along such lines in certain parts of India and in course of time new breeds of improved country stock may be evolved.

If poultry are to be a success a proper system of feeding must be followed, and at the same time a careful record kept of the cost of food used to produce eggs and table birds. The data so far available on the proper feeding of poultry for local conditions in India are, as mentioned above, somewhat scanty, but some feeding trials have been carried out in recent years at the Government Poultry Farm at Gurdaspur in the Punjab, [1940-41], and at the Poultry Research Section of the Indian Veterinary Research Institute, Izatnagar, in the United Provinces [Macdonald 1941].

Some of the findings from these trials will now be described, but the results must be regarded as tentative and further work is necessary in more widely separated places, as those from one locality are not necessarily strictly applicable to another in a different climate and under different management.

Gurdaspur Trials

In the Gurdaspur trials on young stock the following quantities of food were found necessary to produce 1 lb. live weight increase in each of the four breeds named:—

Asil	4·97 pounds,	costing 3·95 annas.
Chittagong	5·99 pounds,	costing 4·65 annas.
Improved Punjab (Brown)	6·47 pounds,	costing 4·99 annas.
Desi (Mongrel) ..	6·47 pounds,	costing 4·87 annas.

The food given consisted of dry mashes, grain and plenty of green stuff shown below for fowls in crates and runs :—

(A) *For fowls in crates*

Fattening ration

Barley meal	50 per cent (by weight).
Wheat meal	25 per cent (by weight).
Yellow maize meal	25 per cent (by weight).

2½ lb. of skimmed milk was added to every pound of the above dry mash, after which a little water was added to produce the required consistency.

The ration was mixed fresh daily and given in bowls to the stock in the fattening crates three times a day, for approximately 15 minutes at a time, after which the bowls were removed and the crates covered by sacking to discourage movement. The covering was lifted at the feeding times only.

The fowls in the crates were subjected to a 24 hour period of starvation before the trial commenced, receiving only water to drink.

(B) *For fowls in runs*

Basal mash

Wheat meal	40 per cent (by weight).
Yellow maize meal	20 per cent (by weight).
Wheat bran	30 per cent (by weight).
Fish meal	10 per cent (by weight).

Supplements

Ground charcoal	2 per cent of basal mash.
Ground linseed	3 per cent of basal mash.

The dry mash was given in metal hoppers for consumption *ad lib.*, and precautions were taken to account for any mash spilt by the fowls.

Grain mixture

Wheat	50 per cent (by weight).
Yellow maize	50 per cent (by weight).

A fixed quantity of ½ oz. per fowl per day was given to all young stock in the runs, early in the morning.

Water was provided in the bowls for consumption *ad lib.* and green grass growing in the runs was available.

The cost of the food will naturally vary according to the relative proportion of mash and grain fed, and other factors, and these costs are perhaps on the high side, as the rate of growth is lower than might have been expected and hence the utilisation of the food not so efficient.

Tentatively, however, these trials indicate that the Asil at Gurdaspur converted food into flesh more effectively than the other breeds, five pounds of food being needed to produce one pound of flesh.

Some preliminary data have also been obtained in regard to the food consumed and the number of eggs laid. These are recorded below:—

	Pounds of food consumed in a year per hen.	Eggs laid.	Average gain or loss per hen in body weight (ounces).
Asil	68	38	+ 3
Chittagong	63	81	− 3
Improved Punjab (Brown)	65	79	− 9
Desi (Mongrel) ..	66	100	− 1

A series of fattening trials carried out with Asil, Chittagong and Improved Punjab (Brown) breeds, showed the following results for live weight increase, per pound of food eaten:—

Breed	Weight in ounces gained in 15 days.	Live weight gained per lb. of food eaten.
Asil	10.69	3.51 ounces.
Chittagong	8.81	3.25 ounces.
Improved Punjab (Brown) ..	7.88	2.77 ounces.

These live weight increases are not as high as might have been expected, and compare unfavourably with data obtained by Macdonald at Izatnagar [1941]. The efficiency of the food during fattening, however, is always low.

The time taken for some of the breeds to reach a certain definite weight was also ascertained and the results which are presented below are not as good as in the Izatnagar trials, which may be due to the rations not being as satisfactory as they might have been:—

Average time taken (in weeks) to reach two pounds weight

	1939-40		1940-41	
	Cockerels.	Pullets.	Cockerels.	Pullets.
Asil	19	20	16	17
Chittagong	18½	21	16	17
Improved Punjab (Brown)	19½	22½	17	17½
Desi (Mongrel) ..	19	21	17	18

The economics of these feeds is given in the following table:

Breed.	Weight gained in 24 weeks.	Mash & grain consumed.	Cost of mash and grain.
	lb.	lb.	Annas.
Asil	3.57	17.72	14.10
Chittagong	3.27	19.46	15.09
Improved Punjab (Brown)	3.09	19.85	15.32
Desi (Mongrel) ..	2.64	17.05	12.83

The Izatnagar Trials

Macdonald [1941] has carried out an interesting series of trials to investigate the growth promoting value of various supplements in the diets of chicks fed on a standard basal diet consisting of locally produced foodstuffs supplemented with green food and crushed limestone.

He used separated milk as the control supplement in each ration and compared the results with those obtained with the standard diet only and with the standard diet supplemented with soybean meal plus salt.

Three breeds were selected for the trials:—

The White Leghorn, the Rhode Island Red, and the *desi*; the White Leghorn and Rhode Island eggs were purchased from reliable dealers, and the *desi* eggs in the local bazar.

The ration

All the chicks produced were fed on mash *ad lib.* in suitable containers, and the grain was strewn in the litter, morning and evening, from the day after hatching. They were also provided with lucerne and berseem twice daily and broken limestone.

The basal mash consisted of 50 parts of wheat bran, 30 parts of yellow maize meal and 20 parts of finely ground oats, and was fed throughout to all the experimental groups. For the first eight weeks the grain ration consisted of equal parts of broken yellow maize, juar and millet, *cheena* (*Panicum miliaceum*), and after this period, of equal parts of cracked yellow maize, wheat and paddy.

Separated milk as a Supplement to the cereal ration for promoting growth.

Eighty-four day old chicks hatched in January were divided into groups each consisting of 28 White Leghorns, and 14 Rhode Island Reds, and each group was fed on the basal standard ration. One group was given separated milk only to drink, in addition to the basal ration, upto 10 weeks of age, and from then till 6 months old, separated milk and water in separate containers. The other group was given water only upto 6 weeks of age, separated milk for the next two weeks, then again water only till 3 months old, and separated milk and water from 3 to 6 months of age.

The average weight of the birds in each group, No. 1 the

milk group, and No. 2 water group, are shown in the following table:—

TALLE I

AVERAGE WEIGHTS IN OUNCE FOR EACH BREED

Breed	White Leghorn.				Rhode Island Red.			
Group	1		2		1		2	
Sex	Male	Female	Male	Female	Male	Female	Male	Female
Day old	1.48	1.46	1.47	1.47	1.42	1.44	1.48	1.41
1 week	2.4	2.2	1.7	1.8	2.3	2.2	1.8	1.7
2 weeks	4.2	3.9	2.2	2.2	4.2	4.1	2.1	2.1
3 weeks	6.1	5.7	2.5	2.6	6.1	5.8	2.6	2.7
4 weeks	9.4	8.5	3.0	3.0	9.1	8.9	2.7	3.0
5 weeks	13.3	12.6	3.1	3.3	13.4	13.1	2.7	3.4
6 weeks	15.8	14.8	3.1	3.8	16.6	16.3	3.0	3.9
7 weeks	21.5	20.5	5.6	5.9	22.4	21.1	5.0	6.4
8 weeks	26.1	24.6	8.0	8.5	27.9	25.2	7.2	9.1
9 weeks	30.7	28.8	9.5	9.6	33.5	28.2	8.6	10.3
10 weeks	37.0	34.6	10.2	10.2	41.3	34.5	9.4	11.2
12 weeks	46.0	40.8	13.1	13.9	49.5	41.2	14.5	16.6
14 weeks	52.0	47.5	21.7	22.0	59.2	49.2	23.3	25.0
16 weeks	—	52.4	—	30.4	—	56.2	—	32.0
18 weeks	—	55.7	—	35.6	—	60.2	—	38.7
20 weeks	—	57.1	—	36.8	—	57.5	—	41.3
22 weeks	—	61.0	—	41.2	—	60.5	—	43.3
24 weeks	—	63.6	—	46.0	—	65.5	—	48.0

This shows the much better growth obtained when milk was fed continuously for 6 months over the water group. Records of food consumption also showed that the milk fed group consumed much more per bird than the water fed group and that the rate of mortality among the milk fed birds was almost negligible compared with a fairly high mortality among the water fed group. Furthermore, the milk fed birds layed eggs at a much earlier date than the second group and in larger numbers and of a greater size as seen below:—

TABLE II

FOOD CONSUMPTION

	Average food consumption per bird (lb.)		Food consumption in lb. per lb. live weight gain.	
	Group 1	*Group* 2	*Group* 1	*Group* 2
1. 0— 4 weeks ..	1.71	1.15	3.70	22.20
2. 4— 8 weeks ..	4.44	1.40	3.74	6.32
3. 8—12 weeks ..	5.53	3.24	5.05	12.38
4. 12—16 weeks ..	4.77	4.01	8.22	5.32
5. 16—20 weeks ..	4.61	4.21	26.50	6.99
6. 20—24 weeks ..	5.32	4.90	8.75	9.50

Note:—The food consumption figures include the separated milk reckoned on a dry matter basis of 9 per cent.

TABLE III

AVERAGE CONSUMPTION (IN LB.) PER CHICK (MILK & WATER GROUPS)

Period.	Group 1		Group 2	
	Average milk consumption per chick	Average water consumption per chick	Average milk consumption per chick	Average water consumption per chick
1. 0— 4 weeks	2.48	nil	nil	Unrecorded
2A. 4— 6 weeks	2.21	nil	nil	—do—
2B. 6— 8 weeks	2.48	nil	0.97	nil
3A. 8— 10 weeks	3.29	nil	nil	Unrecorded
3B. 10— 12 weeks	1.01	3.70	nil	—do—
4. 12— 16 weeks	2.31	9.42	3.82	4.51
5. 16—20 weeks	6.06	17.20	6.71	15.00
6. 20—24 weeks	7.84	14.49	6.56	17.97

Mortality

The value of separated milk in the ration of chickens is well illustrated in Table IV which gives a summary of the number of deaths which occurred in each group at different ages during the trials.

TABLE IV

MORTALITY (0—24 WEEKS)

Period.	Number of deaths. Group 1	Group 2
1. 0— 1 week	1	2
2. 1— 2 weeks	—	3
3. 2— 3 weeks	1	2
4. 3— 4 weeks	—	4
5. 4— 5 weeks	—	4
6. 5— 6 weeks	—	5
7. 6— 7 weeks	—	1
8. 7— 8 weeks	—	1
9. 8—11 weeks	—	—
10. 11—12 weeks	—	2
11. 12—13 weeks	—	1
12. 13—14 weeks	—	1
13. 14—24 weeks	—	—
Total Mortality	2	26
Percentage mortality:	4.8	61.9

Egg production

Table V gives details of the average age of the birds when they laid their first eggs, the percentage of birds laying when 6 months old, the average egg production and the weight of the eggs for each breed on the two experimental rations.

TABLE V

EGG PRODUCTION

Group	1		2	
Breed	White Leghorn.	Rhode Island Red.	White Leghorn.	Rhode Island Red.
Age at first egg, (days)	137	134	159	215
Percentage of birds laying at 24 weeks	42.8	16.6	20.0	nil
Average number of eggs per bird to 24 weeks	10.1	12.9	0.83	nil
Average egg size (oz.)	1.61	1.51	1.55	nil

Similar trials were also conducted with varying quantities of milk fed during 6 months, and with soybean meal plus salt as supplements to the basal ration, and the following conclusions were arrived at as a result of all these trials:—

1. The feeding of chicks on a mixed cereal diet, although supplemented with liberal amounts of green food and calcium, results in very poor growth, excessive mortality and poor food utilization.

2. A ration containing liberal amounts of wheat bran and green food failed to prevent the occurrence of gizzard lesions, which appeared typical of those produced by diets deficient in the gizzard factor.

3. During the early stages of growth (up to 6 weeks), a ration of cereals supplemented by liberal amounts of green food and calcium along with separated milk to drink proved adequate.

4. During the early stages of growth the feeding of milk and water in separate containers gave substantially poorer results than when milk alone was provided as a drink.

5. From 6 weeks onwards the feeding of milk and water in separate containers gave satisfactory results as regards growth etc., and resulted in a considerable saving in cost.

6. The nutritive value of cereal rations is greatly improved by a supplement of soybean meal plus salt.

7. Soybean meal plus salt proved inferior to milk as a supplementary feed for chicks up to 6 weeks of age and to milk and water from 6 weeks onwards.

8. The feeding of cereals plus separated milk to drink from 0—6 weeks and cereals plus soybean meal and salt from 6 weeks onwards gives better results than cereals plus soybean meal and salt from day-old onward. If no milk is available for chicks less than 6 weeks old, soybean meal and salt can be used with fair success.

9. The protein requirements for young chicks during early life are considerably lower than the accepted standard of 20 per cent when the protein supplement is provided in the form of milk.

The Feeding Regime

The nature of the food given to poultry is a matter of prime importance and many poultry keepers fail to make profits through errors in feeding. The food should be palatable and well balanced in regard to proteins, non-proteins, minerals and vitamins, but it should also be cheap. It is, however, bad policy to feed damaged grain and mashes or inferior and bad house and garden refuse.

Hairy or rough cooked foods are not relished by poultry, and they prefer coarsely ground foods to finely ground ones, as the latter when fed as mashes form a sticky mass in the mouth and are not readily swallowed. Granular, loose textured mixtures are far better and more palatable. A certain amount of bulk in the rations is desirable, because too concentrated foods tend to be unpalatable and are not properly digested. The amount of fibre in the food, however, should be moderate as fowls cannot readily digest it, and it does not supply the necessary nutrients either for growth or egg production. From 3 to 10 per cent appears to be a safe limit, but foods of high fibre content should not normally be used, although occasionally they may be included in the ration for special purposes. For example, if the ration is very low in fibre the addition of from 5 to 10 per cent alfalfa (lucerne) meal may improve matters, and similarly the inclusion of a similar amount of a fairly fibrous meal will prove beneficial when an otherwise normal ration is deficient in vitamin A.

Feeding Stuffs for Poultry

The Cereals

The cereals and their by-products, a detailed description of which has been given in Chapter VI, constitute the basis of the rations for all types of poultry, but as they are not well balanced it is better to use a mixture rather than a single type.

Wheat

Wheat is the staple grain of northern India and both the whole grain and bran are widely used as poultry foods.

Wheat and gram are very palatable and the birds will usually select them first from a mixed grain ration. Wheat may be fed either as the whole grain in the grain mixture or ground up with others in the form of a mash, and when cheap may constitute up to 50 per cent of the total feed. Halnan [1928] has shown in digestibility trials on White Leghorn cockerels that it is immaterial whether what are known as 'weak' or 'strong' wheats are fed to poultry, as both are equally suitable. In growing wheat for poultry feeding, therefore, the yield rather than the strength of the grain should be the object to aim at.

Wheat bran

Bran is an excellent food as part of the ration both for young and fully grown birds and may be fed up to 50 per cent of the mash or from 25 to 30 per cent of the total ration.

Barley

Barley has much the same nutritive value as wheat and may be used to replace the latter, although fowls prefer the wheat as a whole grain. When barley is fed it should preferably be crushed and fed with wheat meal and skimmed milk or butter milk.

Ragi

Ragi has a similar protein content to paddy, but less fibre, and is extensively used in poultry rations in south India.

Maize

Maize is excellent for poultry but as it is very fattening it should be used in moderation and not as a sole feed for breeding and growing birds, although it is unrivalled for bringing birds to prime condition for the table. It is perhaps better than wheat for egg production when constituting a part of the ration, and in the U.S.A. as much as 60 per cent in the ration has been used with good results. Yellow maize should be used in preference to white, owing to its high carotene content. Maize may be fed whole, crushed, or as a meal.

Oats

Oats are much more fibrous than either maize or wheat and may be fed either hulled or unhulled. Some poultry keepers do not consider it necessary or desirable to hull them. Oats are not normally consumed as readily as wheat and maize, and those grown in India are usually poorer in quality and less palatable than those grown in more temperate climates. Halnan [1926] has shown that the digestibility of the fibre of oats is very low, averaging 7.6 per cent. Other workers [Kaupp 1904] also have found very low digestible coefficients for the crude fibre of whole oats, varying between 4.2 to 11.2 per cent. These trials all illustrate the fact that poultry cannot digest crude fibre very well, and furthermore, grinding oats does not improve the digestibility of the fibre. If cheap supplies of oats are available, however, they may constitute upto 20 per cent of the mash.

Rice

Rice is the chief grain produced in many parts of India and is extensively used for poultry. Unhusked rice, or paddy, is not so palatable as wheat on account of its rough coat, but it is nevertheless a useful grain.

American workers have obtained satisfactory results using up to 75 per cent of rice products in the ration, but when fed at this level it is necessary to supplement it with a considerable proportion of food rich in proteins and minerals. Whole rice is a more valuable food than polished rice, because polishing removes the vitamin 'B' complex.

Rice bran

Rice bran is richer in protein than the whole rice but has more fibre and is much less palatable than wheat bran. On account of its variable composition and consequent uncertain value in poultry rations it is not advisable to feed more than 40 per cent in the mash or mash and grain ration.

Juar or cholam

This grain is of similar composition to wheat and can be used as a substitute for it.

The Millets

The millets vary somewhat in composition but have a higher fibre and lower protein content than wheat. They can be used as substitutes for other grains and are largely used in chick mixtures.

Other carbohydrate feeds

Tapioca is a carbohydrate rich food and contains but little protein and is not much good as a poultry feed, unless fed with suitable protein rich feeding stuffs. It is nevertheless extensively used on the west coast of southern India.

Potatoes

Potatoes are deficient in protein, and too bulky to use exclusively for poultry and should be fed in conjunction with other foods. When thus employed they form a useful substitute for cereals for all classes of stock, but should not be used too largely for laying hens, which require a considerable amount of protein feeds to produce the yolk of the eggs. When fed to young stock or laying hens it is best to boil the potatoes and mix them properly mashed with an equal quantity of wheat bran.

Protein Supplements

Young stock and laying hens need a considerable amount of proteins of high quality to balance the deficiencies of the cereals in this respect, and it is better to furnish these from animal rather than vegetable sources, as the former are as a rule far better balanced in regard to their amino acid content. It is not generally realised as fully as it should be that milk is an excellent food for poultry and contains proteins of high biological value.

For domestic flocks, refuse from the table and the kitchen should be finely cut and may be fed either as such, or mixed with the mash feed. Halnan [1942] has recorded the results of some trials carried out to ascertain the biological value of proteins on young chickens. He has shown that the method of preparing animal proteins affects their nutritive value and that high temperatures should be avoided in the methods of preparations. He has also shown that both the quantity and quality of the protein fed with a cereal ration affects to a considerable extent the efficiency of the total or gross energy of the ration. It is interesting to note that Halnan [1942] finds as a result of these experiments strong support for the system of expressing the nutritive values of feeding stuffs and nutritive requirements of animals in general in terms of digestible protein and total digestible nutrients, *i.e.* in terms of metabolizable energy rather than in terms of calories or starch equivalents. Whole-meat meal as shown by Halnan possesses a high biological value for chicks, but it has certain disadvantages and tends to cause trouble in the gizzard.

Butter milk and skim milk

Whole milk is usually too expensive to feed to poultry but butter milk and skim milk give, on the whole, just as good results and should be used whenever possible, whether for young or fully grown birds. They may also replace animal food such as meat to a considerable extent. Butter milk contains a certain amount of lactic acid which aids digestion, and it also has valuable laxative qualities. It may be fed either separately in properly protected vessels, or mixed with the soft food, but not to such an extent as to make a slop. Butter milk also has a beneficial action on the digestive tract and may help to ward off many of the diseases which afflict poultry. It is an unrivalled food both for meat and egg production and should be given whenever available.

Skim milk contains a little more fat than butter milk but otherwise has all the valuable properties of the latter. Particular care should be taken in feeding both these to ensure strict cleanliness.

Whey

Whey is useful in supplying the vitamins necessary for growth in young chickens and may be fed in conjunction with protein rich foods. Milk products are not always obtainable and may be costly if fed as sole protein rich supplements, so it is advisable to use them in conjunction with cheaper protein rich supplements of either animal or vegetable origin.

Fish meal

Good quality white fish meal is an excellent protein feed for poultry, but is usually too expensive, except on the sea coast, to be used as a sole protein feed. In coastal areas it is an economical protein supplement to use in conjunction with cereal rations.

Meat and meat and bone meals

Tests carried out by Macdonald [1941] at Izatnagar have shown that although these feeds are, as a rule, poorer in biological value than white fish meals they, nevertheless, give as good results as milk products and should be used whenever they can be bought cheaply.

Vegetable protein supplements

Soybean meal is generally regarded as one of the best vegetable protein rich supplements. The extracted meal is said to give better results than whole soybean meal but the most recent trials at Izatnagar do not support this view. Soybean meal plus salt has a very beneficial effect on the rate of growth of chickens but is not as satisfactory as milk or meat offals. Like other vegetable protein supplements this meal is useless without added salt.

Legumes

Pea, bean and gram meals are moderately satisfactory sources of protein if fed with salt, but the balance of evidence suggests that they are not as good as soybean meal.

The Dals

The dals contain nearly twice as much protein as wheat and give fairly satisfactory results, fed either as grain, or as part of the mash, but they should not be fed as the sole protein supplement.

Groundnut meal

Trials conducted at Izatnagar have shown that groundnut meal is as valuable as soybean meal for growing chickens. As it is usually cheaper than the latter it should be used much more extensively than it is for different classes of poultry.

Ground linseed and linseed meal

Although the proteins of linseed products are of poorer quality than those of the soybean, they are frequently used as protein supplements.

Small quantities of linseed meal are useful in stimulating the growth of feathers in backward chicks and moulting hens, but the amount fed should not exceed 5 per cent of the total ration.

Cottonseed meal

This meal is rich in protein content but the proteins are not as suitable as those of the soybean or groundnut. Hence cotton seed meal should not be used to a greater extent than 5 per cent in the ration for laying birds, as more than this tends to cause discolouration of the yolks, especially of eggs put into cold storage.

MINERALS

Minerals, especially calcium, are as important for poultry as for other classes of stock and when the birds are free to roam at large they are able as a rule to pick up from the ground all they need to supplement the deficiencies of the rations.

Most of the foodstuffs fed to poultry are deficient in one or more of the essential minerals, so when they are confined in runs and there is a danger of mineral deficiency they should be supplied with broken or moderately finely ground limestone or other form of calcium, such as ground shells. No harm will be done if limestone is supplied, even when the birds are able to wander about the fields. Rations containing fish meal as the sole protein supplement will contain sufficient common salt for normal growth and egg production, but insufficient calcium, which is essential for forming the shell of the egg, which is largely composed of calcium carbonate—the same chemical substance as limestone. Absence or deficiency of this substance is likely to cause rickets and the production of soft shelled eggs.

The amount of calcium required in the ration varies according to the age of the birds, the rate of production, and the amount of vitamin D and phosphorus in the diet, but the provision of ground shells and limestone as an invariable adjunct to the rations should be sufficient precaution against calcium deficiency. Mineral supplements should be kept in proper containers so that they are not contaminated, and the birds can have access to them at all times.

It should be remembered, however, that there is nothing which fowls enjoy more than scratching about on fresh uncontaminated ground, or newly ploughed or dug up earth. This provides them with employment, they get the needed minerals, and also a considerable amount of animal food from the worms and insects they find. It is a good plan, therefore, when poultry are kept in confined spaces, to provide them with fresh sods of earth in their runs or to dig up periodically a certain amount of the ground.

Common Salt

Common salt is very necessary for poultry when little or no animal proteins are fed, if the best results are to be obtained, and an amount constituting from ½ to 1 per cent by weight of the total ration should be given finely ground and mixed in the mash. This will guard against either sodium or chlorine deficiency, particularly in warm weather, and act as a preventive against feather picking.

Under normal conditions of feeding it should not be necessary to feed any other mineral supplements such as steamed bone flour, iodine, sulphur, etc., and feeding of charcoal, which is a common practice among many poultry keepers, should be unnecessary when the rations are well balanced. If the inclusion of charcoal gives better results it is usually a sign of improper feeding, and the cause should be removed rather than corrected by including charcoal.

There is still considerable controversy in regard to the value of insoluble grit in the ration. The balance of evidence suggests that this material is not essential but that it may tone up the digestive system and help in the digestion of the food. When birds are kept in confined conditions it may be advisable to feed such materials as finely broken granite, crockery, etc.

Green Foods and Vitamins

Vitamin A

Poultry have relatively high needs for vitamin A, and yet most poultry foods, other than yellow maize, which cannot be fed in sufficiently large quantities to provide the amount needed, are deficient in this vitamin. Poultry should therefore always be given a regular supply of good quality green foods. Any form of green material from the garden or farm, and also young pasture leaves if not rank, are much relished. When poultry do not get sufficient green material both growth and egg production are liable to suffer, and the eggs may be sterile, due to vitamin A deficiency. Recent developments in our knowledge of vitamins have perhaps done more than anything else in the prevention of poultry diseases and uneconomic production. Should there be any difficulty in getting green fodder, sprouted oats may be fed. The oats should be moistened with water and allowed to sprout until the young shoots are just turning green. An alternative is to feed cod or shark liver oil.

Vitamin B or B_1

Vitamin B is not likely to be lacking in any of the ordinary rations, except in cases where polished rice forms a large part of the ration. In polishing the rice the germ containing the vitamin is removed and avian polyneuritis, the counterpart of *beri beri* in man, may result. The only remedy is to improve the rations.

Vitamin D

Under ordinary conditions when fowls get plenty of sunlight, or access to the fields, there should be no danger of vitamin D deficiency. Should they be confined, however, for any length of time in closed-in houses, or behind ordinary glass, there may be real danger of rickets or osteomalacia developing, owing to insufficient calcium-phosphorus metabolism due to lack of vitamin D. It will be necessary in such cases to give about 1 per cent of guaranteed fish liver oil with the ration as a preventative.

Vitamin B_2; (or G, U.S.A.), the growth promoting vitamin

Most of the cereals are deficient in vitamin B_2, and unless the rations are supplemented by materials rich in this vitamin, growth will be poor and the eggs tend to be infertile. Glandular tissues, such as liver, pancreas etc., yeast and milling products are rich in B_2, also young green leaves and pasture, and these should be fed whenever possible to counter the defect of the cereals. (See also Chapter II, Vitamins).

The above constitute the only precautions which need usually be taken against vitamin deficiency—but their observance may make the difference between success and failure.

Introducing Variety into the Ration

It is desirable to change the ration according to the season and temperature. A greater percentage of barley, oats, gram, peas and animal foods should be given in cold weather, while during the hot weather poultry may be given more vegetables, green feeds, butter milk and wheat. Special care should be taken during the monsoon period and when the birds are moulting to ensure the strictest cleanliness and most careful feeding. During such periods plenty of green stuffs and vegetables should be given with wheat, paddy, barley and gram, with a small quantity of meat periodically.

The Amount of Food to be Fed

It is not easy to lay down any fixed standards of how much food should be fed to poultry, but whilst they should not be underfed they should also not be overfed, but given a sufficient and regular amount of food at definite intervals. Over-feeding fowls is not only uneconomic but leads to digestive troubles, which might otherwise be avoided, and diminishes the number of eggs laid rather than increases them. Like most other farm stock different breeds of poultry and different individuals of the same breed will require different amounts of food, and naturally, during the cold weather, more heat producing food is needed than in the hot, and a hen laying may require almost double the amount of food she would need when not laying.

It is a sound plan to give only that amount of feed which the birds will eat with relish, but not more than is necessary to satisfy their appetite. Mashes, whether wet or dry, should be placed in proper containers so that the birds cannot contaminate them, whilst grains should be strewn over the ground.

Fowls enjoy scratching for the grain amongst straw and rubble or from rough overturned earth, but food should not be left about to become stale and dirty.

Just precisely how much food is required per bird depends on a number of factors, such as the extent of run which the birds have, and on the amount of odd scraps which may come from the domestic household. On large pasturage runs, poultry will pick up a considerable proportion of food for themselves and about 2 ounces of additional food per bird will be sufficient. If the runs are small and do not provide much opportunity for the birds to supplement their rations, double this amount, or 4 ounces a day may be needed. Practical experience however will soon indicate whether the birds are having their appetites satisfied and getting enough food.

Protein Supplement

The question arises as to how much protein supplement should be fed to poultry. Rapidly growing chickens need rations rich in well balanced proteins, and the general consensus of opinion is that during the first eight weeks the best results are obtained from rations containing 20 per cent protein, or even 15 per cent, if milk is the sole protein rich supplement. If inferior types of protein are fed a higher percentage will be needed, whilst well grown chickens will do well on considerably less. On the other hand fully grown birds may be fattened on a cereal ration with as little as 3 per cent protein rich supplements.

If the rations are well balanced, good egg production should be secured with rations containing from 12 to 13 per cent protein, although this may be slightly raised.

Young turkeys and quickly growing ducks, such as the Aylesbury, need a somewhat higher protein level than chickens.

Drinking Water

Fowls need a continuous supply of clean drinking water. This should be provided in shallow vessels, which must be thoroughly cleaned every night ready for next day. A better system is an automatic water supply such as may be obtained by

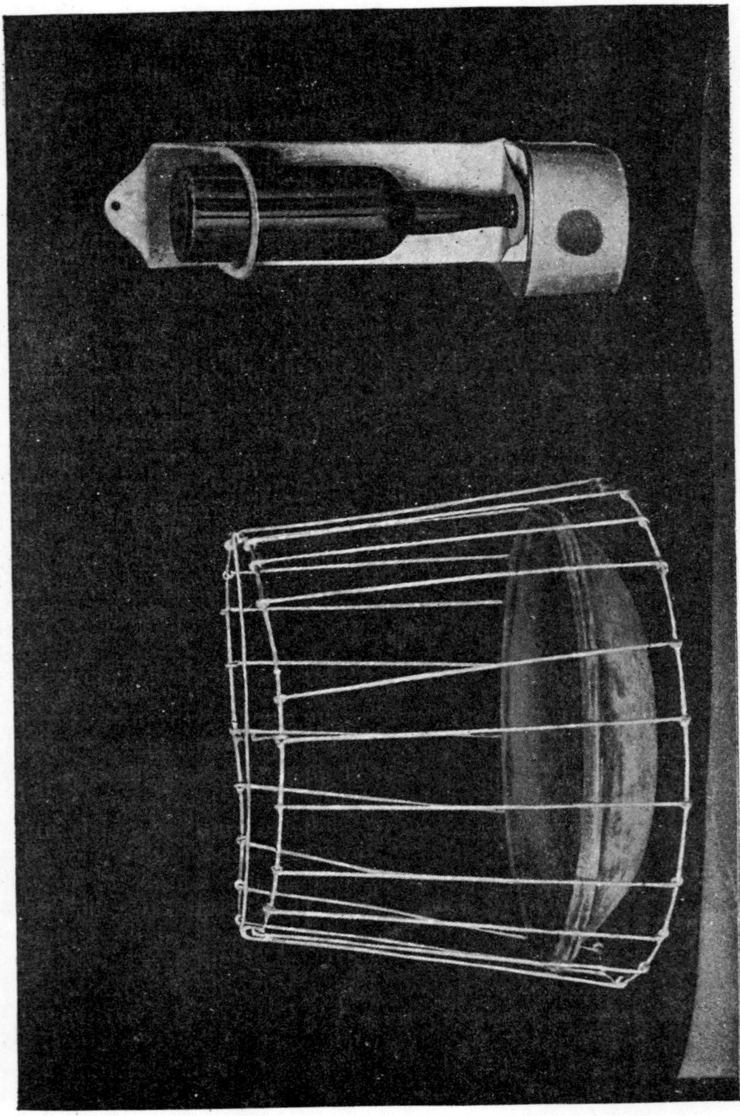

Example of hygienic water supply

inverting a bottle of water in a shallow basin, or a contrivance such as that shown in Plate XLIV, will guard against birds putting their feet into the water. The amount of water poultry need varies with circumstances. Fattening birds need less than growing chicks, and heavy laying birds more than other classes of stock. Naturally more water is required in the hot weather than in the cold.

The water supply should be renewed at least every day in the morning, and again in the evening, and should be placed so as to avoid contamination or exposure to the sun.

Some Systems of Feeding

Different poultry keepers feed their birds according to their own particular systems, which they have found to be satisfactory, and which may conform to one or other of the following:—

1. The all mash system
2. Mash and grain
3. Pellet feeding
4. Grain and milk

The *all mash system* is easy to follow and ensures that all the birds receive a well balanced diet. It is also hygienic as the food is fed from hoppers. It is costly, however, as everything has to be ground and the protein supplements mixed in with the ground cereals. It also tends to become monotonous for the birds and diminishes their natural propensity to scratch and search for the grain part of the food.

The *mash and grain system* gives better results, if properly carried out, as the ration can be varied at will to suit requirements according to the condition of the birds. For example if the birds are becoming too fat the grain is reduced and the mash increased and *vice versa*.

In the *pellet system* the food is all ground and formed into a mash which is then passed through suitable dies under pressure. It is in reality a modification of the mash system but is not practised in India as the machinery needed is expensive and only justifiable on large poultry farms.

The *grain and milk system* is excellent when these feeds can be cheaply bought and good results may be obtained as described above in the Izatnagar trials. Its practicability, however, will depend on the proximity to large dairies and the cost of the milk products. It may be used for all classes of poultry.

Wet versus dry mashes

It is immaterial whether the mash be fed wet or dry, although the former involves a higher degree of skill and more labour than the latter. Most farmers follow the dry method although wet mashes may be fed to backward birds and fattening chickens. The mash should not be so wet as to be sloppy, but in a crumbly condition, and only as much fed at a time as will be finished, as mash left over becomes stale and unpalatable and liable to contamination.

A good combination is to feed dry mash as a morning meal and wet mash and grain in the evening.

How often to Feed Poultry

Fowls are early risers and when confined at night should be let out at dawn and given access to clean drinking water followed by the first meal. This may consist of pollards or well ground wheat, barley or oats mixed with skimmed milk, or butter milk when available, and made into a mash sufficiently damp to hang together and crumble. It should not be made into a slop. Finely chopped onions or garlic may occasionally be mixed with the mash. Domestic refuse, such as finely cut meat, bread, vegetables etc., may also be included in the first mash feed. Periodically a little salt, but not too much, or one of the standard poultry powders may be fed mixed in the mash. In the cold weather fowls should be allowed about an ounce of food per head per day for the first meal, but in the hot weather, and if there is a tendency for laying fowls to become too fat, the first morning meal may be modified by wholly or partially reducing the non-grain part of the feed. Fowls are very partial to soft food and if given too much may find little necessity for their special aptitude in getting other food and minerals by scratching the earth. Where laying birds tend to become fat, egg laying is retarded and they should then be put on to whole grain only. On large runs it may not be necessary to feed again until the last meal in the evening, but in confined spaces the birds should be fed about an ounce per head of whole grain at mid-day strewn about the ground. The final meal in the evening should consist of good grain comprising the balance of the daily ration.

Poultry appreciate variety in their rations and do better if the constitution of the morning meal is frequently changed, and whatever feed happens to be available should be introduced into the ration. For example, boiled roots such as beet, carrots, turnips etc., when finely chopped and mixed with bran and some good quality bone meal and a trace of salt, makes an excellent morning meal. The salt, however, should never be fed as such, but always dissolved in the water or milk used for the morning feed in the amount mentioned above. Wet sloppy food should not be fed to poultry.

Some Typical Feeding Systems

(1) *At the Lyallpur Agricultural Farm*

Newly hatched chickens are given no food for the first two days but are provided with water and some fine grit and charcoal before any solid food is given, and a little sour butter milk to prevent diarrhoea. After two days the chicks are fed according to the following regime:—

3rd day to 3rd week

(1) Crushed grain consisting of three parts wheat, one part maize (2 ounces for 12 chicks at 8 a.m., 12 noon and 6 p.m.)

(2) Wet mash consisting of 2 parts wheat bran and 1 part maize *atta* and 2 parts wheat *atta* (2 ounces for 12 birds, at 10 a.m., 2 p.m., and 4 p.m.)

(3) As much curdled skimmed milk at noon as they can take, and again in the evening with no water.

From about the 4th week up to the age of 6 months the grain and mash rations are increased until the chicks consume approximately 2 ounces of mash and one ounce of grain mixture per head per day. At the age of 3 months the chickens should have increased their weight at birth about 25 times, and therefore the food must be rich in body producing material. This is best attained by a variety of feeds which should not be exclusively grain.

Ground limestone and plenty of green food are also fed to guard against rickets.

Some form of animal food is necessary for chickens, if they are to make the best growth, and their requirements in this respect will be met if they are given all the skim and butter milk that they can drink. The mash fed to chickens in the early stages of growth, however, may contain about 5 per cent finely cut meat, this being gradually raised at the end of the first month to 10 per cent. As with fully grown fowls no definite rules can be laid down as to the actual amounts of food needed, as this will vary with circumstances, but they should always be given as much as they can eat last thing in the evening so as to retire at night with full crops. Chickens should have access to the dry mash throughout the day and after the first two months the best method for feeding this is to place it in hoppers as illustrated in Plate XLV.

From the age of 6 months onwards a satisfactory feed would be 12 lb. mash and 12 lb. grain mixture per 100 birds. This ration is given at the Agricultural College Poultry Farm at Lyallpur and is found satisfactory, but some people may prefer to give no wet feed at all until the chickens are about two months old, when a moist mash may be fed at about 4 o'clock in the afternoon. Except in dry and arid tracts this may be better because, although wet mashes are excellent if properly fed, any slackness or carelessness in feeding on the part of attendants may lead to digestive troubles if wet mashes are improperly fed at too early an age.

Finely chopped up green food may be given after the first week and cooked beets and turnips mixed with the dry mash.

(2) *The Government Poultry Farm, Gurdaspur, Punjab*

A rationing regime which has been in vogue at the Poultry Farm at Gurdaspur in the Punjab for adult and young stock is as follows:—

(a) *Adult stock, i.e., about* 6 *months old and over*

(i) Two ounces of grain per head per day is given to the birds, half in the morning and the other half in the evening. The grain ration is made up as follows:—

Wheat	..	50 parts.	*Winter.*
Yellow maize	..	50 parts.	
Wheat	..	66.6 parts.	*Summer.*
Yellow maize	..	33.3 parts.	

Fig. 1. Waste-proof mash hopper for adult birds.

Fig. 2. Another type of waste-proof mash hopper.

(ii) *Mash*

Wheat *ata*	..	35 parts.	
Yellow maize *ata*	..	20 parts.	*Winter.*
Wheat bran	..	20 parts.	
Barley meal	..	15 parts.	

100 lb. of the above mash are mixed with 75 lb. of beef-offal-extract and fed *ad lib.* Each bird consumes approximately two ounces per day.

Wheat *ata*	..	40 parts.	
Yellow maize *ata*	..	20 parts.	*Summer.*
Wheat bran	..	20 parts.	
Barley meal	..	15 parts.	

100 lb. of the above mash are mixed with 75 lb. of beef-offal-extract and fed *ad lib.* Each bird consumes about two ounces per day.

(b) *Young stock, i.e., from one day old to about* 6 *months*

(i) Grain mixture is fed in the same proportions as for adults, the only difference being that the grains are kibbled and the quantity allowed per day per bird averages about 1 ounce.

(ii) *Mash* (Winter and Summer)

Wheat *ata*	..	35 parts.
Yellow maize *ata*	..	20 parts.
Wheat bran	..	25 parts.
Barley meal	..	10 parts.
Linseed	..	2 parts.

100 lb. of the above mash are mixed with 75 lb. of beef-offal-extract and fed *ad lib.* On an average each bird consumes 1 ounce per day.

In addition to the grain mixture and wet mash, the birds are given any green food available once a day. Rivershell grit is also allowed *ad lib.*

At a time when it became necessary to find alternatives to wheat and maize, when these were difficult to obtain, the writer suggested replacing these grains by bajra, juar and oats. Both the original rations and the alternative ones suggested, however, were deficient in calcium, and so wheat *mamni,* which is very rich in calcium, was suggested as an alternative to any of the above grains, with the further addition of finely ground bone meal or limestone. (Wheat *mamni* is wheat screenings and small foreign seeds.)

Alternative rations proposed by the Officer-in-charge of the Poultry Section of the Imperial Veterinary Research Institute, Izatnagar, were the following:—

Mashes

(1) Wheat bran	..	30 parts.
Rice bran	..	30 parts.
Ground barley	..	30 parts.
Ground gram	..	20 parts.
Salt	..	½ part.

Or,

(2) Wheat bran	..	30 parts.
Ground paddy	..	20 parts.
Ground oats	..	10 parts.
Ground barley	..	20 parts.
Ground gram	..	20 parts.
Salt	..	½ part.

Grains for adults

Paddy	..	1 part.
Juar	..	1 part.
Barley	..	1 part.
Gram	..	1 part.

A mixture of 3 or 4 of the 5 grains mentioned above will do well for chickens, and if meat offals with plenty of green food and rivershell grit and crushed limestone are fed with the above rations, they should prove satisfactory.

(3) *The Poultry Research Section, Imperial Veterinary Research Institute, Izatnagar.* (Details supplied by Mr. A. J. Macdonald.)

The following system of feeding has given good results at the Imperial Veterinary Research Institute, Izatnagar. As the ingredients of the rations used are typical of those found in northern India they are capable of fairly wide application either as formulated, or with slight modifications according to the local supplies of foodstuffs.

System 1.

All the birds are fed dry mash and grain from the day after hatching. During the first week the grain is spread out on paper and fed in small quantities several times daily. From the first to the eighth week the grain is fed in the litter in amounts depending on the chicks' appetite every morning at 7 a.m. and every evening at 4.30 p.m.

All the grain is generally consumed in about 15 minutes. From the age of 8 weeks onwards the grain is fed in the mornings and evenings. Up to 8 weeks of age the birds are given separated milk only to drink and after 8 weeks, separated milk and water are given in separate containers to growing birds and laying stock; liberal amounts of cut succulent green food are fed daily. Broken limestone is fed *ad lib* from the age of one day onwards.

The mash consists of wheat bran 50 parts, yellow maize meal 30 parts, ground oats 20 parts, supplemented with ½ per cent common salt. From 0—8 weeks the grain mixture consists of equal parts of yellow crushed maize, juar and *cheena.* From 8 weeks onward the grain mixture consists of equal parts of yellow maize, wheat and paddy.

System 2.

Another system which has given good results with chickens and laying hens consists in replacing the separated milk with meat offals. In this system the birds are fed the usual grain ration, green food and broken limestone. The meat offals (stomachs and intestines from the slaughter house) are ground up in a mincer and cooked in a small quantity of water for one hour. The cooked offals and the residual water left after cooking are then mixed with dry mash and fed in a crumbly state. Young chickens receive grain in the morning and evening, and a wet mash according to appetite at 9 a.m., 12 noon and 3 p.m. Laying birds receive only one feed of wet mash per day at about 8 a.m., grain at 11 a.m. and 4.30 p.m. Dry mash is fed at 10 a.m. and 3 p.m. Adult birds receive about 1 oz. of meat offals per day. Young chickens from 0—8 weeks receive relatively about 50 per cent more meat offals than the laying stock. From 8 weeks onward chickens can be fed the same proportions of meat offals to mash and grain as laying hens.

Food Consumption

Economy in food utilization is very important from the commercial stand point, both for table purposes and egg production. The efficiency factor, or the pounds of food given per pound of live weight gain depends on various factors, such as the nutritive value of the diet, the breed, climate, etc. Light breeds, such as *desi* fowls, make poorer gains in weight per pound of food consumed than large fast-growing breeds, such as Rhode Island Reds. Similarly, heavy layers utilize their food better for egg production than poor layers. From various reports it would appear that the food consumed per pound of gain, for successive pounds beginning with the first, comes to 4.0, 6.0 and 10.5 for Leghorns, and 3.8, 4.5, 6.5 and 10.0 for Rhode Island Reds. With poor diets these figures would be increased. In the hot weather the rate of growth and efficiency of food utilization are lowered. From 0—24 weeks a Leghorn pullet will consume about 20 lb. of food.

Good laying birds normally consume 3½—4 oz. of food per day or roughly 80—90 lb. of food per year. Though comparatively poor layers, such as the *desi* fowls, consume considerably less food than high producers such as Leghorns, they do not make as efficient use of their food, judged by the pounds of food consumed per pound of eggs produced.

Feeding for Egg Production

The egg laying season varies in different parts of India, and in northern India it is roughly from early October to March. As previously stated the prime essentials for the production of eggs are good breeding and good feeding. For example, a White Leghorn laying on an average 120 eggs during the winter egg laying season, and 20 during the off summer season, may be fed at the rate of 24 lb. of grain and bran mash mentioned on page 478. Other suitable grain mixtures for egg production are as follows:—

1. Equal parts of cracked maize, wheat and oats.
2. Cracked maize .. .. 3 parts.
   Cracked oats .. .. 2 parts.
   Wheat .. .. 1 part.
3. Cracked maize .. .. 2 parts.
   Oats .. .. 1 part.

Alternative dry mash mixtures are:—

1. Maize meal .. .. 2 parts.
   Wheat middlings .. .. 1 part.
   Wheat bran .. .. 1 part.
   Meat scrap .. .. 1 part.
2. Maize meal .. .. 3 parts.
   Meat scrap .. .. 1 part.

As with chickens, skimmed milk, butter milk and meat scraps of good quality, if available, are the best protein feeds to feed, and when available, fowls should be given as much milk as they will consume. All these rations should be supplemented by green stuff, grit, and finely ground shell. The mash should always be fed in hoppers and the grain strewn on the ground which may be littered with straw so that the hens have to scratch for it. More grain should invariably be fed in the evening than in the morning, so that the hens go to roost with their crops full.

Fattening

Birds which are required to be fattened for the market will not need so high a protein content in their rations as laying hens, and, if given good fattening food, should increase in weight from 30 to 50 per cent in two or three weeks, and when fed for this purpose they should be confined in small numbers to fattening pens and fed twice a day. A suitable ration for fattening fowls could be made from the following mixture:—

Maize meal	60 lb.
Middlings	40 lb.
Meat scraps	5 lb.

This may be mixed with sour milk to a moderate though not too soft consistency, but if milk is not available and water is used, the meat scraps must be increased to 10 lb. Drinking water during the fattening period should be reduced to a minimum, or not given at all, depending on climatic conditions.

BIBLIOGRAPHY

1. Annual Report on the Scheme of Research on the Improvement of Poultry for Table Purposes in Northern India, 1940-41.
2. Halnan, E. T. 1926. Jour. Agri. Sci. Vol. 16, pp. 451-458.
3. Halnan, E. T. 1928. Jour. Agri. Sci. Vol. 18, pp. 421-425.
4 & 5. Halnan, E. T. 1942. Jour. Agri. Sci. Vol. 32, pp. 179-193.
6. Kaupp & Ivey, 1923. North Carolina Agr. Exp. Station. Technical. Bul. No. 22, E. W. Brown, U. S. Dept. of Agri. Bureau of Animal Husbandry, Bul. No. 56, 1904.
7, 8 & 9. Macdonald A. J. 1941. Nutritional Experiments with Chickens. Ind. Jour. Vety. Sci. and Animal Husbandry, XI.

CHAPTER XVI

THE ECONOMICS OF FEEDING

The question is sometimes asked "What is the best ration to feed for any particular type of animal ?" The answer is, there is no one ration which can be considered the ration *par excellence* to the exclusion of others.

In the preceding chapters some account has been given of the considerations which must be taken into account in selecting rations for different types of farm stock, and in every case the farmer should be guided by the qualitative and quantitative factors which must be provided for in each particular case.

The computation of rations based on digestible nutrients takes into consideration only the quantitative aspects of the rationing problem, and deals with only some of the factors which govern the nutritive value of rations for different classes of farm stock. Different feeding stuffs have different values for different types of animal, and these values may have little correlation with the amounts of digestible nutrients which the feeding stuffs contain. Certain types of animal may have special requirements for minerals and vitamins which certain rations provide, and their value cannot thus be stated merely in terms of total digestible nutrients and digestible protein. For example, high quality legumes, such as lucerne, either as green fodder or hay, possess special nutritive features for cattle, pigs and growing animals, especially in the dry season and winter months, on account of their richness in calcium and protein. Similarly, dairy by-products are especially valuable for young pigs and poultry, on account of their richness in vitamin B.

There are many other differences in the value of certain feeding stuffs compared with others, which have been brought to light by trial and experience, and for which no adequate explanation based on digestible nutrients alone is forthcoming. For example, ground barley and ground maize have approximately the same value for dairy cattle, but barley has not the same value as maize for fattening cattle, pigs and sheep. Again, oats have a considerably higher relative value for horses and dairy cows than for fattening cattle, sheep and pigs, particularly the latter. Cottonseed and linseed meal when forming protein supplements to otherwise satisfactory rations are of approximately equal value for dairy cows and fattening lambs, but the former is of less value than the latter for fattening cattle when it is the sole protein supplement. Similarly, linseed has a special conditioning value for horses and certain other classes of stock, and hence is of greater value than cottonseed.

Corresponding considerations have already been described in regard to the value of oilseeds and cakes for cattle, sheep and pigs. Soybeans are a good part-protein supplement for dairy cows but are not to be recommended as the sole protein supplement for pigs, as, apart from other reasons they tend to produce soft pork. These and many other considerations which have been touched on in the chapters dealing with the various classes of live stock have to be kept in mind in formulating rations with a view to getting the best ration for any particular purpose.

When all these factors have been taken fully into account, however, one of the most important problems which has to be considered by the farmer is how to provide the cheapest ration consistent with efficiency.

Cost of the ration

A considerable number of feeding stuffs to choose from will be available either on the farm or in the market, and economic considerations demand, other things being equal, that those should be used which can be obtained most cheaply.

It does not follow that because a particular feeding stuff fetches a higher price than another it is necessarily of higher nutritive value. In fact it sometimes happens that higher priced feeding stuffs are of less nutritive value than lower priced ones, as in the case of *desi* and American cotton seeds. It is necessary, therefore, that the farmer should have some approximate guide which will enable him to ascertain the relative values of feeding stuffs in terms of cash.

Various methods have been devised from time to time to ascertain the relative cash values of feeding stuffs in terms of basal food ingredients, and two systems which may be used for this purpose will now be described.

No. 1

A method which takes into account the cost of the food in comparison with its energy value to the animal in terms of calories, calculated from the digestible nutrient content and the cash value of the manure produced.

No. 2

A method based on the total digestible nutrients and digestible protein but applicable only when protein rich feeding stuffs are more expensive than carbonaceous ones.

Method No. 1 based on food units

The three chief constituents of a feeding stuff are proteins, fats and carbohydrates, and one pound of each of these respectively will supply to the body that amount of heat expressed in calories as shown below:—

	Heat value per pound
Proteins	2,130 calories.
Fats	4,000 calories.
Carbohydrates	1,707 calories.

Thus the heat production values provide a method for comparing the values of proteins, fats and carbohydrates. If the heat producing value of one pound of carbohydrate is taken as unity then obviously the corresponding value of one pound of protein will be $\frac{2130}{1707}$ or 1.23 and the corresponding value of one pound of fat will be $\frac{4000}{1707}$ or 2.3.

If these values are applied to any particular feeding stuff, say cottonseed cake, it will be possible to find the percentage of carbohydrate which will have the same heat producing value as the sum of the percentage of the protein, fat and carbohydrate as shown by the digestible nutrients in cottonseed cake.

Thus, taking the digestibility data obtained at Lyallpur for cottonseed cake, the value per digestible food unit for the cake may be calculated as follows on the basis of cost (Rs. 3/- per maund):—

	Digestible Nutrients. %		Starch factor.		Expressed in terms of starch.
Protein	17.97	×	1.23	=	22.10
Fat	8.30	×	2.3	=	19.09

Carbohydrates	20.53	×	1.0	= 20.53
Fibre	16.43	×	1.0	= 16.43
	Digestible food units			= 78.15

At Rs. 3/- per maund the price per digestible food unit will be

$$\frac{\text{Rs. } 3}{78.15} = 7.4 \text{ pies}$$

In a similar way corresponding calculations may be made for any other feeding stuff if the amount of digestible nutrients it contains are known.

A certain amount of manure is produced, however, from the food eaten, the value of which is chiefly determined by the amount of nitrogen which it contains. So every maund of food employed will, in addition to its intrinsic value as a food, yield a certain value in terms of manure. Therefore, the factor 1.23 by which the digestible protein figure was multiplied should be increased in order to account for the manurial value. A reasonably safe but somewhat empirical guide based on the value of the manure would be to increase the protein factor from 1.23 to 1.50. Taking this figure the carbohydrate equivalent of the cottonseed cake will then, with this modification, be calculated thus:—

Protein percentage × 1.50, plus fat percentage × 2.3, plus carbohydrate and fibre percentage.

Protein	17.97	x	1.50	= 26.96
Fat	8.30	x	2.3	= 19.09
Carbohydrates	20.53	x	1.0	= 20.53
Fibre	16.43	x	1.0	= 16.43
				83.01

At Rs. 3 per maund the price per digestible food unit will be

$$\frac{\text{Rs. } 3}{83.01} = 7.0 \text{ pies.}$$

This provides a convenient practical method for comparing the relative values per food unit of different feeding stuffs. It is open to two criticisms, however. Firstly, it takes no account of the fact that the digestible nutrients of different feeding stuffs are not all of the same value, and hence it is to this extent inaccurate. Secondly, it attempts a compromise between the nutritive value of the feeding stuff and the value of the manure produced. Naturally the intrinsic value of manure will depend on the relative amounts of the total digestible nutrients and digestible protein which the feeding stuffs contain, and consequently, although the above method forms a useful guide in regard to purchasing price it is not strictly accurate in terms of rationing value.

Method No. 2 based on total digestible nutrients and digestible protein

The principle underlying this method is based on the estimation of the cost per unit of total digestible nutrients in the food. In its simplest form this method is very easy to employ and it shows which feeding stuffs are cheapest per unit of food contained in them. It is accurate in evaluating feeding stuffs when protein rich feeds are no more expensive than those richer in non-protein digestible nutrients, as is generally the case in India; it has certain drawbacks however, when used for protein rich feeds if the latter are more expensive than carbonaceous ones, as is usually the case in foreign countries, because the method does not take into consideration the differences in the digestible protein content of the various feeding stuffs; a matter of importance when one has to decide which of a number of protein rich supplements to use to balance a ration poor in protein.

This difficulty can be overcome by computing the cost per pound of total digestible nutrients, and also the cost per pound of digestible protein for each feeding stuff, both of which factors must be taken into consideration in deciding which protein supplement to employ.

Considerable experience, however, is necessary to arrive at a correct decision, because one supplement may furnish the cheapest source of total digestible nutrients, while another may be cheapest in regard to digestible protein.

Application of the method

a) When protein rich feeding stuffs are no more expensive than carbonaceous feeds.

Suppose a farmer has to purchase his feeding stuffs, and the concentrates he has to choose from are, maize at Rs. 4 per maund, *bajra* at Rs. 3-12 per maund and oats at Rs. 4-8 per maund. Which of these will be cheapest? The nutritive value of each may be determined from the total digestible nutrients,

DIAGRAM I

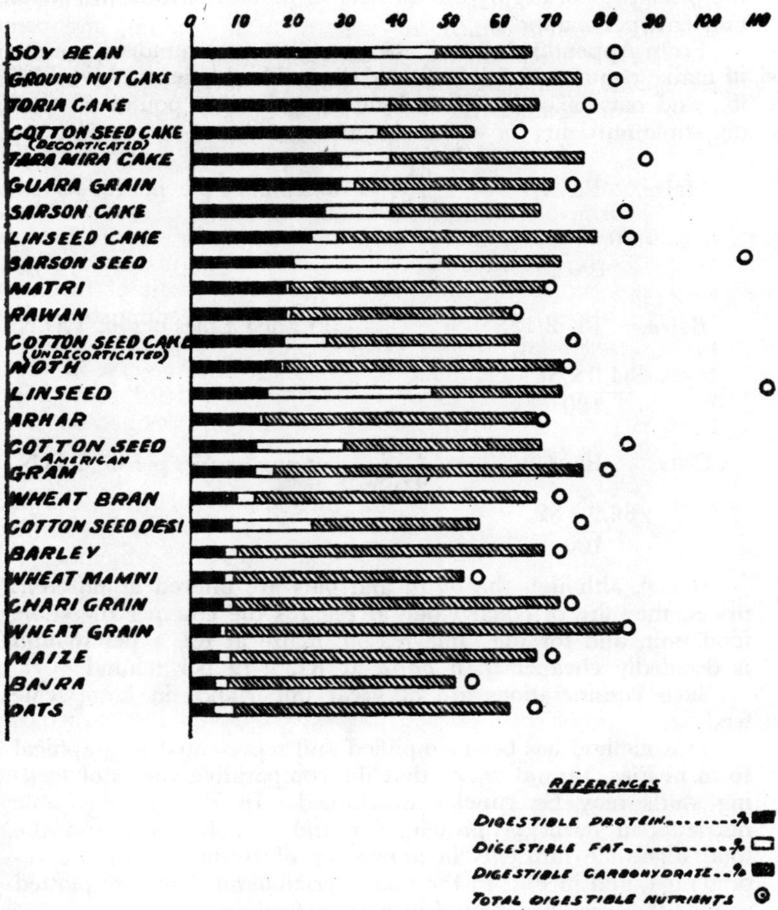

FIG. 14.

and the price per pound of these may be ascertained by dividing the price per maund by the number of pounds of total digestible nutrients per maund.

From Appendix I, it will be seen that one hundred pounds of maize contain 70.5 lb. of total digestible nutrients, *bajra* 54.3 lb., and oats 66.9 lb. Therefore the price per pound of total digestible nutrients for each respectively will be:—

Maize, $\text{Rs. } 4/- = \dfrac{64}{57.81} = \text{1 anna 1 pie per lb. T.D.N.}$

$$\frac{70.5 \times 82}{100}$$

Bajra, $\text{Rs. } 3/12/- = \dfrac{60}{44.53} = \text{1 anna 4 pies per lb. T.D.N.}$

$$\frac{54.3 \times 82}{100}$$

Oats, $\text{Rs. } 4/8/- = \dfrac{72}{54.86} = \text{1 anna 4 pies per lb. T.D.N.}$

$$\frac{66.9 \times 82}{100}$$

Hence, although the *bajra* and oats are offered at different prices, they are of equal value as regards the cost per digestible food unit, and for the same reason, maize at Rs. 4 per maund is decidedly cheaper than *bajra* at Rs. 3-12 per maund.

Such considerations are of great importance in large scale feeding.

This method has been simplified and represented in graphical form in Figs. 14 and 15, so that the comparative values of feeding stuffs may be quickly ascertained. In Fig. 14 digestible nutrients in terms of protein, fat and carbohydrate, and the total digestible nutrients in a number of feeding stuffs are represented, and in Fig. 15 the total digestible nutrients are plotted against the price per maund of various feeding stuffs.

Suppose we wish to compare gram at Rs. 3 per maund with *rawan* (*Dolichos lablab*) at Rs. 2-12 per maund. The total digestible nutrients in 100 lb. of gram and *rawan* are 80 lb. and 63 lb. respectively. It will be seen that curve 'A' strikes the

DIAGRAM II

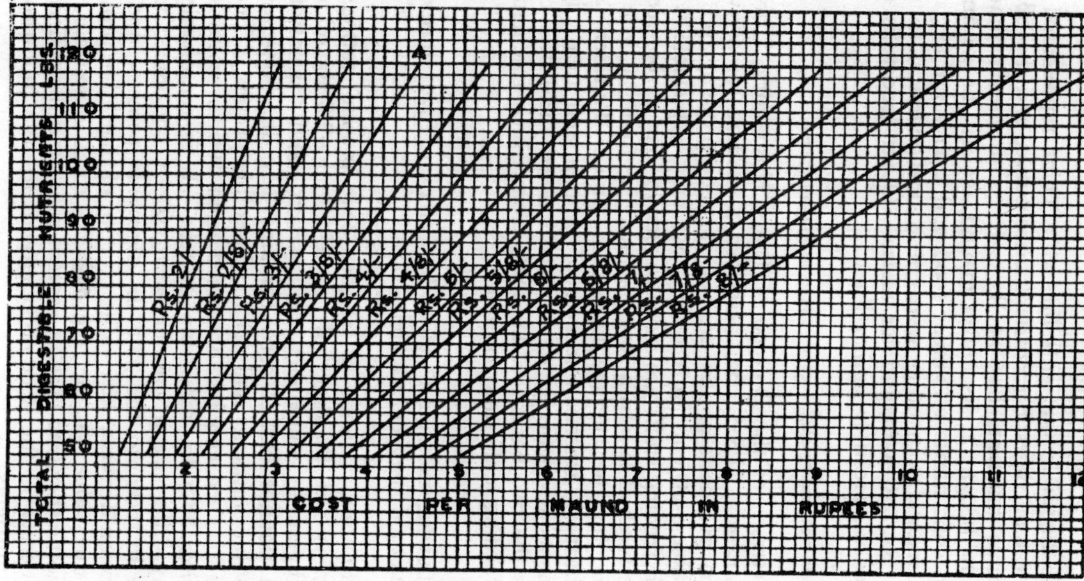

FIG. 15. Each curve represents price per maund where it cuts the 80 lb. total digestible nutrients line; i.e. Rs. 2/-, Rs. 2/8/-, Rs. 3/- and so on.

80 lb. total digestible nutrients line at a point represented by Rs. 3 on the total digestible nutrient price curve, and the 63 lb. total digestible nutrients line at Rs. 2-6. This means that if *rawan* can be bought at Rs. 2-6 a maund, we will be getting a protein rich concentrate of similar food and cash value as gram at Rs. 3 a maund. In other words it is uneconomical to pay more than Rs. 2-6 a maund for *rawan*. Other foodstuffs can be compared on precisely similar lines.

(*b*) *When protein-rich feeding stuffs cost more than carbonaceous ones.*

Wm. Petersen of the Division of Dairy Husbandry, University of Minnesota, U.S.A., has described a method for arriving at the respective costs of the protein, the non-protein, and the total digestible nutrients in a ration, but only applicable when protein rich feeding stuffs are more expensive than carbonaceous ones. The method is satisfactory, but may be found a little too cumbersome for the average farmer in India. Petersen takes the two cheapest American sources of protein and total digestible nutrients respectively, namely, cottonseed meal and corn (maize) as the two basal feeding stuffs from which calculations for other feeding stuffs may be made.

The following example as given by Petersen [1932] illustrates how this method may be applied:—

"One hundred pounds of 43 per cent protein cottonseed meal contains 37.6 lb. digestible protein and 42.6 lb. of non-protein digestible nutrients. Fifty-seven and twenty-four hundredth pounds No. 2 corn contains 42.6 lb. non-protein total digestible nutrients and 4.1 lb. digestible protein. Therefore 57.24 lb. corn can be subtracted from 100 lb. cottonseed meal leaving 33.5 lb. digestible protein. It is reasonable to ascribe any difference in cost between 100 lb. cottonseed meal and 57.24 lb. corn to the 33.5 lb. digestible protein, and by simple division arrive at the value of 1 lb. of digestible protein. To find the value of the non-protein total digestible nutrients, credit either feed with its protein value and the difference from the cost of the feed is charged to the non-protein digestible nutrients.

The formula for determining the cost for a pound of protein is:

$$\frac{\text{Cost of 100 pounds cottonseed meal} - .5724 \text{ (cost of 100 pounds corn)}}{33.5}$$

Example: Substituting with cottonseed meal at \$50.00 per ton and corn at \$25.00 per ton.

$$\frac{\$2.50 - (.5724 \times \$1.25)}{33.5} = 5.328 \text{ cents}$$

The value of a pound of non-protein digestible nutrients may be arrived at by the following formula:

$$\frac{\text{Cost of 57.24 pounds corn} - \text{(value of protein} \times 4.1)}{42.6}$$

Again substituting with corn at \$1.25 per hundredweight and protein at 5.328 cents per pound:

$$\frac{1.25 \times 57.24 - (4.1 \times 5.328)}{42.6}$$

or 1.167 cents per pound.

Should some other feeds be a cheaper source of either protein or non-protein nutrient such feeds may be used to establish the value of either protein or non-protein nutrients. This, however, is seldom necessary as by using corn and cottonseed meal as base feeds other cheaper feeds would be indicated, as selling for less than their value on the basis of corn and cottonseed meal.

By applying these values to the digestible protein and non-protein digestible nutrients of other feeds their values are arrived at. It is apparent that for a given feed both cottonseed meal and corn exert a definite influence on its value and therefore constants can be determined for each feed which when applied to the prices of cottonseed meal and corn should give the correct value of the feed.

By calculating the values of a feed with corn constant and cottonseed meal at two different values a constant can be determined by which cottonseed meal may be multiplied to indicate the influence upon the price of the feed. Likewise constants may be determined for corn.

Example :—Using previous formulae oats have a value of \$19.61 per ton when cottonseed meal is \$40.00 and corn \$20.00 per ton, and \$18.45 per ton when corn is \$20.00 and cottonseed meal $30.00 per ton. Therefore a $10.00 increase in cottonseed meal increases the value of oats $1.16.

Keeping the price of cottonseed meal constant and increasing the price of corn we find that a $10.00 increase in price of corn increases the value of oats $7.49. For any price therefore the value of oats = .116 × price of cottonseed meal + .749 × price of corn.

It is impossible to formulate a method which is applicable in all circumstances, irrespective of what the market price of a particular feeding stuff may be. Each method has its drawbacks, but method No. 2 (a) appears to form a simple and reliable means for ascertaining the approximate price limits, within which it would be economical to interchange protein foods among themselves, or similarly with carbonaceous ones. The farmer must accordingly exercise his ingenuity and experience in deciding which feeding stuffs to use, at the least cost, and with maximum efficiency.

BIBLIOGRAPHY

1. Petersen. Wm. E. 1932, Journal of Dairy Science, Vol. XV. Page 293.

APPENDICES

APPENDIX I

The analytical data relating to feeding stuffs grown in India is, as mentioned in the context, not complete. In compiling the following table all the analytical data which have been obtained at Lyallpur have been included, and also as many as are available from other provinces in India. In order to make the table more complete for reference and the compilation of rations, a number of analyses given by Morrison in his " Feeds and Feeding " have been included. Hitherto, owing to the gaps in our knowledge of the composition of Indian feeding stuffs, rations have in many cases been computed from foreign data. As described earlier in the book, there is considerable variation in the composition of similar feeding stuffs, not only in different years in the same locality, but also in the same year in different localities, according to soil conditions and climatic and other factors. Considerable variations in composition also occur according to the stage of growth, and wherever possible, in the case of growing crops, the stage at which the samples were taken has been indicated.

APPENDIX II

FEEDING STANDARDS

In order to enable farmers to calculate the rations required for different types of animal, various feeding standards based on experimental data have been drawn up in different countries to show how much of the various types of digestible nutrients are required for different classes of animal. Some of these standards are given in terms of starch equivalents and protein equivalents, others are drawn up in terms of digestible protein and total digestible nutrients (see context). In the past, India has been almost entirely dependent on standards drawn up in foreign countries, and those most commonly used, although in somewhat limited circles, have been the Morrison standards. In the last few years, work has been conducted at Lyallpur, as described in Chapter IX, to evaluate standards under Indian conditions for cows and working bullocks. The data regarding them are still incomplete. No feeding standards derived from experimental work on other animals are yet available in India, and in the absence of such it is necessary to use foreign standards. The standards given in this appendix, other than those evaluated at Lyallpur, are those of Morrison and are reproduced by permission.

The experimental work at Lyallpur has been done on cows of 800 pounds weight and giving an average yield of 24 pounds milk containing 5% fat. As described in Chapter IX, both the maintenance and milk production requirements of these animals in terms of dry matter, digestible protein and total digestible nutrients are somewhat lower than the Morrison figures. For example, the maintenance requirements of a cow of 800 pounds weight are :—

	Dry matter.	Digestible protein.	Total digestible nutrients.
	lb.	lb.	lb.
Morrison ..	18.00	0.536	6.53
Lyallpur ..	12.26	0.460	6.20

For the production of each pound of milk containing 5 per cent fat, the following additional nutrients are required :—

	Digestible protein.	Total digestible nutrients.
	lb.	lb.
Morrison ..	0.046–0.056	0.353–0.373
Lyallpur ..	0.048	0.357

APPENDIX II

MAINTENANCE REQUIREMENTS OF DAIRY COWS. LYALLPUR.

Body weight.	Digestible protein (per head daily).	Total digestible nutrients (per head daily).
lb.	lb.	lb.
400	0.280	3.70
500	0.330	4.40
600	0.370	5.00
700	0.420	5.60
800	0.460	6.20
900	0.510	6.75
1,000	0.550	7.30

MAINTENANCE REQUIREMENTS OF BULLOCKS. LYALLPUR.

Body weight.	Dry matter.	Digestible protein.	Total digestible nutrients.
lb.	lb.	lb.	lb.
500	8.66	0.248	4.52
600	9.90	0.284	5.16
700	11.07	0.317	5.78
800	12.20	0.350	6.37
900	13.29	0.381	6.94
1,000	14.35	0.412	7.49

Additional requirements as determined at Lyallpur for each hour's work of ordinary ploughing, etc., are:—

Dry matter.	Digestible protein.	Total digestible nutrients.
lb.	lb.	lb.
0.409	0.0297	0.327

Therefore, the requirements of a working bullock weighing 1,000 pounds and doing an average day's work of 6 hours ploughing are :—

2.45	0.178	1.96

APPENDIX II

MORRISON FEEDING STANDARDS.

	Digestible protein (per head daily).		Total digestible nutrients (per head daily).	
	Minimum allowance advised.	Recommended for good cows under usual conditions.	Minimum allowance advised.	Recommended for good cows under usual conditions.
1. *Dairy Cows.*				
A. *For Maintenance.*	lb.	lb.	lb.	lb.
700 lb. Cow	0.440	0.476	5.13	5.81
750 lb. Cow	0.476	0.506	5.45	6.18
800 lb. Cow	0.494	0.536	5.77	6.53
850 lb. Cow	0.521	0.564	6.08	6.88
900 lb. Cow	0.547	0.593	6.38	7.23
950 lb. Cow	0.574	0.621	6.69	7.58
1,000 lb. Cow	0.600	0.650	7.00	7.93
1,050 lb. Cow	0.626	0.678	7.30	8.27
1,100 lb. Cow	0.652	0.706	7.60	8.61
1,150 lb. Cow	0.677	0.734	7.90	8.95
1,200 lb. Cow	0.703	0.762	8.20	9.29
1,250 lb. Cow	0.730	0.790	8.51	9.64
1,300 lb. Cow	0.754	0.817	8.80	9.97
1,350 lb. Cow	0.779	0.844	9.09	10.29
1,400 lb. Cow	0.805	0.872	9.39	10.63
1,450 lb. Cow	0.829	0.898	9.67	10.96
1,500 lb. Cow	0.854	0.925	9.96	11.28
1,550 lb. Cow	0.878	0.952	10.25	11.61
1,600 lb. Cow	0.904	0.979	10.54	11.94
1,650 lb. Cow	0.928	1.005	10.82	12.26
1,700 lb. Cow	0.952	1.032	11.11	12.58
1,750 lb. Cow	0.976	1.058	11.39	12.90
1,800 lb. Cow	1.001	1.084	11.68	13.23

B. *To be added to the allowance for maintenance given above for producing one pound of milk containing the following fat percentage :—*

2.5 per cent fat ..	0.034	0.040	0.238	0.251
3.0 " " " ..	0.036	0.043	0.261	0.276
3.5 " " " ..	0.038	0.046	0.284	0.300
4.0 " " " ..	0.041	0.049	0.307	0.324
4.5 " " " ..	0.044	0.052	0.330	0.349
5.0 " " " ..	0.046	0.056	0.353	0.373
5.5 " " " ..	0.049	0.059	0.376	0.397
6.0 " " " ..	0.052	0.062	0.399	0.422
6.5 " " " ..	0.054	0.065	0.422	0.446
7.0 " " " ..	0.057	0.068	0.445	0.470

APPENDIX II

MORRISON FEEDING STANDARDS—(Contd.)

	Dry matter.	Digestible protein.	Total digestible nutrients.	Nutritive ratio.
	Requirements per head daily.			
	lb.	lb.	lb.	1:
2. *Growing dairy cattle.*				
Weight 100 lb. ..	1.4— 2.4	0.24—0.40	1.2— 2.0	3.9—4.5
„ 150 „ ..	3.4— 4.0	0.41—0.52	2.3— 3.0	4.4—5.1
„ 200 „ ..	4.6— 5.6	0.52—0.62	3.3— 4.0	5.0—5.5
„ 250 „ ..	5.9— 6.9	0.61—0.71	4.1— 4.8	5.7—6.2
„ 300 „ ..	7.2— 8.0	0.67—0.78	4.9— 5.5	6.3—6.8
„ 400 „ ..	9.0—10.0	0.80—0.90	6.1— 6.6	6.5—7.0
„ 500 „ ..	10.6—11.8	0.87—0.98	6.9— 7.7	6.9—7.4
„ 600 „ ..	12.0—13.6	0.94—1.06	7.7— 8.7	7.2—7.7
„ 700 „ ..	13.4—15.5	1.00—1.13	8.4— 9.7	7.4—7.9
„ 800 „ ..	14.8—17.4	1.06—1.20	9.1—10.7	7.6—8.1
„ 900 „ ..	16.1—19.2	1.11—1.27	9.8—11.7	7.8—8.3
„ 1,000 „ ..	17.5—21.0	1.16—1.33	10.4—12.6	8.0—8.4
3. *Fattening yearling cattle.*				
Weight 600 lb. ..	13.2—16.3	1.20—1.41	10.3—12.7	7.0—8.0
„ 700 „ ..	15.2—18.3	1.41—1.60	12.0—14.4	7.0—8.0
„ 800 „ ..	17.0—20.0	1.59—1.79	13.5—16.1	7.0—8.0
„ 900 „ ..	18.5—21.8	1.74—1.94	14.8—17.4	7.0—8.0
„ 1,000 „ ..	19.7—22.9	1.87—2.06	15.9—18.5	7.0—8.0
„ 1,100 „ ..	20.8—24.0	1.99—2.17	16.9—19.5	7.0—8.0
4. *Fattening 2 years old cattle.*				
Weight 900 lb. ..	18.7—22.3	1.62—1.83	14.6—17.4	7.5—8.5
„ 1,000 „ ..	20.0—23.5	1.78—1.98	16.0—18.8	7.5—8.5
„ 1,100 „ ..	20.9—24.1	1.87—2.07	17.0—19.6	7.5—8.5
„ 1,200 „ ..	21.8—24.7	1.95—2.12	17.7—20.1	7.5—8.5
5. *Horses, idle.*				
Weight 1,000 lb. ..	13.0—18.0	0.6—0.8	7.0— 9.0	10.0—12.0
„ 1,100 „ ..	13.9—19.3	0.6—0.9	7.5— 9.7	10.0—12.0
„ 1,200 „ ..	14.8—20.6	0.7—0.9	8.0—10.3	10.0—12.0
„ 1,300 „ ..	15.7—21.8	0.7—1.0	8.5—10.9	10.0—12.0
„ 1,400 „ ..	16.6—23.0	0.8—1.0	8.9—11.5	10.0—12.0
„ 1,500 „ ..	17.5—24.2	0.8—1.1	9.4—12.1	10.0—12.0
„ 1,600 „ ..	18.3—25.4	0.8—1.1	9.9—12.7	10.0—12.0
„ 1,700 „ ..	19.1—26.5	0.9—1.2	10.3—13.3	10.0—12.0
„ 1,800 „ ..	20.0—27.6	0.9—1.2	10.8—13.8	10.0—12.0

APPENDIX II

MORRISON FEEDING STANDARDS—(Contd.)

	Dry matter.	Digestible protein.	Total digestible nutrients.	Nutritive ratio.
	Requirements per head daily.			
	lb.	lb.	lb.	1:
6. *Horses at light work.*				
Weight 1,000 lb. ..	15.0—20.0	0.8—1.0	9.0—11.0	9.0—11.0
,, 1,100 ,, ..	16.2—21.6	0.9—1.1	9.7—11.9	9.0—11.0
,, 1,200 ,, ..	17.4—23.1	0.9—1.2	10.4—12.7	9.0—11.0
,, 1,300 ,, ..	18.5—24.7	1.0—1.2	11.1—13.6	9.0—11.0
,, 1,400 ,, ..	19.6—26.2	1.0—1.3	11.8—14.4	9.0—11.0
,, 1,500 ,, ..	20.8—27.7	1.1—1.4	12.5—15.2	9.0—11.0
,, 1,600 ,, ..	21.9—29.2	1.2—1.5	13.1—16.0	9.0—11.0
,, 1,700 ,, ..	23.0—30.6	1.2—1.5	13.8—16.8	9.0—11.0
,, 1,800 ,, ..	24.0—32.0	1.3—1.6	14.4—17.6	9.0—11.0
7. *Horses at medium work.*				
Weight 1,000 lb. ..	16.0—21.0	1.0—1.2	11.0—13.0	9.0—11.0
,, 1,100 ,, ..	17.4—22.8	1.1—1.3	11.9—14.1	9.0—11.0
,, 1,200 ,, ..	18.8—24.6	1.2—1.4	12.9—15.2	9.0—11.0
,, 1,300 ,, ..	20.1—26.4	1.3—1.5	13.8—16.3	9.0—11.0
,, 1,400 ,, ..	21.5—28.2	1.3—1.6	14.8—17.4	9.0—11.0
,, 1,500 ,, ..	22.8—29.9	1.4—1.7	15.7—18.5	9.0—11.0
,, 1,600 ,, ..	24.1—31.6	1.5—1.8	16.6—19.6	9.0—11.0
,, 1,700 ,, ..	25.4—33.3	1.6—1.9	17.5—20.6	9.0—11.0
,, 1,800 ,, ..	26.7—35.0	1.7—2.0	18.3—21.7	9.0—11.0
8. *Horses at hard work.*				
Weight 1,000 lb. ..	18.0—22.0	1.2—1.4	13.0—16.0	9.0—11.0
,, 1,100 ,, ..	19.7—24.0	1.3—1.5	14.2—17.5	9.0—11.0
,, 1,200 ,, ..	21.3—26.1	1.4—1.7	15.4—19.0	9.0—11.0
,, 1,300 ,, ..	23.0—28.1	1.5—1.8	16.6—20.5	9.0—11.0
,, 1,400 ,, ..	24.7—30.2	1.6—1.9	17.8—21.9	9.0—11.0
,, 1,500 ,, ..	26.3—32.2	1.8—2.0	19.0—23.4	9.0—11.0
,, 1,600 ,, ..	28.0—34.2	1.9—2.2	20.2—24.8	9.0—11.0
,, 1,700 ,, ..	29.6—36.2	2.0—2.3	21.4—26.3	9.0—11.0
,, 1,800 ,, ..	31.2—38.1	2.1—2.4	22.5—27.7	9.0—11.0

APPENDIX II

MORRISON FEEDING STANDARDS—(Contd.)

	Requirements per head daily.			
	Dry matter.	Digestible protein.	Total Digestible nutrients.	Nutritive ratio.
	lb.	lb.	lb.	1:
9. *Brood mares nursing foals, but not at work.*				
Weight 1,000 lb. ..	15.0—22.0	1.2—1.5	9.0—12.0	6.5—7.5
„ 1,100 „ ..	16.2—23.8	1.3—1.6	9.7—13.0	6.5—7.5
„ 1,200 „ ..	17.4—25.5	1.4—1.7	10.4—13.9	6.5—7.5
„ 1,300 „ ..	18.5—27.1	1.5—1.9	11.1—14.8	6.5—7.5
„ 1,400 „ ..	19.6—28.8	1.6—2.0	11.8—15.7	6.5—7.5
„ 1,500 „ ..	20.8—30.4	1.7—2.1	12.5—16.6	6.5—7.5
„ 1,600 „ ..	21.9—32.1	1.7—2.2	13.1—17.5	6.5—7.5
„ 1,700 „ ..	23.0—33.7	1.8—2.3	13.8—18.4	6.5—7.5
„ 1,800 „ ..	24.0—35.2	1.9—2.4	14.4—19.2	6.5—7.5
10. *Growing draft colts after weaning.*				
Weight 400 lb. ..	9.2—11.3	0.8—0.9	5.6— 7.2	6.5—7.0
„ 500 „ ..	10.9—13.3	0.9—1.0	6.6— 8.4	6.6—7.1
„ 600 „ ..	12.4—15.2	1.0—1.2	7.6— 9.6	6.7—7.2
„ 700 „ ..	13.9—17.0	1.1—1.3	8.5—10.8	6.8—7.3
„ 800 „ ..	15.3—18.7	1.2—1.4	9.4—11.9	6.9—7.4
„ 900 „ ..	16.7—20.4	1.3—1.5	10.2—13.0	7.0—8.0
„ 1,000 „ ..	18.0—22.0	1.4—1.6	11.0—14.0	7.0—8.0
„ 1,100 „ ..	19.3—23.6	1.5—1.6	11.8—15.0	7.2—8.2
„ 1,200 „ ..	20.6—25.1	1.5—1.7	12.6—16.0	7.5—8.5
11. *Pregnant ewes, up to 4 to 6 weeks before lambing.*				
Weight 100 lb. ..	2.0—2.3	0.16—0.18	1.5—1.8	7.5—8.5
„ 110 „ ..	2.2—2.4	0.17—0.20	1.6—1.9	7.5—8.5
„ 120 „ ..	2.3—2.6	0.18—0.21	1.7—2.0	7.5—8.5
„ 130 „ ..	2.4—2.8	0.19—0.22	1.8—2.1	7.5—8.5
„ 140 „ ..	2.6—2.9	0.20—0.23	1.9—2.2	7.5—8.5
„ 150 „ ..	2.7—3.1	0.21—0.24	2.0—2.4	7.5—8.5

APPENDIX II

MORRISON FEEDING STANDARDS—(Contd.)

	Requirements per head daily.			
	Dry matter.	Digestible protein.	Total Digestible nutrients.	Nutritive ratio.
	lb.	lb.	lb.	1:
12. *Pregnant ewes, last 4 to 6 weeks before lambing.*				
Weight 100 lb. ..	2.5—2.8	0.21—0.23	1.9—2.2	7.2—8.2
,, 110 ,, ..	2.7—2.9	0.22—0.25	2.0—2.3	7.2—8.2
,, 120 ,, ..	2.8—3.1	0.23—0.26	2.1—2.4	7.2—8.2
,, 130 ,, ..	2.9—3.3	0.24—0.27	2.2—2.5	7.2—8.2
,, 140 ,, ..	3.1—3.4	0.25—0.28	2.3—2.6	7.2—8.2
,, 150 ,, ..	3.2—3.6	0.26—0.29	2.4—2.8	7.2—8.2
13. *Ewes nursing lambs.*				
Weight 100 lb. ..	2.9—3.2	0.27—0.29	2.3—2.6	6.7—7.7
,, 110 ,, ..	3.1—3.3	0.28—0.31	2.4—2.7	6.7—7.7
,, 120 ,, ..	3.2—3.5	0.29—0.32	2.5—2.8	6.7—7.7
,, 130 ,, ..	3.3—3.7	0.30—0.33	2.6—2.9	6.7—7.7
,, 140 ,, ..	3.5—3.8	0.31—0.34	2.7—3.0	6.7—7.7
,, 150 ,, ..	3.6—4.0	0.32—0.35	2.8—3.2	6.7—7.7
14. *Fattening lambs.*				
Weight 50 lb. ..	1.9—2.3	0.16—0.19	1.2—1.5	6.5—7.0
,, 60 ,, ..	2.0—2.5	0.20—0.23	1.5—1.8	6.7—7.2
,, 70 ,, ..	2.2—2.7	0.21—0.24	1.7—2.0	6.9—7.4
,, 80 ,, ..	2.3—2.8	0.22—0.25	1.8—2.1	7.1—7.6
,, 90 ,, ..	2.4—2.9	0.23—0.26	1.9—2.2	7.3—7.8
,, 100 ,, ..	2.5—3.0	0.25—0.27	2.0—2.3	7.5—8.0
15. *Growing and fattening pigs.*				
Weight 30 ,, ..	1.3—1.9	0.25—0.32	1.2—1.7	4.0—4.5
,, 50 ,, ..	2.1—2.8	0.35—0.43	1.9—2.5	4.5—5.0
,, 75 ,, ..	2.9—3.9	0.43—0.52	2.6—3.5	5.3—5.8
,, 100 ,, ..	3.6—4.8	0.50—0.60	3.2—4.3	5.8—6.2
,, 150 ,, ..	4.8—6.2	0.65—0.75	4.3—5.6	6.2—6.5
,, 200 ,, ..	5.8—7.1	0.73—0.83	5.2—6.4	6.4—6.7
,, 250 ,, ..	6.5—7.8	0.80—0.90	5.9—7.0	6.5—6.8
,, 300 ,, ..	7.1—8.4	0.85—0.95	6.4—7.6	6.6—7.0

APPENDIX II

MORRISON FEEDING STANDARDS—(Contd.)

	Dry matter.	Digestible protein	Total Digestible nutrients.	Nutritive ratio.
	Requirements per head daily.			
	lb.	lb.	lb.	1:
16. *Wintering pregnant gilts.*				
Weight 200 lb. ..	3.3—4.0	0.43—0.47	3.0—3.6	6.0—7.0
„ 250 „ ..	3.9—4.7	0.50—0.55	3.5—4.2	6.0—7.0
„ 300 „ ..	4.4—5.4	0.57—0.63	4.0—4.8	6.0—7.0
17. *Wintering pregnant older sows.*				
Weight 300 lb. ..	3.7—4.5	0.43—0.49	3.2—4.1	6.5—7.5
„ 400 „ ..	4.6—5.6	0.53—0.60	4.0—5.0	6.5—7.5
„ 500 „ ..	5.4—6.6	0.63—0.71	4.7—5.9	6.5—7.5
„ 600 „ ..	6.2—7.6	0.72—0.81	5.4—6.8	6.5—7.5
18. *Brood sows nursing litters.*				
Weight 300 lb. ..	8.9—10.9	1.16—1.23	8.1— 9.5	6.0—7.0
„ 400 „ ..	9.4—11.5	1.22—1.29	8.5—10.0	6.0—7.0
„ 500 „ ..	9.8—12.7	1.28—1.35	9.0—10.5	6.0—7.0
„ 600 „ ..	11.2—13.8	1.34—1.42	9.4—11.0	6.0—7.0

APPENDIX IV

Our knowledge concerning the vitamin content of feeding stuffs for animals is not as extensive as that concerning articles of human dietaries. As described in the context the only feeding stuffs which are likely to cause symptoms of vitamin deficiency are those deficient in vitamins A & D, and indications have been given as to how such deficiencies may be met. The data in the following table have been compiled from various sources including some analyses conducted at Lyallpur.

The indications of vitamin content given must not be regarded as absolute, but rather as a general guide, as very considerable variations occur in different samples of the same feeding stuff. For example, hay may show very wide variations due to the condition of the grass when cut and the manner in which the hay has been prepared. Similarly for many of the other items, and any specific analysis for vitamin content may show some deviation one way or the other from that recorded.

APPENDIX IV

VITAMIN CONTENT OF SOME FEEDING STUFFS.

The relative amounts of the different vitamins in each feed are indicated by the following symbols :—

- **x** indicates the presence of only a small amount of the vitamin.
- **xx** means the presence of the vitamin in satisfactory quantity.
- **xxx** means that the feeding stuff is a very good source of the vitamin.
- **xxxx** indicates that the feed is exceptionally rich in the vitamin.
- **O** means entire absence of the vitamin.
- **–** indicates that the information concerning the amount of the vitamin is lacking or is not conclusive.

Feeding stuff.	Vitamins.					
	A	B_1	B_2	C	D	E
Green roughages, roots etc.						
Alfalfa	xxx	x	xx	xxx	O	xxx
Beets, common	O	O	–	x	O	–
Beets, sugar	O	O	–	x	O	–
Cabbage, green leaves	xx	x	x	xxx	O	–
Cabbage, white portion	O to x	x	–	xxx	O	–
Carrots, yellow	xxx	x	–	xx	O	–
Clover	xxx	x	xx	xxx	O	xxx
Grasses, growing actively	xxx	x	xx	xxx	O	xxx
Kale	xxx	x	–	xxx	O	–
Lucerne	xxx	x	x	xxx	–	–
Mangels	O	O	–	O	O	–
Potatoes	O	x	–	xx	O	–
Sweet potatoes, yellow	xx	x	–	xx	O	–
Berseem	xxx	–	–	xxx	–	–
Maize	xxx	–	–	xxx	–	–
Oats	xxx	–	–	xxx	–	–
Grass hay, good quality	x to xx	x	x	O	x	xx
Grass hay, poor quality	O to x	–	–	O	–	–
Dry roughages.						
Lucerne hay, good quality	xxx	x	xx	O	xx	xxx
Lucerne hay, poor quality	O to x	x	x	O	–	–
Maize fodder, well cured	x	x	–	O	x	–
Cotton seed hulls	O	–	–	O	–	–
Legume hay in general, good quality	xx	x	xx	O	xx	xxx
Millet straw	O	–	–	O	–	–
Sorghum fodder	x	x	–	O	x	–

APPENDIX IV

VITAMIN CONTENT OF SOME FEEDING STUFFS.

Feeding stuff.	Vitamins.					
	A	B_1	B_2	C	D	E
Concentrates.						
Barley	O to x	xx	x	O	O	xx
Beans	O	xx	–	O	O	–
Beet pulp, dried ..	O	–	–	O	O	O
Buttermilk ..	x	x	xx	O to x	O	x
Cotton seed meal ..	O	–	–	O	O	–
Fish meal ..	O to x	O	O to xx	O	O to x	–
Groundnuts ..	O to x	xx	x	O	–	–
Linseed meal ..	O	–	–	O	O	x
Maize, white ..	O	xxx	x	O	O	xx
Maize, yellow ..	xx	xxx	x	O	O	xx
Maize gluten feed ..	xx	–	O	O	O	O
Maize gluten meal ..	xxx	–	O	O	O	O
Millet seed ..	O to x	xx		O	O	–
Milk, whole ..	xx	x	xx	x	O to x	x
Molasses, cane ..	O	xx	–	O	O	xx
Oats ..	O	xx	x	O	O	xx
Peas, green ..	xx	xx	x	O	O	xx
Rice, whole grain, unpolished	O	xx	x	O	O	xx
Rice, whole grain, polished ..	O	O	O	O	O	O
Rye ..	O	xx	x	O	O	xx
Skim milk ..	O	x	xx	O to x	O	x
Sorghum grain ..	O	xx	–	O	O	xx
Soybeans ..	O	xx	x	O	O	–
Soybean meal, whole pressed	O	xx	x	O	O	–
Tankage ..	O	O	x	O	O	–
Wheat ..	O	xx	x	O	O	xx
Wheat bran ..	O	xx	x	O	O	xx
Wheat germ ..	O	xxxx	x	O	O	xxxx
Wheat middlings, standard ..	O	xxx	x	O	O	xxx
Whey ..	O	x	xx	O to x	O	x
Whey, dried ..	O	x	xxx	O to x	O	x
Silage.						
Maize silage, good quality, from green maize at silage stage ..	x to xx	x	–	O to x	O to x	x
Sorghum silage, good quality	x to xx	x	–	O to x	O to x	x

APPENDIX IV

VITAMIN CONTENT OF SOME FEEDING STUFFS.

Feeding stuff.	Vitamins.					
	A	B_1	B_2	C	D	E
Miscellaneous.						
Apples ..	x	x	x	x to xx	–	–
Butter ..	xxx	O	–	O	x	x
Cheese ..	xx	O	x	–	–	–
Cream ..	x to xx	xx	–	O to x	x	–
Cod liver oil ..	xxxx	O	O	O	xxxx	O
Eggs ..	xx	x	x to xx	xx	xx	–
Fresh fruit in general ..	–	x	O to xxx	–	–	x
Lean meat ..	–	x	xx	x	–	x
Orange juice ..	x	x,	–	xxx	–	x
Tomatoes ..	xx	x	x	xxx	–	–
Yeast ..	O	xxxx	xxx	O	O	O

APPENDIX V

This appendix shows detailed rationing schemes followed by the Military Dairy Farms, India, and is reproduced by kind permission of the Director of Military Farms, India.

APPENDIX V

BASIC RATION TABLE (A) FOR CALVES UPTO SIX MONTHS

Figures under all heads, except fodder, denote the quantity to be fed for every 10 pounds body weight. (Five pounds and over to be counted as 10 pounds, under five pounds to be ignored).

The figures under fodder show the actual proportion of the standard fodder ration to be fed.

Age.	Cow calves.				Buffalo calves.				Cow and Buffalo calves.	
	Dam's colostrum. lb.	Milk. lb.	Separated milk. lb.	Concentrates. lb.	Dam's colostrum. lb.	Milk. lb.	Separated milk. lb.	Concentrates. lb.	Fodder.	No. of times rations fed daily.
1st-3rd day	1.0	..	..	..	1.0	..	..	..	..	
4th-7th day	..	1.0	..	..	..	1.0	..	..	..	4
2nd week	..	1.25	..	..	..	0.8	0.1	..	..	3
3rd ,,	..	1.25	..	..	..	0.7	0.2	..	..	2
4th ,,	..	0.75	0.25	1/20	..	0.6	0.2	1/20	..	2
5th ,,	..	0.75	0.25	1/20	..	0.5	0.4	1/20	1/20	2
6th ,,	..	0.75	0.25	1/20	..	0.4	0.4	1/20	1/10	2
7th ,,	..	0.50	0.40	1/12	..	0.3	0.4	1/10	1/8	2
8th ,,	..	0.50	0.40	1/12	..	0.2	0.5	1/10	1/8	2
9th ,,	..	0.50	0.40	1/12	..	0.15	0.5	1/10	1/6	2
10th-12th week	..	0.30	0.50	1/12	..	..	0.5	1/10	1/5	2
13th-16th ,,	..	0.10	0.50	1/12	..	..	0.5	1/10	1/5	2
17th-19th ,,	..	..	0.40	1/12	..	..	0.5	1/10	1/4	2
20th-23rd ,,	..	..	0.25	1/12	..	..	0.5	1/10	1/3	2
24th-26th ,,	..	..	0.10	1/12	..	..	0.5	1/10	1/2	2

COMPOSITION OF CONCENTRATE MIXTURE. (*As for animals over six tmonhs*).

NOTES—1. Rations will be worked out to the nearest pound in the case of colostrum, milk, separated milk and fodder, and to the nearest half pound in the case of concentrates.

2. Managers have the authority to alter the ration of an individual animal if considered desirable.

3. The milk and separated milk required for calves rations have prior claim on all issues. The Calves Incharge must report in the Cattleyard Report Book every time that the full rations of milk or separated milk have not been supplied.

4. At Young Stock Farms when separated milk is not available, it will be replaced by 50 per cent. whole milk.

5. The quality of fodder fed to calves will be the best available. As a rule the mixed fodder should contain about 25 per cent. dry matter.

6. When fodders 1 and 2 (see fodder table) are fed, the concentrate ration for cow calves over 12 weeks old will be increased to 1/10.

APPENDIX V

BASIC RATION TABLE (B) FOR CATTLE OVER SIX MONTHS.

(Except Horses and Mules for which see Ration Table (C).

Weight in pounds.	Concentrates.								Fodder.
	A		B		C		D		Proportion of standard fodder to be fed.
	1	2	1	2	1	2	1	2	
	lb.	lb.	lb.	lb.	lb.	lb.	lb.	lb.	
Upto 400	3	3	3	2	2	2	2	2	$\frac{1}{2}$
400-500	4	3	3	2	2	1	2	1	$\frac{2}{3}$
500-600	4	3	3	2	2	1	1	0	$\frac{3}{4}$
600-700	4	3	2	1	1	0	0	0	$\frac{5}{6}$
700-800	4	3	2	1	1	0	0	0	1
800-1000 (a)*	4	3	2	1	1	0	0	0	$1\frac{1}{6}$
1000-1200 (b)*	5	4	2	1	1	0	0	0	$1\frac{1}{3}$
1200-1400 (c)*	5	4	2	2	1	0	0	0	$1\frac{2}{3}$
Over 1400	6	5	3	2	2	0	0	0	$1\frac{3}{4}$

NOTES:—1. Category 1 is for cattle containing over 50 per cent. foreign blood.

2. Category 2 is for indigenous cattle and cattle containing 50 per cent. or less foreign blood.

3. Bulls in service and bullocks at work must not be fed under a category below 'C'.

4. Bulls, Bullocks and all pure Friesians will be fed on actual weight basis.

*To simplify calculations in feeding concentrates and fodder to mature female stock only, it has been assumed that all indigenous cows and those containing upto 25 per cent. foreign blood come in category (a) (800-1000 lb.), those with over 25 per cent. foreign blood (except pure Friesians) in category (b) (1000-1200 lb.) and buffaloes in category (c) (1200-1400 lb.)

APPENDIX V

ADDITIONAL CONCENTRATES TO BE FED FOR EXTRA STRAIN ON CONSTITUTION.

[*Add to Basic Ration Table (B)*].

1. *For Milk* :—

A and B scales :—One pound for 2½ lb. milk.

C scale :—One pound for 3 lb. milk upto two gallons and one pound for 2½ lb. milk for yields beyond that.

D scale :—One pound for 3 lb. milk upto three gallons and 1 pound for 2½ lb. milk for yields beyond that.

NOTE— 1. Concentrates and fodder will be worked out to the nearest pound.

2. The full concentrate ration for the expected maximum yield will not be given until 14 days after calving as follows :—

1—7 days..........50 %. 8—14 days............75%.

After 14 days.............100%.

2. *For low Temperature* :—

NOTE—To animals in milk only :—45°—55°F.......1 lb. Below 45°F.......2 lb.

Temperature will be that inside the shed.

3. *For Pregnancy* :—

During the last 3 months only. Category 1........2 lb.

Category 2........1 lb. (See note 3 under " Reductions ").

4. *For work to Bullocks* :—

6 lb. concentrates for hard work, 4 lb. for light work or no work.

5. *For Service to Bulls* :—

7 lb. concentrates. Mature bulls not in regular service to be given 4 lb. concentrates.

6. *For Growth to Friesian Bulls and Cows* :—

Friesian Young Stock below 1400 lb. body weight will be given 1 lb. extra concentrates.

APPENDIX V

REDUCTIONS IN NORMAL CONCENTRATE AND FODDER ALLOWANCE IN THE CASE OF INDIVIDUAL ANIMALS.

1. *Animals in Milk.*

Certain individual animals (category 2 only) maintain excessive condition on normal concentrate scales. The concentrate ration of such animals will be reduced by 50 to 100%. The condition and milk yield will be carefully watched when making the reduction.

2. *Dry adults and heifers in excessive condition and not holding to service or not coming into heat.*

Maintenance concentrate ration to be reduced by 50 to 100%. In the case of fat animals in category 2, the fodder allowance may also be reduced by 25%.

3. *Animals that will be dry for over 6 months prior to next calving.*

No pregnancy allowance to be given unless condition is unsatisfactory. Those in good condition will have even their maintenance concentrate ration reduced by 50 to 100% so long as they remain in good condition.

BASIC RATION TABLE (C) FOR HORSES AND MULES

Body weight in pounds.	Concentrates. lb.				Proportion of standard fodder to be fed.
	A	B	C	D	
Upto 600 ..	5	4	3	3	$\frac{5}{6}$
600-800 ..	6	5	4	4	$\frac{11}{12}$
800-1000 ..	7	6	5	5	1
Over 1000 ..	8	6	5	5	$1\frac{1}{12}$

Note.—1. Add 4 lb. concentrates for work ; if an animal is idle, feed according to the basic ration table on weight basis.

2. Horses and mules at work should not be fed under a category below " C "

APPENDIX V

COMPOSITION OF CONCENTRATE MIXTURES.

Mixtures.	Bran.	Crushed gram.	Crushed barley.	Crushed cotton seed cake.	Crushed ground nut cake.
Cattle.					
(a) ..	4	1	1	..	2
(b) ..	4	..	2	..	2
(c) ..	4	1	1	2	..
(d) ..	5	1	..	..	2
(e) ..	4	2	..	2	..
(f) ..	4	1	2	..	1
(g) ..	3	..	2	3	..
(h) ..	3	1	2	2	..
(i) ..	5	..	1	..	2
Horses and Mules.					
(a) ..	4	4	..	..	..
(b) ..	3	2	3	..	..

Managers will feed the most economical mixture made up of stocks in hand. The cheapest mixture according to current prices should be the basis for indenting for future requirements of concentrates.

INDICES

INDEX TO AUTHORS

INDEX TO AUTHORS

SUBJECT INDEX

SUBJECT INDEX

SUBJECT INDEX

SUBJECT INDEX

SUBJECT INDEX

SUBJECT INDEX

SUBJECT INDEX

SUBJECT INDEX

SUBJECT INDEX

ERRATA

Page	Line	For	Read
3	2 from bottom	an inevitable	inevitable
7	5 from top	that vast	vast
14	12 from bottom	the bacteria	bacteria
15	9 from top	The Soil	Soil
17	18 from top	is	are
72	15 from bottom	million	μ on
77	13 from bottom	constitution	constitutions
82	17 from bottom	that	the
82	17 from bottom	digestive	digestion
113	14 from top	ma	may
118	12 from top	of calories	or calories
127	3 from top	15	1·5
133	18 from bottom	vary	varies
140	3 from bottom	cheifly	chiefly
201	3 from bottom	skim	skim milk
223	Table III Column IV	Digestibility Protein	Digestible Protein
226	17 from bottom	and	which
293	last line	orisson	Morrison
351	13 from bottom	Automatically	Anatomically
370	4 from top	Sxffer	suffer
374	20 from bottom	feed	food
378	8 from bottom	feed	food
421	6 from top	as important	important
422	16 from top	though	trough
438	Table I. B. Column 1	D. P. I	T. D. N.
438	Table I. B. Column 2	D. P. I.	D. P
441	7 from top	Service	Services
441	2 from bottom	wider	narrower